Graduate Texts in Mathematics 187

Springer
New York
Berlin
Heidelberg
Barcelona
Budapest
Hong Kong
London
Milan
Paris
Singapore
Tokyo

Graduate Texts in Mathematics

continued after index

Joe Harris
Ian Morrison

Moduli of Curves

Springer

Joe Harris
Department of Mathematics
Harvard University
Cambridge, MA 02138
USA

Ian Morrison
Department of Mathematics
Fordham University
Bronx, NY 10458-5165
USA

Mathematics Subject Classification (1991): 14H10

Library of Congress Cataloging-in-Publication Data
Harris, Joe.
    Moduli of curves / Joe Harris, Ian Morrison.
      p.   cm. — (Graduate texts in Mathematics; 187)
    Includes bibliographical references and index.
    ISBN 0-387-98438-0 (hardcover : alk. paper). — ISBN 0-387-98429-1
(pbk. : alk. paper)
      1. Moduli theory.   2. Curves, Algebraic.   I. Morrison, Ian, 1950–
II. Title.
QA564.H244   1998
516.3′5—dc21   98-13036

Printed on acid-free paper.

Production managed by Allan Abrams; manufacturing supervised by Jeffrey Taub.
Photocomposed copy prepared from the authors' TeX files.
Printed and bound by R.R. Donnelley and Sons, Harrisonburg, VA.
Printed in the United States of America.

9 8 7 6 5 4 3 2 1

ISBN 0-387-98438-0 Springer-Verlag New York Berlin Heidelberg   SPIN 10659835 (hardcover)
ISBN 0-387-98429-1 Springer-Verlag New York Berlin Heidelberg   SPIN 10659843 (softcover)

To Phil Griffiths
and David Mumford

Preface

Aims

The aim of this book is to provide a guide to a rich and fascinating subject: algebraic curves, and how they vary in families. The revolution that the field of algebraic geometry has undergone with the introduction of schemes, together with new ideas, techniques and viewpoints introduced by Mumford and others, have made it possible for us to understand the behavior of curves in ways that simply were not possible a half-century ago. This in turn has led, over the last few decades, to a burst of activity in the area, resolving long-standing problems and generating new and unforeseen results and questions. We hope to acquaint you both with these results and with the ideas that have made them possible.

The book isn't intended to be a definitive reference: the subject is developing too rapidly for that to be a feasible goal, even if we had the expertise necessary for the task. Our preference has been to focus on examples and applications rather than on foundations. When discussing techniques we've chosen to sacrifice proofs of some, even basic, results — particularly where we can provide a good reference — in order to show how the methods are used to study moduli of curves. Likewise, we often prove results in special cases which we feel bring out the important ideas with a minimum of technical complication.

Chapters 1 and 2 provide a synopsis of basic theorems and conjectures about Hilbert schemes and moduli spaces of curves, with few or no details about techniques or proofs. Use them more as a guide to the literature than as a working manual. Chapters 3 through 6 are, by contrast, considerably more self-contained and approachable. Ultimately, if you want to investigate fully any of the topics we discuss, you'll have to go beyond the material here; but you *will* learn the techniques fully enough, and see enough complete proofs, that when you finish a section here you'll be equipped to go exploring on your own.

If your goal is to work with families of curves, we'd therefore suggest that you begin by skimming the first two chapters and then tackle the later chapters in detail, referring back to the first two as necessary.

Contents

As for the contents of the book: Chapters 1 and 2 are largely expository: for the most part, we discuss in general terms the problems associated with moduli and parameter spaces of curves, what's known about them, and what sort of behavior we've come to expect from them. In Chapters 3 through 5 we develop the techniques that have allowed us to analyze moduli spaces: deformations, specializations (of curves, of maps between them and of linear series on them), tools for making a variety of global enumerative calculations, geometric invariant theory, and so on. Finally, in Chapter 6, we use the ideas and techniques introduced in preceding chapters to prove a number of basic results about the geometry of the moduli space of curves and about various related spaces.

Prerequisites

What sort of background do we expect you to have before you start reading? That depends on what you want to get out of the book. We'd hope that even if you have only a basic grounding in modern algebraic geometry and a slightly greater familiarity with the theory of a fixed algebraic curve, you could read through most of this book and get a sense of what the subject is about: what sort of questions we ask, and some of the ways we go about answering them. If your ambition is to work in this area, of course, you'll need to know more; a working knowledge with many of the topics covered in *Geometry of algebraic curves, I* [7] first and foremost. We could compile a lengthy list of other subjects with which some acquaintance would be helpful. But, instead, we encourage you to just plunge ahead and fill in the background as needed; again, we've tried to write the book in a style that makes such an approach feasible.

Navigation

In keeping with the informal aims of the book, we have used only two levels of numbering with arabic for chapters and capital letters for sections within each chapter. All labelled items in the book are numbered consecutively within each chapter: thus, the orderings of such items by label and by position in the book agree.

There is a single index. However, its first page consists of a list of symbols, giving for each a single defining occurrence. These, and other, references to symbols also appear in the main body of the index where they are alphabetized "as read": for example, references to $\overline{\mathcal{M}}_g$ will be found under Mgbar; to κ_i under kappai. Bold face entries in the main body index point to the defining occurrence of the cited term. References to all the main results stated in the book can be found under the heading theorems.

Production acknowledgements

This book was designed by the authors who provided Springer with the PostScript file from which the plates were produced. The type is a very slightly modified version of the Lucida font family designed by Chuck Bigelow and Kristin Holmes. (We added swashes to a few characters in the \mathcal alphabet to make them easier to distinguish from the corresponding upper-case \mathit character. These alphabets are often paired: a \mathcal character is used for the total space of a family and the \mathit version for an element.) It was coded in a customized version of the LaTeX2e format and typeset using Blue Sky Research's Textures TeX implementation with EPS figures created in Macromedia's Freehand7 illustration program.

A number of people helped us with the production of the book. First and foremost, we want to thank Greg Langmead who did a truly wonderful job of producing an initial version of both the LaTeX code and the figures from our earlier WYSIWYG drafts. Dave Bayer offered invaluable programming assistance in solving many problems. Most notably, he devoted considerable effort to developing a set of macros for overlaying text generated within TeX onto figures. These allow precise one-time text placement independent of the scale of the figure and proved invaluable both in preparing the initial figures and in solving float placement problems. If you're interested, you can obtain the macros, which work with all formats, by e-mailing Dave at bayer@math.columbia.edu.

Frank Ganz at Springer made a number of comments to improve the design and assisted in solving some of the formatting problems he raised. At various points, Donald Arseneau, Berthold Horn, Vincent Jalby and Sorin Popescu helped us solve or work around various difficulties. We are grateful to all of them.

Lastly, we wish to thank our patient editor, Ina Lindemann, who was never in our way but always ready to help.

Mathematical acknowledgements

You should not hope to find here the sequel to *Geometry of algebraic curves, I* [7] announced in the preface to that book. As we've already noted, our aim is far from the "comprehensive and self-contained account" which was the goal of that book, and our text lacks its uniformity. The promised second volume is in preparation by Enrico Arbarello, Maurizio Cornalba and Phil Griffiths.

A few years ago, these authors invited us to attempt to merge our then current manuscript into theirs. However, when the two sets of material were assembled, it became clear to everyone that ours was so far from meeting the standards set by the first volume that such a merger made little sense. Enrico, Maurizio and Phil then, with their usual generosity, agreed to allow us to withdraw from their project and to publish what we had written here. We cannot too strongly acknowledge our admiration for the kindness with which the partnership was proposed and the grace with which it was dissolved nor our debt to them for the influence their ideas have had on our understanding of curves and their moduli.

The book is based on notes from a course taught at Harvard in 1990, when the second author was visiting, and we'd like to thank Harvard University for providing the support to make this possible, and Fordham University for granting the second author both the leave for this visit and a sabbatical leave in 1992-93. The comments of a number of students who attended the Harvard course were very helpful to us: in particular, we thank Dan Abramovich, Jean-Francois Burnol, Lucia Caporaso and James McKernan. We owe a particular debt to Angelo Vistoli, who also sat in on the course, and patiently answered many questions about deformation theory and algebraic stacks.

There are many others as well with whom we've discussed the various topics in this book, and whose insights are represented here. In addition to those mentioned already, we thank especially David Eisenbud, Bill Fulton and David Gieseker.

We to thank Armand Brumer, Anton Dzhamay, Carel Faber, Bill Fulton, Rahul Pandharipande, Cris Poor, Sorin Popescu and Monserrat Teixidor i Bigas who volunteered to review parts of this book. Their comments enabled us to eliminate many errors and obscurities. For any that remain, the responsibility is ours alone.

Finally, we thank our respective teachers, Phil Griffiths and David Mumford. The beautiful results they proved and the encouragement they provided energized and transformed the study of algebraic curves — for us and for many others. We gratefully dedicate this book to them.

Contents

Chapter 1

Parameter spaces: constructions and examples

A Parameters and moduli

Before we take up any of the constructions that will occupy us in this chapter, we want to make a few general remarks about moduli problems in general.

What is a moduli problem? Typically, it consists of two things. First of all, we specify a class of objects (which could be schemes, sheaves, morphisms or combinations of these), together with a notion of what it means to have a family of these objects over a scheme B. Second, we choose a (possibly trivial) equivalence relation $\sim$ on the set $S(B)$ of all such families over each B. We use the rather vague term "object" deliberately because the possibilities we have in mind are wide-ranging. For example, we might take our families to be

1. smooth flat morphisms $C \to B$ whose fibers are smooth curves of genus g, or

2. subschemes C in $\mathbb{P}^r \times B$, flat over B, whose fibers over B are curves of fixed genus g and degree d,

and so on. We can loosely consider the elements of $S(\mathrm{Spec}(\mathbb{C}))$ as the objects of our moduli problem and the elements of $S(B)$ over other bases as families of such objects parameterized by the complex points of B.[1]

The equivalence relations we will wish to consider will vary considerably even for a fixed class of objects: in the second case cited above, we might wish to consider two families equivalent if

---

[1] More generally, we may consider elements of $S(\mathrm{Spec}(k))$ for any field k as objects of our moduli problem defined over k.

1. the two subschemes of $\mathbb{P}^r \times B$ are equal,

2. the two subcurves are projectively equivalent over B, or

3. the two curves are (biregularly) isomorphic over B.

In any case, we build a functor **F** from the category of schemes to that of sets by the rule
$$F(B) = S(B)/ \sim$$
and call **F** the moduli functor of our moduli problem.

The fundamental first question to answer in studying a given moduli problem is: to what extent is the functor **F** representable? Recall that **F** is *representable* in the category of schemes if there is a scheme $\mathcal{M}$ and an isomorphism Ψ (of functors from schemes to sets) between **F** and the *functor of points* of $\mathcal{M}$. This last is the functor $\text{Mor}_{\mathcal{M}}$ whose value on B is the set $\text{Mor}_{\text{sch}}(B, \mathcal{M})$ of all morphisms of schemes from B to $\mathcal{M}$.

DEFINITION (1.1) *If **F** is representable by $\mathcal{M}$, then we say that the scheme $\mathcal{M}$ is a* fine moduli space *for the moduli problem* **F**.

Representability has a number of happy consequences for the study of **F**. If $\varphi : \mathcal{D} \to B$ is any family in (i.e., any element of) $S(B)$, then $\chi = \Psi(\varphi)$ is a morphism from B to $\mathcal{M}$. Intuitively, (closed) points of $\mathcal{M}$ classify the objects of our moduli problem and the map χ sends a (closed) point b of B to the moduli point in $\mathcal{M}$ determined by the fiber $\mathcal{D}_b$ of $\mathcal{D}$ over b. Going the other way, pulling back the identity map of $\mathcal{M}$ itself via Ψ constructs a family $1 : \mathcal{C} \to \mathcal{M}$ in $S(\mathcal{M})$ called the *universal* family. The reason for this name is that, given any morphism $\chi : B \to \mathcal{M}$ defined as above, there is a commutative fiber-product diagram

(1.2)

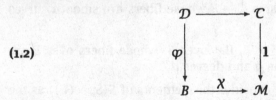

with $\varphi : \mathcal{D} \to B$ in $S(B)$ and $\Psi(\varphi) = \chi$. In sum, every family over B is the pullback of $\mathcal{C}$ via a *unique* map of B to $\mathcal{M}$ and we have a perfect dictionary enabling us to translate between information about the geometry of families of our moduli problem and information about the geometry of the moduli space $\mathcal{M}$ itself. One of the main themes of moduli theory is to bring information about the objects of our moduli problem to bear on the study of families and vice versa: the dictionary above is a powerful tool for relating these two types of information.

Unfortunately, few natural moduli functors are representable by schemes: we'll look at the reasons for this failure in the next chapter. One response to this failure is to look for a larger category (e.g., algebraic spaces, algebraic stacks, ...) in which F can be represented: the investigation of this avenue will also be postponed until the next chapter. Here we wish to glance briefly at a second strategy: to find a scheme $\mathcal{M}$ that captures enough of the information in the functor F to provide us with a "concise edition" of the dictionary above.

The standard way to do this is to ask only for a natural transformation of functors $\Psi = \Psi_{\mathcal{M}}$ from F to $\mathrm{Mor}(\cdot, \mathcal{M})$ rather than an isomorphism. Then, for each family $\varphi : \mathcal{D} \to B$ in $S(B)$, we still have a morphism $\chi = \Psi(\varphi) : B \to \mathcal{M}$ as above. Moreover, these maps are still natural in that, if $\varphi' : \mathcal{D}' = \mathcal{D} \times_B B' \to B'$ is the base change by a map $\xi : B' \to B$, then $\chi' = \Psi(\varphi') = \Psi(\varphi) \circ \xi$. This requirement, however, is far from determining $\mathcal{M}$. Indeed, given any solution $(\mathcal{M}, \Psi)$ and any morphism $\pi : \mathcal{M} \to \mathcal{M}'$, we get another solution $(\mathcal{M}', \pi \circ \Psi)$. For example, we could *always* take $\mathcal{M}'$ to equal $\mathrm{Spec}(\mathbb{C})$ and $\Psi(\varphi)$ to be the unique morphism $B \to \mathrm{Spec}(\mathbb{C})$ and then our dictionary would have only blank pages; or, we could take the disjoint union of the "right" $\mathcal{M}$ with any other scheme. We can rule such cases out by requiring that the complex points of $\mathcal{M}$ correspond bijectively to the objects of our moduli problem. This still doesn't fix the scheme structure on $\mathcal{M}$: it leaves us the freedom to compose, as above, with a map $\pi : \mathcal{M} \to \mathcal{M}'$ as long as π itself is bijective on complex points. For example, we would certainly want the moduli space $\mathcal{M}$ of lines through the origin in $\mathbb{C}^2$ to be $\mathbb{P}^1$ but our requirements so far don't exclude the possibility of taking instead the cuspidal rational curve $\mathcal{M}'$ with equation $y^2 z = x^3$ in $\mathbb{P}^2$ which is the image of $\mathbb{P}^1$ under the map $[a, b] \to [a^2 b, a^3, b^3]$. This pathology can be eliminated by requiring that $\mathcal{M}$ be universal with respect to the existence of the natural transformation Ψ: cf. the first exercise below. When all this holds, we say that $(\mathcal{M}, \Psi)$, or more frequently $\mathcal{M}$, is a *coarse moduli space* for the functor F. Formally,

DEFINITION (1.3) *A scheme $\mathcal{M}$ and a natural transformation $\Psi_{\mathcal{M}}$ from the functor F to the functor of points $\mathrm{Mor}_{\mathcal{M}}$ of $\mathcal{M}$ are a coarse moduli space for the functor F if*

1) *The map $\Psi_{\mathrm{Spec}(\mathbb{C})} : F(\mathrm{Spec}(\mathbb{C})) \to \mathcal{M}(\mathbb{C}) = \mathrm{Mor}(\mathrm{Spec}(\mathbb{C}), \mathcal{M})$ is a set bijection.[2]*

2) *Given another scheme $\mathcal{M}'$ and a natural transformation $\Psi_{\mathcal{M}'}$ from $F \to \mathrm{Mor}_{\mathcal{M}'}$, there is a unique morphism $\pi : \mathcal{M} \to \mathcal{M}'$ such that*

---

[2] Or more generally require this with $\mathbb{C}$ replaced by any algebraically closed field.

the associated natural transformation $\Pi : \mathrm{Mor}_{\mathcal{M}} \to \mathrm{Mor}_{\mathcal{M}'}$ *satisfies* $\Psi_{\mathcal{M}'} = \Pi \circ \Psi_{\mathcal{M}}$.

EXERCISE (1.4) Show that, if one exists, a coarse moduli scheme $(\mathcal{M}, \Psi)$ for F is determined up to canonical isomorphism by condition 2) above.

EXERCISE (1.5) Show that the cuspidal curve $\mathcal{M}'$ defined above is *not* a coarse moduli space for lines in $\mathbb{C}^2$. Show that $\mathbb{P}^1$ is a fine moduli space for this moduli problem. What is the universal family of lines over $\mathbb{P}^1$?

EXERCISE (1.6) 1) Show that the j-line $\mathcal{M}_1$ is a coarse moduli space for curves of genus 1.
2) Show that a j-function J on a scheme B arises as the j-function associated to a family of curves of genus 1 only if all the multiplicities of the zero-divisor of J are divisible by 3, and all multiplicities of $(J - 1728)$ are even. Using this fact, show that $\mathcal{M}_1$ is not a fine moduli space for curves of genus 1.
3) Show that the family $y^2 - x^3 - t$ over the punctured affine line $\mathbb{A}^1 - \{0\}$ with coordinate t has constant j, but is not trivial. Use this fact to give a second proof that $\mathcal{M}_1$ is not a fine moduli space.

The next exercise gives a very simple example which serves two purposes. First, it shows that the second condition on a coarse moduli space above doesn't imply the first. Second, it shows that even a coarse moduli space may fail to exist for some moduli problems. All the steps in this exercise are trivial; its point is to give some down-to-earth content to the rather abstract conditions above and working it involves principally translating these conditions into English.

EXERCISE (1.7) Consider the moduli problem F posed by "flat families of reduced plane curves of degree 2 up to isomorphism". The set $F(\mathrm{Spec}(\mathbb{C}))$ has two elements: a smooth conic and a pair of distinct lines.
1) Show (trivially) that there is a natural transformation Ψ from F to $\mathrm{Mor}(\cdot, \mathrm{Spec}(\mathbb{C}))$.
Now fix any pair (X, Ψ') where X is a scheme and Ψ' is a natural transformation from F to $\mathrm{Mor}(\cdot, X)$.
2) Show that, if $\varphi : \mathcal{C} \to B$ is any family of smooth conics, then there is a unique $\mathbb{C}$-valued point $\pi : \mathrm{Spec}(\mathbb{C}) \to X$ of X such that $\Psi'(\varphi) = \pi \circ \Psi(\varphi)$.
3) Let $\varphi : \mathcal{C} \to \mathbb{A}^1_t$ be the family defined by the (affine) equation $xy - t$ and φ' be its restriction to $\mathbb{A}^1 - \{0\}$. Use the fact that φ' is a family of smooth conics to show that $\Psi'(\varphi) = \pi \circ \Psi(\varphi)$.

4) Show that the pair $(\text{Spec}(\mathbb{C}),\ \Psi)$ has the universal property in 2) above but does *not* satisfy 1). Use Exercise (1.4) to conclude that there is no coarse moduli space for the functor **F**.

We conclude by introducing one somewhat vague terminological dichotomy which is nonetheless quite useful in practice. We would like to distinguish between problems that focus on purely intrinsic data and those that involve, to a greater or lesser degree, extrinsic data. We will reserve the term *moduli space* principally for problems of the former type and refer to the classifying spaces for the latter (which until now we've also been calling moduli spaces) as *parameter spaces*. In this sense, the space $\mathcal{M}_g$ of smooth curves of genus g is a moduli space while the space $\mathcal{H}_{d,g,r}$ of subcurves of $\mathbb{P}^r$ of degree d and (arithmetic) genus g is a parameter space. The extrinsic element in the second case is the g_d^r that maps the abstract curve to $\mathbb{P}^r$ and the choice of basis of this linear system that fixes the embedding. Of course, this distinction depends heavily on our point of view. The space G_d^r classifying the data of a curve plus a g_d^r (without the choice of a basis) might be viewed as either a moduli space or a parameter space depending on whether we wish to focus primarily on the underlying curve or on the curve plus the g_d^r. One sign that we're dealing with a parameter space is usually that the equivalence relation by which we quotient the geometric data of the problem is trivial; e.g., for $\mathcal{M}_g$ this relation is "biregular isomorphism" while for $\mathcal{H}_{d,g,r}$ it is trivial.

Heuristically, parameter spaces are easier to construct and more likely to be fine moduli spaces because the extrinsic extra structure involved tends to rigidify the geometric data they classify. On the other hand, *complete* parameter spaces can usually only be formed at the price of allowing the data of the problem to degenerate rather wildly while complete — even compact — moduli spaces can often be found for fairly nice classes of objects. In the next sections, we'll look at the Hilbert scheme, a fine parameter space, which provides the best illustration of the parameter space side of this philosophy.

B Construction of the Hilbert scheme

The Hilbert scheme is an answer to the problem of parameterizing subschemes of a fixed projective space $\mathbb{P}^r$. In the language of the preceding section, we might initially look for a scheme $\mathcal{H}$ which is a fine parameter space for the functor whose "data" for a scheme B consists of all proper, connected, families of subschemes of $\mathbb{P}^r$ defined over B. This functor, however, has two drawbacks. First, it's too large to give us a parameter space of finite type since it allows hypersurfaces of all degrees. Second, it allows families whose fibers vary so wildly that, like the example in Exercise (1.7), it cannot even be coarsely represented. To solve the first problem, we would like to fix the principal numerical invariants of the subschemes. We can solve the second by restricting our attention to flat families which, loosely, means requiring that the fibers vary "continuously". Both problems can thus be resolved simultaneously by considering only families with constant Hilbert polynomial.

Recall that the Hilbert polynomial of a subscheme X of $\mathbb{P}^r$ is a numerical polynomial characterized by the equations $P_X(m) = h^0(X, \mathcal{O}_X(m))$ for all sufficiently large m. If X has degree d and dimension s, then the leading term of $P_X(m)$ is $dm^s/s!$: cf. Exercise (1.13). This shows both that P_X captures the main numerical invariants of X, and that fixing it yields a set of subschemes of reasonable size. Moreover, if a proper connected family $\mathcal{X} \to B$ of such subschemes is flat, then the Hilbert polynomials of all fibers of $\mathcal{X}$ are equal, and, if B is reduced, then the converse also holds. Thus, fixing P_X also forces the fibers of the families we're considering to vary nicely.

Intuitively, the Hilbert scheme $\mathcal{H}_{P,r}$ parameterizes subschemes X of $\mathbb{P}^r$ with fixed Hilbert polynomial P_X equal to P: More formally, it's a fine moduli space for the functor $\mathbf{Hilb}_{P,r}$ whose value on B is the set of proper *flat* families

$$
(1.8) \qquad
\begin{array}{ccccc}
\mathcal{X} & \xrightarrow{\;i\;} & \mathbb{P}^r \times B & \xrightarrow{\;\pi_{\mathbb{P}^r}\;} & \mathbb{P}^r \\
& \searrow{\varphi} & \downarrow{\pi_B} & & \\
& & B & &
\end{array}
$$

with $\mathcal{X}$ having Hilbert polynomial P. The basic fact about it is:

THEOREM (1.9) (GROTHENDIECK [67]) *The functor* $\mathbf{Hilb}_{P,r}$ *is representable by a projective scheme* $\mathcal{H}_{P,r}$.

The idea of the proof is essentially very simple. We'll sketch it, but we'll only give statements of the two key technical lemmas whose proofs are both somewhat nontrivial. For more details we refer you to the recent book of Viehweg [148], Mumford's notes [120] or Grothendieck's original Seminaire Bourbaki talk [67]. First some notation: it'll be convenient to let $S = \mathbb{C}[x_0, \dots, x_r]$ and to let $O_r(m)$ denote the Hilbert polynomial of $\mathbb{P}^r$ itself (i.e.,

$$
(1.10) \qquad O_r(m) = \binom{r+m}{m} = \dim(S_m)
$$

is the number of homogeneous polynomials of degree m in ($r + 1$) variables) and to let $Q(m) = O_r(m) - P(m)$. For large m, $Q(m)$ is then the dimension of the degree m piece $I(X)_m$ of the ideal of X in $\mathbb{P}^r$.

The subscheme X is determined by its ideal $I(X)$ which in turn is determined by its degree m piece $I(X)_m$ for any sufficiently large m. The first lemma asserts that we can choose a single m that has this property uniformly for every subscheme X with Hilbert polynomial P.

LEMMA (1.11) (UNIFORM m LEMMA) *For every P, there is an m_0 such that if $m \geq m_0$ and X is a subscheme of $\mathbb{P}^r$ with Hilbert polynomial P, then:*

1) *$I(X)_m$ is generated by global sections and $I(X)_{l \geq m}$ is generated by $I(X)_m$ as an S-module.*

2) *$h^i(X, I_X(m)) = h^i(X, \mathcal{O}_X(m)) = 0$ for all $i > 0$.*

3) *$\dim(I(X)_m) = Q(m)$, $h^0(X, \mathcal{O}_X(m)) = P(m)$ and the restriction map $r_{X,m} : S_m \rightarrow H^0(X, \mathcal{O}_X(m))$ is surjective.*

The key idea of the construction is that the lemma allows us to associate to *every* subscheme X with Hilbert polynomial P the point $[X]$ of the Grassmannian $G = \mathbb{G}(P(m), O_r(m))$ determined by $r_{X,m}$.[3] More formally again, if $\varphi : X \rightarrow B$ is any family as in (1.8), then from the sheafification of the restriction maps

$$(\pi_\mathbb{P})^*(\mathcal{O}_{\mathbb{P}^r}(m)) \longrightarrow (\pi_\mathbb{P})^*(\mathcal{O}_{\mathbb{P}^r}(m) \otimes \mathcal{O}_X) \longrightarrow 0$$

we get a second surjective restriction map

$$(\pi_B)_*(\pi_\mathbb{P})^*(\mathcal{O}_{\mathbb{P}^r}(m)) \longrightarrow (\pi_B)_*(\pi_\mathbb{P})^*(\mathcal{O}_{\mathbb{P}^r}(m) \otimes \mathcal{O}_X) \longrightarrow 0 .$$
$$\|$$
$$\mathcal{O}_B \otimes S_m$$

The middle factor is a locally free sheaf of rank $P(m)$ on B and therefore yields a map $\Psi(\varphi) : B \rightarrow G$. Since these maps are functorial in B, we have a natural transformation Ψ to the functor of points of some subscheme $\mathcal{H} = \mathcal{H}_{P,r}$ of G.

It remains to identify $\mathcal{H}$ and to show it represents the functor **Hilb**$_{P,r}$. The key to doing so is provided by the universal subbundle $\mathcal{F}$ whose fiber over $[X]$ is $I(X)_m$ and the multiplication maps

$$\times_k : \mathcal{F} \otimes S_k \rightarrow S_{k+m}.$$

---

[3] Or, equivalently, for those who prefer their Grassmannians to parameterize *subspaces* of the ambient space, the point in $G = \mathbb{G}(Q(m), O_r(m))$ determined by $I(X)_m$.

LEMMA (1.12) *The conditions that* $\mathrm{rank}(\times_k) \le Q(m + k)$ *for all* $k \ge 0$ *define a determinantal subscheme* $\mathcal{H}$ *of* G *and a morphism* $\psi : B \rightarrow G$ *arises by applying the construction above to a family* $\varphi : X \rightarrow B$ *(i.e.,* $\psi = \Psi(\varphi)$*) if and only if* ψ *factors through this subscheme* $\mathcal{H}$.

Grothendieck's theorem follows immediately. By definition, $\mathcal{H}$ is a closed subscheme of G (and hence in particular projective). The second sentence of the lemma is just another way of expressing the condition that the transformation Ψ is an isomorphism of functors between **Hilb**$_{P,r}$ and the functor of points of $\mathcal{H}$.

A few additional remarks about the lemmas are nonetheless in order. When we feel that no confusion will result, we'll often elide the words "the Hilbert point of". Most commonly this allows us to say that "the variety X lies in" a subscheme of a Hilbert scheme when we mean that "the Hilbert point $[X]$ of the variety X lies in" this locus. More generally, we'll use the analogous elision when discussing loci in other parameter and moduli spaces. In our experience, everyone who works a lot with such spaces soon acquires this lazy but harmless vice.

For a fixed X, the existence of an m_0 with the properties of the Uniform m lemma is a standard consequence of Serre's FAC theorems [138]. The same ideas, when applied with somewhat greater care, yield the uniform bound of the lemma. A natural question is: what is the minimal value of m_0 that can be taken for a given P and r? The answer is that the worst possible behavior is exhibited by the combinatorially defined subscheme X_{lex} defined by the *lexicographical ideal*. With respect to a choice of an ordered system of homogeneous coordinates $(x_0, \ldots, x_r)$ on $\mathbb{P}^r$, this is the ideal whose degree m piece is spanned by the $Q(m)$ monomials that are greatest in the lexicographic order. This ideal exhibits many forms of extreme behavior. For example, its Hilbert function $h^0(X, \mathcal{O}_X(m))$ attains the maximum possible value in every (and not just in every sufficiently large) degree. For more details, see [13].

Second, we may also ask what values of k it is necessary to consider in the second lemma. A priori, it's not even clear that the *infinite* set of conditions $\mathrm{rank}(\times_k) \le Q(m + k)$ define a scheme. A key step in the proof of the lemma is to show that the *supports* of the ideals I_K generated by the conditions $\mathrm{rank}(\times_k) \le Q(m + k)$ for $k \le K$ stabilize for large K. This is done by using the first lemma to show that, if enough of these equalities hold, then $\mathrm{rank}(\times_k)$ is itself represented by a polynomial of degree r which can only be $Q(m+k)$. It then follows by noetherianity that for some possibly larger K the ideals I_K stabilize and hence that $\mathcal{H}$ is a scheme. A more careful analysis shows that if m is at least the m_0 of the first lemma and J is *any* $Q(m)$-dimensional subspace of S_m, then the dimension of the subspace $\times_k(J \otimes S_k)$ of S_{k+m} is *at least* $Q(k + m)$. Moreover, equality can hold for *any* $k > 0$

only if J is actually the degree m piece of the ideal of a variety X with Hilbert polynomial P. So $\mathcal{H}$ is actually defined by the equations $\text{rank}(\times_1) \le Q(m+1)$. For details, see [63].

The next three exercises show that Hilbert schemes of hypersurfaces and of linear subspaces are exactly the familiar parameter spaces for these objects. For concreteness, the exercises treat special cases but the arguments generalize in both cases.

EXERCISE (1.13) 1) Use Riemann-Roch to show that, if $X \subset \mathbb{P}^r$ has degree d and dimension s, then the leading term of $P_X(m)$ is $(\frac{d}{s!})m^s$.
2) Fix a subscheme $X \subset \mathbb{P}^r$. Show, by taking cohomology of the exact sequence of $X \subset \mathbb{P}^r$, that X is a hypersurface of degree d if and only if

$$P_X(m) = \binom{r+m}{m} - \binom{r+m-d}{m-d}.$$

3) Show that X is a linear space of dimension s if and only if

$$P_X(m) = \binom{s+m}{m}.$$

EXERCISE (1.14) Show that the Hilbert scheme of lines in $\mathbb{P}^3$ (that is, the Hilbert scheme of subschemes of $\mathbb{P}^3$ with Hilbert polynomial $P(m) = m+1$) is indeed the Grassmannian $G = \mathbb{G}(1,3)$. *Hint*: Recall that G comes equipped with a universal rank 2 subbundle $S_G \subset \mathcal{O}_G^4$. The universal line over G is the projectivization of S_G. Conversely, given any family $\varphi : \mathcal{X} \rightarrow B$ of lines in $\mathbb{P}^3$, we get an analogous subbundle $S_B \subset \mathcal{O}_B^4$ by $S_B = \varphi_*(\mathcal{O}_X(1))^\vee \subset H^0(\mathbb{P}^3, \mathcal{O}_{\mathbb{P}^3}(1)) \otimes \mathcal{O}_B \cong \mathcal{O}_B^4$. Check, on the one hand, that the projectivization of this inclusion yields the original family $\varphi : \mathcal{X} \rightarrow B$ in $\mathbb{P}^3$ and, on the other, that the standard universal property of G realizes this subbundle as the pullback of the universal subbundle by a unique morphism $\chi : B \rightarrow G$. Then apply Exercise (1.4).

EXERCISE (1.15) This exercise checks that the Hilbert scheme of plane curves of degree d is just the familiar projective space of dimension $N = d(d+3)/2$ whose elements correspond to polynomials f of degree d up to scalars.
1) Show that the incidence correspondence

$$\mathcal{C} = \{(f,P)\,|\,f(P) = 0\} \subset \mathbb{P}^N \times \mathbb{P}^2$$

is flat over $\mathbb{P}^N$.

The plan of attack is clear: to show that the projection $\pi : \mathcal{C} \rightarrow \mathbb{P}^N$ is the universal curve. To this end, let $\varphi : \mathcal{X} \rightarrow B$ be a flat family of plane curves over B and $\mathcal{I}$ be the ideal sheaf of $\mathcal{X}$ in $\mathbb{P}^2 \times B$.

2) Show that $\mathcal{I}$ is flat over B. *Hint*: Apply the fact that a coherent sheaf $\mathcal{F}$ on $\mathbb{P}^r \times B$ is flat over B if and only if, for large m, $(\pi_B)_*(\mathcal{F}(m))$ is locally free to the twists of the exact sheaf sequence of X in $\mathbb{P}^2 \times B$.

3) Show that $(\pi_B)_*(\mathcal{I}(d))$ is a line bundle on B and that the associated linear system gives a morphism $\chi : B \longrightarrow \mathbb{P}^N$.

4) Show that $\varphi : \mathcal{X} \longrightarrow B$ is the pullback via χ of the universal family $\pi : \mathcal{C} \longrightarrow \mathbb{P}^N$. Then use the universal property of projective space to show that χ is the unique map with this property.

We should warn you that these two examples are rather misleading: in both cases, the Hilbert schemes parameterize only the "intended" subschemes (linear spaces in the first case, and hypersurfaces in the second). Most Hilbert schemes largely parameterize projective schemes that you would prefer to avoid. The reason is that, in contrast to the conclusions in Exercise (1.13), the Hilbert polynomial of a "nice" (e.g., smooth, irreducible) subscheme of $\mathbb{P}^r$ is usually also the Hilbert polynomial of many nasty (nonreduced, disconnected) subschemes too. The *twisted cubics* — rational normal curves in $\mathbb{P}^3$ that have Hilbert polynomial $P_X(m) = 3m+1$ — give the simplest example: a plane cubic plus an isolated point has the same Hilbert polynomial. We will look, in more detail, at this example and many others in the next few sections.

A natural question is: what is the relationship between the Hilbert scheme and the more elementary Chow variety which parameterizes cycles of fixed degree and dimension in $\mathbb{P}^r$? The answer is that they are generally very different. The most important difference is that the Hilbert scheme has a natural scheme structure whereas the Chow variety does not.[4] This generally makes the Hilbert scheme more useful. It is the source of the universal properties on which we'll rely heavily later in this book and one reflection is that the Hilbert scheme captures much finer structure. Here is a first example.

EXERCISE (1.16) Let $C \subset \mathbb{P}^3$ be the union of a plane quartic and a noncoplanar line meeting it at one point. Show that C is not the flat specialization of a smooth curve of degree 5. What if C is the union of the quartic and a noncoplanar conic meeting it at two points?

---

[4]We should note that several authors have produced scheme structures on the Chow variety: the most complete treatment is in Sections I.3-5 of [100] which gives an overview of alternate approaches. However, the most natural scheme structures don't represent functors in positive characteristics. This means many aspects of Hilbert schemes have no analogue for Chow schemes, most significantly, the characterization of the tangent space in Section C and the resulting ability to work infinitesimally on it.

There are a number of useful variants of the Hilbert scheme whose existence can be shown by similar arguments.[5]

DEFINITION (1.17) (Hilbert schemes of subschemes) *Given a subscheme Z of* $\mathbb{P}^r$, *we can define a closed subscheme* $\mathcal{H}^Z_{P,r}$ *of* $\mathcal{H}_{P,r}$ *parameterizing subschemes of Z that are closed in* $\mathbb{P}^r$ *and have Hilbert polynomial P.*

DEFINITION (1.18) (Hilbert schemes of maps) *If* $X \subset \mathbb{P}^r$ *and* $Y \subset \mathbb{P}^s$, *there is a Hilbert scheme* $\mathcal{H}_{X,Y,d}$ *parameterizing polynomial maps* $f : X \to Y$ *of degree at most d. This variant is most easily constructed as a subscheme of the Hilbert scheme of subschemes of* $X \times Y$ *in* $\mathbb{P}^r \times \mathbb{P}^s$ *using the Hilbert points of the graphs of the maps f.*

DEFINITION (1.19) (Hilbert schemes of projective bundles) *From a* $\mathbb{P}^r$ *bundle* $\mathcal{P}$ *over Z, we can construct a Hilbert scheme* $\mathcal{H}_{P,\mathcal{P}/Z}$ *parameterizing subschemes of* $\mathcal{P}$ *whose fibers over Z all have Hilbert polynomial P.*

DEFINITION (1.20) (Relative Hilbert schemes) *Given a projective morphism* $\pi : X \to Z \times \mathbb{P}^r \to Z$, *we have a relative Hilbert scheme* $\mathcal{H}$ *parameterizing subschemes of the fibers of* π. *Explicitly,* $\mathcal{H}$ *represents the functor that associates to B the set of subschemes* $\mathcal{Y} \subset B \times \mathbb{P}^r$ *and morphisms* $\alpha : B \to Z$ *such that* $\mathcal{Y}$ *is flat over B with Hilbert polynomial P and* $\mathcal{Y} \subset B \times_Z X$.

The following is an application of the fact that Hilbert schemes of morphisms exist and are quasiprojective.

EXERCISE (1.21) Show that for any $g \geq 3$ there is a number $\varphi(g)$ such that any smooth curve C of genus g has at most $\varphi(g)$ nonconstant maps to curves B of genus $h \geq 2$.

One warning about these variants is in order: the notion of scheme "of type X" needs to be handled with caution. For example, look at the following types of subschemes of $\mathbb{P}^2$:

1. Plane curves of degree d;

2. Reduced and irreducible plane curves of degree d;

3. Reduced and irreducible plane curves of degree d and geometric genus g; and,

---

[5]Perhaps, more accurately, in view of our omissions, by citing similar arguments.

    4. Reduced and irreducible plane curves of degree d and geometric genus g having only nodes as singularities.

The first family is parameterized by the Hilbert scheme $\mathcal{H}$, which we have seen in the second exercise above is simply a projective space $\mathbb{P}^N$. The second is parameterized by an open subset $W_d \subset \mathbb{P}^N$. The last one also may be interpreted in such a way that it has a fine moduli space, which is a closed subscheme $U_{d,g} \subset W_d$.

    The third, however, does not admit a nice quasiprojective moduli space at all. It is possible to define the notion of a family of curves with δ nodes over an arbitrary base — so that, for example, the family $xy - \varepsilon$ has no nodes over $\mathrm{Spec}(\mathbb{C}[\varepsilon]/\varepsilon^2)$ — but it's harder to make sense of the notion of geometric genus over nonreduced bases. For families of nodal curves, we can get around this by using the relation $g + \delta = (d-1)(d-2)/2$. One way out is to first define the moduli space $V_{d,g}$ to be the *reduced* subscheme of W_d whose support is the set of reduced and irreducible plane curves of degree d and geometric genus g, and to then consider *only* families of such curves with base B that come equipped with a map $B \to V_{d,g}$. In other words, we could let the moduli space define the moduli problem rather than the other way around. Unfortunately, this approach is generally unsatisfactory because we'll almost always want to consider families that don't meet this condition.

C   Tangent space to the Hilbert scheme

Let $\mathcal{H}$ be the Hilbert scheme parameterizing subschemes of $\mathbb{P}^r$ with Hilbert polynomial P. One significant virtue of the fact that $\mathcal{H}$ represents a naturally defined functor is that it's relatively easy to describe the tangent space to $\mathcal{H}$. Before we do this, we want to set up a few general notions. Recall that the tangent space to any scheme X at a closed point p is just the set of maps $\mathrm{Spec}(\mathbb{C}[\varepsilon]/\varepsilon^2) \to X$ centered at p (that is, mapping the unique closed point $\mathbf{0}$ of $\mathrm{Spec}(\mathbb{C}[\varepsilon]/\varepsilon^2)$ to p). We will write $\mathbb{I}$ for $\mathrm{Spec}(\mathbb{C}[\varepsilon]/\varepsilon^2)$. More generally, we let $\mathbb{I}_k = \mathrm{Spec}(\mathbb{C}[\varepsilon]/(\varepsilon^{k+1}))$ and more generally still

(1.22) $$\mathbb{I}_k^{(l)} = \mathrm{Spec}(\mathbb{C}[\varepsilon_1, \ldots, \varepsilon_l]/(\varepsilon_1, \ldots, \varepsilon_l)^{k+1}),$$

with the convention, already used above, that k and l are suppressed when they are equal to 1.

    If you're unused to this scheme-theoretic formalism, you may wonder: if a tangent vector to a scheme X corresponds to a morphism $\mathbb{I} \to X$, how do we add them? The answer is that two morphisms $\mathbb{I} \to X$ that agree on the subscheme $\mathrm{Spec}(\mathbb{C}) \subset \mathbb{I}$ (i.e., both map it to the same point p) give a morphism from the fibered sum of $\mathbb{I}$ with itself over $\mathrm{Spec}(\mathbb{C})$ to X. But this fibered sum is just $\mathbb{I}^{(2)}$, and we have a sort of "diagonal" inclusion Δ of $\mathbb{I}$ in $\mathbb{I}^{(2)}$ induced by the map of rings $\mathbb{C}[\varepsilon_1, \varepsilon_2]/(\varepsilon_1, \varepsilon_2)^2 \to \mathbb{C}[\varepsilon]/(\varepsilon^2)$ sending both ε_1 and ε_2 to ε; the composition $\pi \circ \Delta$ shown in diagram (1.23) is the sum of the tangent vectors.

(1.23)

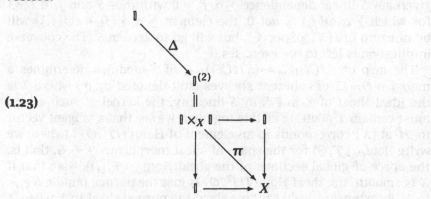

We're now ready to unwind these definitions for Hilbert schemes. Most directly, if $\mathcal{H}$ is a Hilbert scheme and $[X] \in \mathcal{H}$ corresponds to the subscheme $X \subset \mathbb{P}^r$, then by the universal property of $\mathcal{H}$ a map from $\mathbb{I}$ to $\mathcal{H}$ centered at $[X]$ corresponds to a flat family $\mathcal{X} \to \mathbb{I}$ of subschemes of $\mathbb{P}^r \times \mathbb{I}$ whose fiber over $0 \in \mathrm{Spec}(\mathbb{C}[\varepsilon]/\varepsilon^2)$ is X. Such a family is called a *first-order deformation* of X. We will look at such deformations in more detail in Chapter 3.

For the time being, however, there is another way to view its tangent space that is much more convenient for computations. This approach is based on the fact that $\mathcal{H}$ is naturally a subscheme of the Grassmannian G of codimension $P(m)$-dimensional quotients of S_m. Recall that any tangent vector to G at the point $[Q]$ corresponding to the quotient Q of S_m by a subspace L of codimension $P(m)$ in S_m can be identified with a $\mathbb{C}$-linear map $\varphi : L \to S_m/L$. If $\tilde{\varphi} : L \to S_m$ is any lifting of φ, then the collection $\{f + \varepsilon \cdot \tilde{\varphi}(f)\}_{f \in I(X)_m}$ yields the map from $\mathbb{I}$ to G associated to φ. Suppose that $L = I(X)_m$ or, in other words, that the point $[Q]$ is the Hilbert point $[X]$ of a subscheme of $\mathbb{P}^r$ with Hilbert polynomial P and φ is given by a map $I(X)_m \to (S/I(X)_m)$. Then we may view the collection $\{f + \varepsilon \cdot \tilde{\varphi}(f)\}_{f \in I(X)_m}$ as polynomials defining a subscheme $\mathcal{X} \subset \mathbb{I} \times \mathbb{P}^r$. The universal property of the Hilbert scheme implies that such a tangent vector to G will lie in the Zariski tangent space to the subscheme $\mathcal{H}$ if and only if $\mathcal{X}$ is flat over $\mathbb{I}$.

What does the condition of flatness mean in terms of the linear map φ? This is also easy to describe and verify: $\mathcal{X}$ will be flat over $\mathbb{I}$ if and only if the map φ extends to an S-module homomorphism $I(X)_{l \geq m} \to (S/I(X))_{l \geq m}$ (which we will also denote φ). For example, if this condition is *not* satisfied, we claim that the exact sequence of $S \otimes \mathbb{C}[\varepsilon]/\varepsilon^2$ modules

$$0 \longrightarrow I(X) \longrightarrow S \otimes \mathbb{C}[\varepsilon]/\varepsilon^2 \longrightarrow A(X) \longrightarrow 0$$

will fail to be exact after we tensor with the $\mathbb{C}[\varepsilon]/\varepsilon^2$-module $\mathbb{C}$. Indeed, given any S-linear dependence $\sum \alpha_i f_i = 0$ with $\alpha_i \in S$ and $f_i \in I(X)$ for which $\sum \alpha_i \varphi(f_i)$ is *not* 0, the element $\sum \alpha_i \cdot (f_i + \varepsilon \varphi(f_i))$ will be nonzero in $I(X) \otimes \mathrm{Spec}(\mathbb{C})$, but will go to zero in S. The converse implication is left to the exercises.

The map $\varphi : I(X)_{l \geq m} \rightarrow (S/I(X))_{l \geq m}$ of S-modules determines a map $\mathcal{I} \rightarrow \mathcal{O}_{\mathbb{P}^n}/\mathcal{I}$ of coherent sheaves (still denoted by φ) where $\mathcal{I}$ is the ideal sheaf of X in $\mathbb{P}^n$. By S-linearity, the kernel of such a map must contain $\mathcal{I}^2$. Putting all this together, we see that a tangent vector to $\mathcal{H}$ at $[X]$ corresponds to an element of $\mathrm{Hom}(\mathcal{I}/\mathcal{I}^2, \mathcal{O}_X)$ (where we write $\mathrm{Hom}_{\mathcal{O}_C}(\mathcal{F}, \mathcal{G})$ for the space of sheaf morphisms $\mathcal{F} \rightarrow \mathcal{G}$, that is, the space of global sections of the sheaf $\underline{\mathrm{Hom}}_{\mathcal{O}_C}(\mathcal{F}, \mathcal{G})$). Note that if X is smooth, the sheaf $\underline{\mathrm{Hom}}(\mathcal{I}/\mathcal{I}^2, \mathcal{O})$ is just the normal bundle $N_{X/\mathbb{P}^r}$ to X. By extension, we'll call this sheaf the normal sheaf to X when X is singular (or even nonreduced). With this convention, the upshot is that the *Zariski tangent space to the Hilbert scheme at a point X is the space of global sections of the normal sheaf of X*:

(1.24) $$T_{[X]}\mathcal{H} = H^0(X, N_{X/\mathbb{P}^r}).$$

EXERCISE (1.25) Verify that the family $\mathcal{X} \subset \mathbb{P}^r \times \mathrm{Spec}(\mathbb{C}[\varepsilon]/\varepsilon^2)$ induced by an S-linear map $\varphi : I(X)_{l \geq m} \rightarrow (S/I(X))_{l \geq m}$ is indeed flat as claimed.

EXERCISE (1.26) Determine the normal bundle to the rational normal curve $C \subset \mathbb{P}^r$ and show, by computing its h^0, that the Hilbert scheme parameterizing such curves is smooth at any point corresponding to a rational normal curve.

EXERCISE (1.27) Similarly, show that the Hilbert scheme parameterizing elliptic normal curves is smooth at any point corresponding to an elliptic normal curve.

Warning. As we remarked in the last section, the Hilbert scheme, by definition, parameterizes a lot of things you weren't particularly eager to have parameterized. The examples that we'll look at in the next sections will make this point painfully clear. For now, let's return to the example of twisted cubics. These form a twelve-dimensional family parameterized by a component $\mathcal{D}$ of the Hilbert scheme $\mathcal{H}_{3m+1,3}$ of curves in $\mathbb{P}^3$ with Hilbert polynomial $3m + 1$. But $\mathcal{H}$ also has a second irreducible component $\mathcal{E}$, whose general member is the union of a plane cubic and an isolated point: this component has dimension 15. A general point of the intersection corresponds to a nodal plane cubic with an embedded point at the node, and at such a point the dimension of the Zariski tangent space to $\mathcal{H}$ is necessarily larger than 15. In particular, it's hard to tell whether the component $\mathcal{D} \subset \mathcal{H}$ whose general member is a twisted cubic — the component we're most likely to be interested in — is smooth at such a point. That both components are, in fact, smooth, has only recently been established by Piene and Schlessinger [130]. We will return to this point in Chapter 3. The exercises that follow establish some easier facts which will be needed then.

EXERCISE (1.28) Verify that the tangent space to $\mathcal{H}$ at a general point $[X]$ of intersection of the two components of $\mathcal{H}$ has dimension 16. *Hint*: In this example, the minimum degree m that has the properties needed in the construction of $\mathcal{H}$ is 4 and it's probably easiest to explicitly calculate the space of $\mathbb{C}$-linear maps $\varphi : I(X)_4 \to (S/I(X))_4$ that kill $I(X)^2$.

A theme that will be important in later chapters is the use of the natural $\mathrm{PGL}(r+1)$-action on Hilbert schemes of subschemes of $\mathbb{P}^r$. In the Hilbert scheme $\mathcal{H}$ of twisted cubics, this can be used to considerable effect because each component has a single open orbit, namely, that of the generic element. Hence there are only finitely many orbits. Since, by construction, the Hilbert scheme is invariant for the natural $\mathrm{PGL}(r+1)$-action on $\mathcal{G}$, its singular loci are also invariant (i.e., unions of orbits) and can be analyzed completely.

EXERCISE (1.29) 1) Use the Borel-fixed point theorem to show that every subscheme of $\mathbb{P}^r$ has a flat specialization that is fixed by the standard Borel subgroup of upper triangular matrices. Conclude that every component of a Hilbert scheme $\mathcal{H}$ contains a point parameterizing a Borel-fixed subscheme.

2) Show that there are exactly three Borel-fixed orbits in $\mathcal{H} = \mathcal{H}_{3m+1,3}$:

- a spatial double line in $\mathbb{P}^3$ (that is, the scheme C defined by the square of the ideal of a line in $\mathbb{P}^3$);
- a planar triple line plus an embedded point lying in the same plane as the line;
- a planar triple line plus an embedded point *not* lying in the same plane as the line.

3) Show also that these orbits lie in $\mathcal{D}$ only, in $\mathcal{E}$ only and in $\mathcal{D} \cap \mathcal{E}$ respectively. Conclude that $\mathcal{H}$ has exactly two components.

4) Show that the tangent space to $\mathcal{H}$ at points of each of the three orbits in 2) is of dimension 12, 15 and 16 respectively and that in each case the normal sheaf has vanishing h^1.

5) Show that the Hilbert scheme $\mathcal{H}$ of twisted cubics contains finitely many PGL(4)-orbits. How many lie in $\mathcal{D}$ alone? in $\mathcal{E}$ alone? in $\mathcal{D} \cap \mathcal{E}$?

A few remarks about this example are in order. First, the lexicographic ideal of $\mathcal{H}$ (whose degree m piece consists of the first $\dim(S_m) - P(m)$ monomials in the lexicographic order) defines a planar triple line plus a coplanar embedded point. Note that this scheme isn't a specialization of the twisted cubic and that the minimal m_0 satisfying the hypotheses of the Uniform m lemma (1.11) for this scheme is 4. On the other hand, an inspection of the ideals of curves in the list from 2) of the preceding exercise shows that $m_0 = 3$ works for every orbit in the "good" component of $\mathcal{D}$. In general, the least m_0 that can be used in the construction will be much greater than the least m_0 that works for ideals of smooth (or even reduced) subschemes with the given Hilbert polynomial.

This annoying discrepancy is unfortunately just about the only way in which $\mathcal{H}$ is a typical Hilbert scheme. The existence of *any* smooth component of a Hilbert scheme (even those parameterizing complete intersections) is extremely rare.

EXERCISE (1.30) Generalize the scheme C in the preceding exercise to a multiple line which is a flat specialization of a rational normal curve in $\mathbb{P}^r$ and show that for $r > 3$ the corresponding Hilbert scheme is *not* smooth at $[C]$.

How else is the twisted cubic example misleadingly simple? Components of the Hilbert scheme whose general member isn't connected (let alone irreducible) are in fact the rule rather than the exception. For example, in the Hilbert scheme $\mathcal{H}_{d,g,r}$ of curves of degree d and genus g in $\mathbb{P}^r$, there will be component(s) $\mathcal{C}_{d,g',r}$ whose general element C consists of a curve of geometric genus $g' > g$ plus $(g' - g)$ points (so that $p_a(C) = g$ and C has the "correct" Hilbert polynomial $P(m) = md - g + 1$). Worse yet, for large enough d the Hilbert scheme of zero-dimensional subschemes of $\mathbb{P}^3$ of degree d will have, in addition to the "standard" component whose general member consists of d distinct points, components whose general member is nonreduced — though no one knows how many such components the Hilbert scheme will have, or what their dimensions might be. So, for large d, there will be component(s) $\mathcal{C}_{d,g',r}$ whose general element C consists of a curve of geometric genus $g' > g$ plus a subscheme of dimension 0 and degree $(g' - g)$ lying on one of these "exotic" components. As in the twisted cubic example, such components will often (always?) have dimension greater than that of the components that parameterize honest curves of genus g.

To avoid having to rule out such components repeatedly, it'll be convenient to make the

DEFINITION (1.31) *The* restricted Hilbert scheme $\mathcal{R}$ *is the open subscheme of $\mathcal{H}$ consisting of those points* $[X]$ *such that every component $\mathcal{D}$ of $\mathcal{H}$ on which $[X]$ lies has* smooth, nondegenerate, irreducible *general element. In other words, the restricted Hilbert scheme is the complement of those irreducible components of $\mathcal{H}$ every point of which corresponds to a curve that is singular, degenerate or reducible.*

What we would really like to do is to take the (closed) union $\overline{\mathcal{R}}$ of all the components $\mathcal{D}$ so as to have a projective scheme but unfortunately there is no natural scheme structure on $\mathcal{D}$ at points where it meets components outside of $\mathcal{R}$. We can, of course, speak of the *restricted Hilbert variety* $\overline{\mathcal{R}}$ by giving this set its reduced structure but then maps to $\overline{\mathcal{R}}$ will no longer correspond to families of subschemes of $\mathbb{P}^r$.

One further warning: it's almost never possible to analyze all Borel-fixed subschemes explicitly. As a result, even when it is possible to list the components of a Hilbert scheme — restricted or not — it usually requires considerable effort to verify that no others exist. The discussion of Mumford's example in the next section will illustrate this point.

One of the very few positive results about the global geometry of Hilbert scheme is Hartshorne's

THEOREM (1.32) (CONNECTEDNESS THEOREM [83]) *For any P and r, the Hilbert scheme $\mathcal{H}_{P,r}$ is connected.*

Hartshorne's proof involves first showing that every X specializes flatly to a union Y of linear subspaces that he calls a *fan*. In fact, there is an explicit procedure for translating between the coefficients of P and the number of subspaces of each dimension in Y. Next, Hartshorne characterizes those Y whose ideals have maximal Hilbert *function*: these are the *tight* fans for which the i-dimensional subspaces lie in a common $(i+1)$-dimensional subspace. He then shows that all tight fans lie on a common component of $\mathcal{H}$. Finally, he shows that, if Y is a fan that isn't tight, then there is a fan Y' whose Hilbert function majorizes that of Y and a sequence of generalizations and specializations connecting Y and Y'.

The next exercise uses Hartshorne's theorem to characterize Hilbert polynomials of projective schemes; we should point out that this characterization, due to Macaulay [111] (see also, [144]), came first and is a key element of Hartshorne's proof.

EXERCISE (1.33) 1) Calculate the Hilbert polynomial $P_{(n_0,n_1,...,n_r)}(m)$ of a generic (reduced) union $\bigcup_{i=0}^{r}(L_{i1} \cup \cdots \cup L_{in_i})$ where each L_{ij} is an i-plane in $\mathbb{P}^r$.

2) Define

$$Q_{(a_0,a_1,...,a_s)} = \sum_{i=0}^{s} \left[\binom{m+i}{i+1} - \binom{m+i-a_i}{i+1} \right].$$

Show that any rational numerical polynomial $P(m)$ — i.e., an element of $\mathbb{Q}[m]$ that takes integer values for integer m — can be expressed as

$$Q_{(a_0,a_1,...,a_s)}(m)$$

for unique nonnegative integers a_i with $a_s \neq 0$.

3) Define a mapping $(n_0, n_1, \ldots, n_r) \mapsto (a_0, a_1, \ldots, a_s)$ by requiring that

$$P_{(n_0,n_1,...,n_r)}(m) = Q_{(a_0,a_1,...,a_s)}(m).$$

Show that the image of this map is exactly the set of $(a_0, a_1, \ldots, a_s)$ for which $a_0 \geq a_1 \geq \cdots \geq a_s$.

4) Use the first step of Hartshorne's proof to deduce Macaulay's Theorem [111]: a numerical polynomial is the Hilbert polynomial of a projective variety if and only if the sequence $(a_0, a_1, \ldots, a_s)$ of 2) is non-increasing.

There is little convincing evidence either for or against the connectedness of the restricted Hilbert variety or its closure $\overline{\mathcal{R}}$: known examples have so far provided neither a counterexample nor a plausible replacement for the class of fans used in Hartshorne's proof. See, however, Exercise (1.41) on $\mathcal{H}_{9,10,3}$ in Section D.

D   Extrinsic pathologies

The difficulties we've discussed above are relatively minor annoyances. We will see much nastier behavior in the examples that follow. The gist of these examples can be summed in:

LAW (1.34) (MURPHY'S LAW FOR HILBERT SCHEMES) *There is no geometric possibility so horrible that it cannot be found generically on some component of some Hilbert scheme.*

To illustrate the application of this law, and as an example of a tangent-space-to-the-Hilbert-scheme calculation, we now wish to recall Mumford's famous example [118] of a component $\mathcal{J}$ of the (restricted) Hilbert scheme of space curves that is everywhere nonreduced. This example also serves to justify the somewhat technical construction of the Hilbert scheme. Most of the work there was devoted to producing, not the underlying subvariety of the Grassmannian $G = G(P(m), S_m)$, but a natural scheme structure on this subvariety. Mumford's example shows that this scheme structure can be far from reduced. Moreover, since the general point of J is a perfectly innocent-looking (i.e., smooth, irreducible, reduced, nondegenerate) curve in $\mathbb{P}^3$, it shows that we cannot hope to avoid these complications simply by restricting ourselves to subschemes of $\mathbb{P}^r$ that are sufficiently geometrically nice. The point is that the behavior of families X of subschemes of $\mathbb{P}^r$ can exhibit many pathologies even when the individual members X of the family exhibit none. These phenomena are usually caused by constraints imposed by the particular models of the fibers that the Hilbert scheme in question parameterizes. In the examples dealing with space curves that follow, this constraint typically takes the form of a condition that the curve C corresponding to any point on some component of the relevant Hilbert scheme $\mathcal{H}$ lies on a surface of some small fixed degree. One of the motivations for the study of intrinsic moduli space is the possibility of eliminating such extrinsic pathologies.

Mumford's example

The curves we want to look at are those lying on smooth cubic surfaces S, having class $4H + 2L$ where H is the divisor class of a plane section of S and L that of a line on S. (Recall that, on S, $H^2 = 3$, that $H \cdot L = -L^2 = 1$, and that $K_S = -H$.) We immediately see that the degree of such a curve is $d = H \cdot (4H + 2L) = 14$ and that its arithmetic genus is $g = \frac{1}{2}C \cdot (C + K_S) + 1 = 24$. We are therefore going to be working with the Hilbert scheme $\mathcal{H}_{14,24,3}$ or, in practice, with the restricted Hilbert scheme $\mathcal{R}_{14,24,3}$.

Note that the linear series $|H + L|$ is base point free since it's cut out by quadrics containing a conic curve $C \subset S$ coplanar with L. Hence $|4H + 2L|$ is also base point free and its general member is indeed a smooth curve (even, as we leave you to verify, irreducible). Finally, the dimension of the family of such curves isn't hard to compute. On a particular cubic S, the linear system $|4H + 2L|$ has dimension predicted by Riemann-Roch on S as $h^0(\mathcal{O}_S(C)) = \frac{1}{2}C \cdot (C - K_S) = 37$. Since the family of cubic surfaces has dimension 19 and each curve C of this type lies on a unique cubic ($d = 14$), the dimension of the sublocus J_3 of $\mathcal{H}_{14,24,3}$ cut out by such C's is $37 + 19 = 56$.

The family J_3 of curves C that arises in this way is irreducible. This can be proved in two ways. The first is via the monodromy of the family of all cubic surfaces in $\mathbb{P}^3$. In this approach, one first shows that the monodromy group of this family is E_6 and in particular acts transitively on the set of lines on a given S. For details, we refer you to [77]. The second, more elementary, approach is to construct this family as a tower of projective bundles imitating the argument for the irreducibility of the family $\mathcal{J}'_3$ given preceding Exercise (1.37). We leave the details to you, as in that exercise.

The key question is: is $\mathcal{J}_3$ (open and) dense in a component of the Hilbert scheme? To answer this, let C now be any curve of degree 14 and genus 24 in $\mathbb{P}^3$. We ask first: does C have to lie on a cubic? Now, the dimension of the vector space of cubics in $\mathbb{P}^3$ is 20. On the other hand, by Riemann-Roch on C, the dimension of $H^0(C, \mathcal{O}_C(3))$ is

$$h^0(\mathcal{O}_C(3)) = d - g + 1 + h^1(\mathcal{O}_C(3)) = 19 + h^0(K_C(-3)),$$

and since $\deg(K_C(-3)) = 2G - 2 - 3D = 4$, this last term could very well be positive. Indeed, it is for the curves C constructed above: for those, $K_C = \mathcal{O}_C(K_S + C) = \mathcal{O}_C(C - H)$ so $K_C(-3) = \mathcal{O}_C(2L)$ which has $h^0 = 1$. Thus, dimensional considerations alone don't force C to lie on a cubic.

Suppose C doesn't lie on a cubic. We have $h^0(\mathcal{O}_{\mathbb{P}^3}(4)) = 35$, while $h^0(\mathcal{O}_C(4)) = 56 - 24 + 1 = 33$, so C must lie on at least a pencil of quartics. Moreover, an element T of such a pencil must intersect the other elements in the union of C and a curve D of degree 2. Since K_T is trivial, $(C \cdot C)_T = 2(g_C - 1) = 46$. From the linear equivalence of $C + D$ and $4H$, we first obtain $C \cdot D = C \cdot (4H - C) = 56 - 46 = 10$, then $D^2 = (4H - C)^2 = 64 - 112 + 46 = -2$, and finally $g_D = 0$. This is only possible if C is a plane conic. To count the dimension of the family of such curves, then, we reverse this analysis, starting with a conic D, which moves with 8 degrees of freedom. The projective space Λ of quartics containing D has dimension 25. An open subset of the 48-dimensional Grassmannian $\mathbb{G}(1, 25)$ of pencils in Λ will have base locus the union of D and a curve C not lying on any cubic. The dimension of the family $\mathcal{J}_4$ of all such C is thus 56. Since the loci $\mathcal{J}_3$ and $\mathcal{J}_4$ have the *same* dimension, we deduce that *a general curve of class $4H + 2L$ on a smooth cubic surface is* not *the specialization of a curve not lying on a cubic*. This assertion together with the irreducibility of $\mathcal{J}_3$ imply that $\mathcal{J}_3$ is dense in a component of the Hilbert scheme.

We return to the examination of a curve $C \sim 4H + 2L$ in $\mathcal{J}_3$ lying on a smooth cubic S. It's easy to calculate the dimension of the space of sections of the normal bundle of C: the standard sequence

$$0 \longrightarrow N_{C/S} \longrightarrow N_{C/\mathbb{P}^3} \longrightarrow N_{S/\mathbb{P}^3} \otimes \mathcal{O}_C \longrightarrow 0$$

reads

$$0 \longrightarrow K_C(1) \longrightarrow N \longrightarrow \mathcal{O}_C(3) \longrightarrow 0,$$

and since $K_C(1)$ is nonspecial, it follows that:

$$h^0(N) = h^0(K_C(1)) + h^0(\mathcal{O}_C(3))$$

$$= 37 + 20$$
$$= 57.$$

Thus the Hilbert scheme is singular at C, and, since C is generic in $\mathcal{J}_3$, even nonreduced.

What is going on here? It's not hard to see where the extra dimension of $h^0(N)$ is coming from: if $h^0(\mathcal{O}_C(3))$ really is 20 for curves near C, then, at least infinitesimally, deformations of C don't have to lie on cubics. Naively, you might expect that near C the locus in the Hilbert scheme of curves C_λ lying on cubics was the divisor in the Hilbert scheme given by the determinant of the 20×20 matrix associated to the restriction map $H^0(\mathbb{P}^3, \mathcal{O}(3)) \rightarrow H^0(C, \mathcal{O}(\mathbb{Z}))$; thus the local dimension of $\mathcal{H}$ near C should be 57. Of course, it doesn't turn out this way, but this analysis is nonetheless correct *to first order*. There do, in fact, exist first-order deformations of C that don't lie on any cubic, and these account for the extra dimension in the tangent space to $\mathcal{H}$. If you've seen some deformation theory before you may attempt:

EXERCISE (1.35) Make the analysis above precise. What does it mean to say that a first-order deformation of C doesn't lie on a cubic? Find such a deformation.

Deformation theory is discussed in Chapter 3. Until then, even if you're unfamiliar with the subject, you should be able to understand our occasional references to deformations by viewing them as algebraic analogues of perturbations which themselves are parameterized by various schemes.

We've shown above that there is a unique component $\mathcal{J}_4$ of $\mathcal{R}$ whose general member doesn't lie on a cubic surface. Are there other components besides $\mathcal{J}_3$ whose general member does lie on a cubic surface S? The answer is yes: there is exactly one other. Suppose that C is a curve in $\mathcal{R}$ lying on a *smooth* cubic surface S. The key observation is that C must lie on a sextic surface T not containing S: we have $h^0(\mathbb{P}^3, \mathcal{O}(6)) = 84$, while $h^0(C, \mathcal{O}(6)) = 61$ and the space of sextics containing S has vector space dimension 20. We can thus describe C as residual to a curve B or degree 4 in the intersection of S with a sextic.[6] (Note that the curves in Mumford's example are residual to a disjoint union of two conics in such a complete intersection.)

What does B look like? First off, we can tell its arithmetic genus from the liaison formula[7]: if two curves C and D, of degrees d and e

---

[6]Similar dimension counting shows that the generic C lies on no surface of degree less than 6 *not* containing S.

[7]To see this formula, use adjunction on S to write

$$2g - 2 = (K_S + C) \cdot C = ((m-4)H + C) \cdot C = (m-4)d + C^2$$

and genera g and h respectively, together comprise a complete intersection of surfaces S and T of degrees m and n, then

(1.36) $$h - g = \frac{(m + n - 4)(e - d)}{2}.$$

In the present case, this says that B has arithmetic genus (-1) and self-intersection 0 on S; in particular B is reducible. One possibility is that B consists of two disjoint conics; in this case the two conics must be residual to the same line in plane sections of S and we get the Mumford component. Otherwise, B must contain a line. For example, B might consist of the disjoint union of a line L and a twisted cubic E and, unless B has a multiple line, any other configuration must be a specialization of this. In this case, the class of C in the Neron-Severi group of $NS(S)$ will not equal $4H + 2L$. Since $NS(S)$ is discrete, the class of C in it must be constant on any component of $\mathcal{R}$. We therefore conclude that B's of this type give rise to component(s) of $\mathcal{R}$ distinct from $\mathcal{J}_3$.

To see that just one component $\mathcal{J}_3'$ arises in this way, it's simplest to use a liaison-theoretic approach.[8] We will simply list the steps, leaving the verifications as an exercise. First, the set of all pairs (L, E) is irreducible since the locus of E's and L's are $\mathrm{PGL}(4)$-orbits in their respective Hilbert schemes. Second, over a dense open set in this base, the set of triples (S, L, E) such that S is a cubic surface containing $L \cup E$ forms a projective bundle, hence is again irreducible. Third, over a dense set of these triples, the set of quadruples, (T, S, L, E) such that T is a sextic surface containing $L \cup E$ but not S is a dense open set in the fiber of a second projective bundle. Finally, these quadruples map onto a dense subset of $\mathcal{J}_3'$.

EXERCISE (1.37) Verify the four assertions in the preceding paragraph.

It remains to deal with the case when B has a multiple line. If B has a multiple line L, then it must have the form $2L + D$, where D is a conic meeting L once.

EXERCISE (1.38) Let C be a curve in $\mathcal{R}_{14,24,3}$ that lies on the intersection of a cubic surface S and a sextic surface T. Suppose, further, that

---

and conclude that $C^2 = g - 2 - (m - 4)d$. Then plug this into the equation $nd = C \cdot (C + D) = C^2 + C \cdot D$, to obtain $C \cdot D = (m + n - 4)d - (2g - 2)$. By symmetry, $C \cdot D = (m + n - 4)e - (2h - 2)$, from which the formula as stated is immediate.

[8]The same result can also be obtained by showing that the monodromy group E_6 of the family of smooth cubic surfaces acts transitively on the 432 pairs (E, L) as above on a fixed S.

C is residual in this intersection to a quartic B of the form $2L + D$ with L a line and D a conic meeting L once. Show that $L + D$ is the specialization of a twisted cubic disjoint from L and hence that C is a specialization of the generic element of $\mathcal{J}_3'$.

A few additional remarks about this third component are in order. The first is that calculations like those carried out for $\mathcal{J}_3$ show that the dimension of $\mathcal{J}_3'$ is again 56 and that for general $[C]$ in $\mathcal{J}_3'$, $\mathcal{O}_C(3)$ is nonspecial. We therefore conclude that this component of $\mathcal{R}$ is at least generically reduced.

The analysis above shows that $\mathcal{R}_{14,24,3}$ has three 56-dimensional components: the generic elements of $\mathcal{J}_3$ and $\mathcal{J}_3'$ lie on *smooth* cubic surfaces, and any curve C not lying on any cubic surface is parameterized by a point of $\mathcal{J}_4$. In principle, there might exist other components $\mathcal{J}_3^*$ of $\mathcal{R}_{14,24,3}$ whose general elements lie only on a *singular* cubic surface.

EXERCISE (1.39) Complete the analysis of $\mathcal{R}_{14,24,3}$ by showing that, in fact, no such $\mathcal{J}_3^*$ exists.

Here are a few more exercises dealing with ideas that arise in Mumford's example.

EXERCISE (1.40) Make up your own examples of components of the Hilbert scheme of space curves that are everywhere nonreduced. Hartshorne feels that, in some sense, "most" components of the Hilbert scheme are of this type. Do you agree?

EXERCISE (1.41) 1) Use an analysis like that above to show that the restricted Hilbert scheme $\mathcal{R}_{9,10,3}$ of space curves of degree 9 and genus 10 has exactly two components $\mathcal{J}_2$ and $\mathcal{J}_3$.
2) Show further that the general element of $\mathcal{J}_2$ is a curve of type $(3, 6)$ on a quadric surface while the general element of $\mathcal{J}_3$ is the complete intersection of two cubic surfaces, and that both components have dimension 36.
3) Let C be any smooth curve. Show that if the Hilbert point $[C]$ of C lies in $\mathcal{J}_3$, then $K_C = \mathcal{O}_C(2)$ and hence C is *not* trigonal while if $[C] \in \mathcal{J}_2$, then $K_C \neq \mathcal{O}_C(2)$ and hence C *is* trigonal.
4) Conclude that any curve in the intersection of these components is necessarily singular. Find such a curve.

In particular, this last exercise shows that the locus of smooth curves in a Hilbert scheme can form a disconnected subvariety, and shows that there are, in general, limits to how nice we can make the elements of a restricted Hilbert scheme before it becomes disconnected.

Other examples

Exercise (1.39) might tempt you to suppose that if every curve on a
component of $\mathcal{R}$ lies on a hypersurface S of degree d then, for general
C, we can choose S to be smooth. This, heuristically, should not be
true since it would violate Murphy's Law of Hilbert Schemes (1.34). We
would like to exhibit next an explicit counterexample.

Our example uses double lines in $\mathbb{P}^3$. A double line supported on
the reduced line with equations $z = w = 0$ is a scheme X whose ideal
has the form

$$I_X = (z^2, zw, w^2, F(x,y)z + G(x,y)w)$$

where F and G are homogeneous of degree m. If F and G have no common zeros, then X has degree 2 and arithmetic genus $p_a(X) = -m$. If
T is a smooth surface of degree $(m+1)$ and L is a line lying on T, then
the class $2L$ on T will define a double line of arithmetic genus $-m$.

In our example, we want to take $m = 2$ so X is twice the class of a
line L on a smooth cubic. Such an X lies on many quartic surfaces S.
Indeed, the general such S will have equation

$$f = \alpha x(Fz + Gw) - \beta y(Fz + Gw) + h$$

with $h \in \langle z, w \rangle^2$ and α and β suitable constants. A short calculation
shows that this S has a double point at the point $(\beta, \alpha, 0, 0)$. Geometrically, X is a *ribbon*: i.e., a line L with a second-order thickening along a
normal direction at each point. Because these normal directions wind
twice around L, X cannot lie on any smooth surface of degree greater
than 3.

Let C be the curve residual to X in a complete intersection $S \cap T$,
where T is a surface of degree n. Then C has degree $4n - 2$ and the
liaison formula (1.36) shows that its genus is $2n^2 - 2n - 2$. Now a
theorem of Halphen [71] asserts that whenever the degree d and genus
g of a smooth space curve satisfy $g > (d^2 + 5d + 10)/10$, then the
curve lies on a quartic surface. A little arithmetic shows that our C
(and hence any flat deformation of it) satisfy these hypotheses for
all $n \geq 7$. Thus, any deformation C' of our C still lies on a quartic
surface S'.

We next claim that: *such a C' remains residual to a double line in a
complete intersection of S' with a surface T' of degree n not containing
S'.* By the argument above, S' must also be singular, and we conclude
that for $n \geq 7$, the generic curve in the component of $\mathcal{H}_{n-2,2n^2-2n,3}$
containing C lies on a quartic but that this quartic is always singular.

To see the claim, first note that $K_{C'} = \mathcal{O}_{C'}(n)(-X)$ and hence, since
X meets C' positively, that $\mathcal{O}_{C'}(n)$ is nonspecial. By Riemann-Roch,
$h^0(C', \mathcal{O}_{C'}(n)) = n(4n - 2) - (2n^2 - 2n - 2) + 1 = 2n^3 + 3$. The

dimension of the space of surfaces of degree n in $\mathbb{P}^3$ containing C' is thus at least

$$\binom{n+3}{3} - (2n^2 - 3) = \binom{n-1}{3} + 1.$$

Since the binomial coefficient on the right is the dimension of the space of degree n surfaces containing the quartic S', C' continues to lie in the complete intersection of S' and a surface T' of degree n. Reversing the liaison formula, the curve X' residual to C' in $S' \cap T'$ again has degree 2 and genus (-2). Since the curve X has no embedded points and is a specialization of X', X' can have no embedded points itself. This next exercise asks you to show that X' must then be a double line and completes the proof of the claim above.

EXERCISE (1.42) Check that the only X' with no embedded points, degree 2 and genus (-2) is a double line.

We will cite only one more pathological example. But to really grasp the force of Murphy's Law, we suggest that you make up for yourself examples of curves exhibiting other bizarre forms of behavior.

Modulo a number of verifications left to the exercises, we'll construct a smooth, reduced and irreducible curve C lying in the intersection of two components of the Hilbert scheme — so that, in particular, its deformation space (as a subscheme of $\mathbb{P}^r$) is reducible. To do this, let S be a cone over a rational normal curve in $\mathbb{P}^{r-1}$, let $L_1, \ldots, L_{r-2} \subset S$ be lines on S, let $T \subset \mathbb{P}^r$ be a general hypersurface of degree m containing $L_1, \ldots, L_{r-2}$ and let C be the residual intersection of T with S. Assuming m is sufficiently large, C will then be a smooth curve (it'll pass once through the vertex of S).

Such a C is a *Castelnuovo curve*, that is, a curve of maximum genus among irreducible and nondegenerate curves of its degree $m(r-1) - (r-2) = (m-1)(r-1) + 1$ in $\mathbb{P}^r$. Now, Castelnuovo theory [21] tells us that a Castelnuovo curve of that degree in $\mathbb{P}^r$ must lie on a rational normal scroll X on which it must have class either $mH - (r-2)F$ or $(m-1)H + F$. On the singular scroll S, $H \sim (r-1)F$ and these coincide, but in general they are distinct; it follows (at least as long as $r \geq 4$) that there are two components of the Hilbert scheme of curves of the given degree and genus whose general members are Castelnuovo curves.

EXERCISE (1.43) 1) Show that the curve C discussed above can be deformed to a curve on a smooth scroll having either of the classes $mH - (r-2)F$ or $(m-1)H + F$ and hence that $[C]$ lies on both components of the Hilbert scheme of Castelnuovo curves.

2) Find the dimension of the component of the Hilbert scheme parameterizing curves of each type and the dimension of their intersection.
3) Find the dimension of the Zariski tangent space to the Hilbert scheme at the point $[C]$.

E   Dimension of the Hilbert scheme

We will be returning to the Hilbert scheme later on in the book, and will do more with it then. We should mention here, though, some of the principal open questions with regard to $\mathcal{H}$. With an eye to our intended applications, in the remainder of this chapter we'll deal only with Hilbert schemes of *curves*.

The first issue is dimension. To begin with, the description of the tangent space to the Hilbert scheme of curves in $\mathbb{P}^r$ at a point $[C]$ as the space of global sections of the normal bundle to C gives us an a priori guess as to its dimension: we may naively expect that

$$\dim \mathcal{H} = h^0(C, N_C) = \chi(N_C).$$

This number is readily calculated from the sequence

$$0 \to T_C \to T_{\mathbb{P}^3} \otimes \mathcal{O}_C \to N_C \to 0.$$

We see that the degree of the normal bundle is

$$\deg(N_C) = (r+1)d + 2g - 2;$$

and then by Riemann-Roch we have

$$\chi(N_C) = \deg(N_C) - (r-1)(g-1)$$
$$= (r+1)d - (r-3)(g-1).$$

This number we'll call the *Hilbert number* $h_{d,g,r}$.

Of course, neither of the equalities above necessarily holds always — nor even, unfortunately, that often. Even worse, the naive inequalities associated to these estimates ($\dim(\mathcal{H}) \leq h^0(N_C)$ and $h^0(N_C) \geq \chi(N_C)$) go in *opposite* directions. It is nonetheless the case that the *dimension inequality*

(1.44) $\dim(\mathcal{H}) \geq h_{d,g,r} := (r+1)d - (r-3)(g-1)$

always holds at points of $\mathcal{H}$ parameterizing smooth curves, or more generally curves that are locally complete intersections. This follows from a less elementary fact of deformation theory, which we will discuss in Chapter 3. We can also see it from an alternate derivation of the Hilbert number based on a study of tangent spaces to W_d^r's. This topic belongs to the theory of special linear series which we'll take up in Chapter 5. For now, we recall from [7, IV.4.2.i] that, in any family of line bundles of degree d on curves of genus g, the locus of those line bundles having $r + 1$ or more sections has codimension at most $(r + 1)(g - d + r) = g - \rho$ in the neighborhood of a line bundle with exactly $r + 1$ sections.[9] Applying this to the family of all line bundles of degree d on all curves of genus g, we conclude that the family of linear series of degree d and dimension r on curves of genus g has local dimension at least $(3g - 3) + g - (r + 1)(g - d + r)$[10] everywhere. Since such a linear series determines a map of a curve to $\mathbb{P}^r$ up to the $(r^2 + 2r)$-dimensional family $\text{PGL}(r + 1)$ of automorphisms of $\mathbb{P}^r$, we may conclude that

$$\dim(\mathcal{H}) \geq 4g - 3 - (r + 1)(g - d + r) + r^2 + 2r$$
$$= (r + 1)d - (r - 3)(g - 1).$$

so the dimension of $\mathcal{H}$ is at least the Hilbert number. By way of terminology, we'll call a component of $\mathcal{H}$ *general* if its dimension is equal to the Hilbert number, and *exceptional* if its dimension is strictly greater. Note one aspect of the Hilbert number: when $r = 3$, $h_{d,g,3} = 4d$ is independent of the genus, while for $r \geq 4$ it decreases with g.

There is another approach to this estimate which is worth mentioning since in some cases it yields additional local information. Assume for the moment that C is smooth, nondegenerate and irreducible and that $\mathcal{O}_C(1)$ is nonspecial. Then $r \leq d - g$. (We don't necessarily have equality since we aren't assuming that C is linearly normal in $\mathbb{P}^r$.) We can count parameters: the curve C depends on $3g - 3$ and the line bundle $\mathcal{L} \in \text{Pic}^d(C)$ determined by $\mathcal{O}_C(1)$ on g. Moreover, close to our initial choices we continue to have the inequality $h^0(C, \mathcal{O}_C(1)) \leq d - g + 1$. Hence the choice of the linear subsystem of $H^0(C, \mathcal{L})$ of dimension $(r + 1)$ determines a point in a Grassmannian $\mathbb{G}(r, d - g)$ whose dimension is $(r + 1)(d - g + r)$. Finally, we must add $(r^2 + 2r)$ parameters coming from the $\text{PGL}(r + 1)$-orbit of each linear system. The total is exactly $h_{d,g,r}$. Note that this argument actually proves that $\chi(N_{C/\mathbb{P}^r}) = h^0(N_{C/\mathbb{P}^r}) = \dim(\mathcal{H}_{d,g,r})$ and hence leads to the:

COROLLARY (1.45) *If C is a smooth, irreducible, nondegenerate curve of degree d and genus g in $\mathbb{P}^r$ with $\mathcal{O}_C(1)$ nonspecial, then*

---

[9]Here ρ is the *Brill-Noether number* $\rho = \rho_{g,r,d} := g - (r + 1)(g - d + r)$.

[10]In this sum the first term expresses the moduli of the curve C, the second the moduli of the line bundle L of degree d and the third the codimension of the set of pairs (C, L) with at least $(r + 1)$ sections. Note that this postulation also equals $3g - 3 + \rho$.

$H^1(C, N_{C/\mathbb{P}^r}) = 0$ *and*

$$\dim_{[C]} \mathcal{H}_{d,g,r} = \dim T_{[C]} \mathcal{H}_{d,g,r}$$
$$= h_{d,g,r}$$
$$= (r+1)d - (r-3)(g-1).$$

EXERCISE (1.46) Give an alternate proof that the basic estimate $\dim(\mathcal{H}) \geq h_{d,g,r}$ for the dimension of $\mathcal{H}_{d,g,r}$ holds without the assumption that $\mathcal{L} = \mathcal{O}_C(1)$ is nonspecial as follows:

1) Let C be smooth and irreducible of genus g, let $\{\omega_1, \ldots, \omega_g\}$ be a basis of the holomorphic differentials on C and let $\mathcal{L}$ be a line bundle on C. If $D = p_1 + \cdots + p_d$ is an effective divisor on C with line bundle $\mathcal{O}_C(D) \cong \mathcal{L}$, we may define a map of $\varphi_D : H^0(C, \mathcal{L}) \to \mathbb{C}^d$ by taking principal parts of sections at the points of the support of D. Show that the image of this map is the annihilator in $\mathbb{C}^d$ of the $g \times d$ matrix M_D whose ij^{th} entry is

$$\int_{p_0}^{p_j} \omega_i$$

2) Let $\xi = \{(\Lambda, M) | \Lambda \subset \ker(M)\}$ where Λ is an r-dimensional subspace of $\mathbb{C}^d$ and M is a $g \times d$ matrix. The space of quadruples $\mathcal{F} = (C, D, V, \mathcal{L})$ with D an effective divisor on C such that $\mathcal{L} \cong \mathcal{O}_C(D)$ and V an $(r+1)$-dimensional subspace of $H^0(C, \mathcal{L})$ maps onto ξ by taking Λ to be the image of V under the map φ_D and $M = M_D$. Show that this map is dominant and, by calculating $\dim(\xi)$, conclude that $\dim(\mathcal{F}) \geq h_{d,g,r} + r$.

3) Show also that the map from $\mathcal{F}$ to $\mathcal{H}_{d,g,r}$ given by forgetting the choice of D is onto with fiber of dimension r and conclude that $\dim(\mathcal{H}_{d,g,r}) \geq h_{d,g,r}$.

If we start to compute dimensions of components of $\mathcal{H}$, we see, in the low-degree examples, only general components. For example, in $\mathbb{P}^3$ the lines form a four-dimensional family, conics an eight-dimensional family, twisted cubics a twelve-dimensional family, etc. It becomes clear fairly soon, however, that this state of affairs is temporary. For example, we find *only* exceptional components when we look at the following: complete intersections of high degree; curves of high degree on quadric or cubic surfaces; determinantal varieties associated to $n \times (n+1)$ matrices with entries of high degree, etc. The general question of what the dimensions of the components of $\mathcal{H}$ may be remains very much open. Four questions in particular may be asked:

QUESTION (1.47) 1) For fixed d and r, but possibly varying g, what is the largest dimension of a component of the restricted Hilbert scheme $\mathcal{R}_{d,g,r}$ whose general elements are smooth, irreducible and nondegenerate?

2) For any d, g and r, what is the smallest dimension of a component of $\mathcal{H}_{d,g,r}$? Of course, for $r = 3$ the answer is $4d$, but for $r \geq 4$, the Hilbert number will be negative for many values of d, g and r, and it's very much unknown what the smallest dimension may be. In particular, it's conjectured that the only *rigid* curves — that is, curves that admit no deformations other than projectivities of $\mathbb{P}^r$ — are rational normal curves; but this remains open.

3) Can we find a function $\sigma(g)$ such that the basic estimate $\dim(\mathcal{H}) = h_{d,g,r}$ holds for any component of $\mathcal{H}$ whose image in $\mathcal{M}_g$ has codimension less than $\sigma(g)$? This last question is motivated by the empirical evidence that the expected dimension is correct when this codimension is small. It's even possible that $\sigma(g)$ could be taken to be roughly equal to g.

4) Does the inequality $\dim(\mathcal{H}_{d,g,r}) \geq h_{d,g,r}$ hold for *any* component of a Hilbert scheme of curves? Our motivation for suggesting this question is the empirical observation that families of singular curves, which might provide counterexamples, seem to have dimension equal to the Hilbert number exactly when the curves do *not* smooth in the ambient projective space. Consider, for example, the union in $\mathbb{P}^3$ of a line and a plane curve of degree d meeting at a point. If $d = 3$, the family of such curves has dimension 15 (4 for the line plus 2 for the plane plus 10 for the cubic minus 1 so that the line and cubic meet) which is less than the Hilbert number 16; however, such curves smooth in $\mathbb{P}^3$ to elliptic normal curves. If $d = 4$, the family has dimension 20 $(4 + 2 + 15 - 1)$ which equals the Hilbert number: such curves are classic examples of curves that do *not* smooth in $\mathbb{P}^3$.

F   Severi varieties

If we stick to Hilbert schemes parameterizing subcurves of $\mathbb{P}^r$ then, as we've seen in Exercise (1.15), the case $r = 2$ is trivial: the Hilbert point of a plane curve of degree d is just given by the equation of the curve. In this case, we can hope to understand much more precisely the subloci of curves with various geometrically significant properties. In this section, we'll take some first steps in this direction.

We've seen that the space of all plane curves of a given degree is simply a projective space $\mathbb{P}^N$ where $N = \left(\frac{d(d+3)}{2} \right)$. What we wish to do here is to look at the locally closed subvariety of this $\mathbb{P}^N$ consisting of curves of degree d and geometric genus g. We introduce three loci:

DEFINITION (1.48) *In the space of plane curves of degree d, define:*

1) $V_{d,g}$ *to be the locus of reduced and irreducible curves of degree d and genus g;*

2) $U_{d,g}$ to be the locus of reduced and irreducible curves of degree d and genus g having only nodes as singularities;

3) $\overline{V}_{d,g}$ to be the closure of $V_{d,g}$ in $\mathbb{P}^N$.

These are often referred to as *Severi varieties*. Note that, as with the restricted Hilbert scheme, $V_{d,g}$ and $\overline{V}_{d,g}$ don't have a natural "parametric" scheme structure: that is, there is no known way to define a scheme structure on them so that they represent the functor of families of plane curves with the appropriate geometric properties.

On the other hand, the underlying spaces of these varieties are much better behaved than the Hilbert schemes $\mathcal{H}_{d,g,r}$ for $r \geq 3$. To sum up the state of our knowledge, we have the:

THEOREM (1.49) (ZARISKI; HARRIS [80]) *For all d and g,*

1) $U_{d,g}$ is smooth of dimension $3d + g - 1 = h_{d,g,2}$;

2) $U_{d,g}$ is dense in $V_{d,g}$;

3) $V_{d,g}$ is irreducible.

With the tools we have available at this point, we can't prove the irreducibility now or even the fact that the nodal curves are dense in the curves of genus g, but we can at least verify that the locus of nodal curves is smooth of the expected dimension (the proof of the second part will be given in Section 3.B as Corollary (3.46) and of the third part in Section 6.E). To do this, we look first at the variety

$$\Sigma = \{(C, p_1, \ldots, p_\delta) | p_i \in C_{sing}\} \subset \mathbb{P}^N \times (\mathbb{P}^2)^\delta.$$

If we fix (affine) coordinates (x, y) on $\mathbb{P}^2$, let a_{ij} denote the coefficient of $x^i y^j$ in the equation of C and let (x_α, y_α) be the coordinates of the node p_α, then Σ is given by the α triples of equations

$$F_\alpha(a_{ij}, x_\alpha, y_\alpha) = \sum a_{ij}(x_\alpha)^i(y_\alpha)^j = 0,$$

$$G_\alpha(a_{ij}, x_\alpha, y_\alpha) = \sum i \cdot a_{ij}(x_\alpha)^{i-1}(y_\alpha)^j = 0, \qquad \text{and}$$

$$F_\alpha(a_{ij}, x_\alpha, y_\alpha) = \sum j \cdot a_{ij}(x_\alpha)^i(y_\alpha)^{j-1} = 0 \qquad \text{for all } \alpha.$$

(The first equation just says p_α is on C and the last two that it's a singular point. Just the ability to write down such a simple set of explicit equations already distinguishes the analysis of Severi varieties from that of Hilbert schemes in higher dimensions.)

From these equations, we might expect naively that $\dim(\Sigma) = N + 2\delta - 3\delta = N - \delta$. In fact, we'll show that in a neighborhood of a nodal curve C with nodes $p_1, \ldots, p_\delta$, the variety Σ

maps one-to-one to $V_{d,g}$ by showing that the differential of this map is injective at C.

Consider, for example, the case $\delta = 1$. Suppose we're at a point $(C,p)' \in \Sigma$ normalized so that $p = (0,0)$. The matrix of partials of F, G and H with respect to x, y and a_{00} looks like

	F	G	H
$\frac{\partial}{\partial x}$	0	a_{20}	a_{11}
$\frac{\partial}{\partial y}$	0	a_{11}	a_{02}
$\frac{\partial}{\partial a_{00}}$	1	0	0

TABLE (1.50)

The corresponding minor $(a_{20}a_{02} - a_{11}^2)$ is nonzero exactly when C has an ordinary node at p. Note also that all the missing entries in the first column of this matrix are 0 at (C,p). We deduce that at (C,p), Σ_1 is smooth of codimension 3 in $\mathbb{P}^N \times \mathbb{P}^2$. Moreover we see that the projection map $\pi : \Sigma \to \mathbb{P}^N$ is an immersion at (C,p), with the tangent space to Σ_1 at (C,p) mapping isomorphically to the space of polynomials of degree d vanishing at p (i.e., having $a_{00}=0$).

Now, in the general case, if C is a curve with exactly δ nodes $p_1,\ldots,p_\delta$ as singularities, the map from Σ to $\mathbb{P}^n$ factors through Σ_1 in δ ways by distinguishing each of the p_α's in turn. We can therefore represent the locus $V_{d,g}$ in an analytic neighborhood of C as the intersection of the images of analytic neighborhoods of the points $(C,p_i) \in \Sigma_1$. The tangent space to $V_{d,g}$ at C is thus the linear space of polynomials of degree d vanishing at the points p_i. But we know that the p_i impose independent conditions on curves of any degree $m \geq d - 3$ (cf. [7, p. 54, Exercise 11]); it follows that

(1.51)
$$\dim(T_{[C]}V_{d,g}) = N - \delta$$

and hence that $V_{d,g}$ is smooth of this dimension. For emphasis, we again note that

$$N - \delta = \left(\frac{d(d+3)}{2}\right) - \left(\frac{(d-1)(d-2)}{2} - g\right)$$

$$= 3d + g - 1$$

$$= h_{d,g,2}.$$

While we have much better control over Severi varieties than over more general Hilbert schemes, there are many open problems. We might ask for a description of the tangent space $T_{[C]}V_{d,g}$ near a curve

C having other than nodal singularities. As an example, consider the variety $V = V_{3,0}$ of cubics of geometric genus 0 which is a hypersurface in the $\mathbb{P}^9$ of plane cubics. In the affine plane slice $\mathbb{A}$ given by $y^2 = x^3 + ax + b$, the equation of V is $\Delta = 4a^3 + 27b^2 = 0$. For $(a, b) \neq (0, 0)$, the cubic $C_{(a,b)}$ corresponding to a point of Δ is nodal and $\mathbb{A} \cap V$ is smooth at $[C_{(a,b)}]$. But $C_{(0,0)}$ is a cuspidal cubic and $\mathbb{A} \cap V$ itself has a cusp at $[C_{(0,0)}]$.

The upshot is that, while the locus $U_{d,g}$ of nodal curves can be compactified in a natural way to $V_{d,g}$, most of its desirable geometric properties are lost in the process. This leads to the problem: find a better (partial) compactification of $U_{d,g}$. We would like, at least, a parametric compactification whose tangent space at a point $[C]$ has some natural description as a linear space of curves of degree d.

There is one known way to improve $V_{d,g}$ somewhat. Define $T_{d,g}$ to be the space of pairs (C, π) where C is a smooth curve of genus g and $\pi : C \dashrightarrow \mathbb{P}^2$ is a birational map of degree d. This change of point-of-view from subvarieties of $\mathbb{P}^2$ to maps to $\mathbb{P}^2$ may seem to be a somewhat irrelevant one since $T_{d,g}$ and $V_{d,g}$ are bijective and both contain $U_{d,g}$ as a dense open subvariety. However, these two spaces are not isomorphic as varieties. Essentially, $T_{d,g}$ normalizes $V_{d,g}$ at points $[C]$ corresponding to curves with cusps. All these observations generalize: if you're interested you'll find a longer discussion in [33].

EXERCISE (1.52) In the example above show that for (a, b) in Δ but different from $(0, 0)$ the curve $C_{a,b}$ has a node at the double root $(x, y) = (\frac{-3b}{2a}, 0)$ and that the composition of the normalization map $\pi : \tilde{C}_{a,b} \to C_{a,b} \subset \mathbb{P}^2$ with the projection to the x-axis is simply branched over $(x, y) = (0, \frac{3b}{a})$. Show further that the normalization $\tilde{C}_{a,b} \cong \mathbb{P}^1$ has equation $y^2 = x - \frac{3b}{a}$ and is the fiber of a family of $\mathbb{P}^1$'s over $\Delta - \{(0, 0)\}$ which does *not* extend to all of Δ but which *does* have an extension over the normalization $\tilde{\Delta}$ of Δ.

G Hurwitz schemes

The last parameter spaces we wish to discuss are perhaps the most classical ones: the *Hurwitz schemes* $\mathcal{H}_{d,g}$, which are parameter spaces for "maps of curves to $\mathbb{P}^1$", i.e., for branched covers of $\mathbb{P}^1$. There are many variations on their construction in a fairly tight analogy to the Severi varieties. We could simply try to parameterize pairs (C, π) consisting of a smooth curve C of genus g and a finite map $\pi : C \to \mathbb{P}^1$. Alternately, we can associate to a branched cover its branch divisor B, which, by the Riemann-Hurwitz formula, is a divisor of degree $b = 2d + 2g - 2$ in $\mathbb{P}^1$. Since the set of such divisors is canonically a projective space $\mathbb{P}^b$ by associating to B the equation of degree b, unique up to rescaling, that has B as its cycle of zeros, this might seem almost trivial. Of course, divisors with points of multiplicity greater than d cannot correspond to any cover so we cannot hope to get a complete parameter space in this way. A more essential difficulty stems from the fact that the cover C depends on B and the $\mathfrak{S}_d$-conjugacy class of the monodromy homomorphism from $\pi_1(\mathbb{P}^1 - B)$ into the symmetric group $\mathfrak{S}_d$ on a general fiber of π: the number of covers with a given branch locus B and combinatorics of the description of their monodromies thus both depend on the multiplicities of the points in B itself.

The intersection of all these approaches is the case when all the branch points of π are simple, that is, when B consists of b distinct points and hence corresponds to a point in the dense open subset $\widetilde{\mathcal{B}}_{d,g}$ of $\mathbb{P}^b$ isomorphic to the quotient by $\mathfrak{S}_b$ of the complement of all diagonals in $(\mathbb{P}^1)^b$. (Such covers thus form a locus analogous to the locus $U_{d,g}$ of nodal curves in the case of the Severi variety.) A straightforward local analysis then yields:

THEOREM (1.53) *Let $\widetilde{\mathcal{H}}_{d,g}$ be the set of branched covers of $\mathbb{P}^1$ of degree d and genus g having $b = 2d + 2g - 2$ simple branch points. Then,*

1) *$\widetilde{\mathcal{H}}_{d,g}$ is an unramified cover of $\widetilde{\mathcal{B}}_{d,g}$ and, hence, is naturally a smooth quasiprojective variety of dimension*

$$b = 2d + 2g - 2 = h_{d,g,1}.$$

2) *There is a smooth universal family of curves $\mathcal{C}_{d,g}$: i.e., a diagram*

$$
\begin{array}{c}
\mathcal{C}_{d,g} \\
\downarrow \\
\mathbb{P}^1 \times \widetilde{\mathcal{H}}_{d,g} \\
\downarrow \\
\widetilde{\mathcal{H}}_{d,g}
\end{array}
$$

whose fiber over a point $[\pi]$ of $\widetilde{\mathcal{H}}_{d,g}$ is the covering $\pi : C \rightarrow \mathbb{P}^1$ parameterized by $[\pi]$.

The key point is that $\widetilde{\mathcal{H}}_{d,g}$ is a covering space of $\widetilde{\mathcal{B}}_{d,g}$. We will return to the topic of branched covers in G, and give only a precis here. First, for any B, we can choose small loops γ_i around the points b_i in B that generate $\pi_1(\mathbb{P}^1 - B)$ modulo the single relation $\prod_i \gamma_i = 1$. Since each branch point is simple, the γ_i must map to simple transpositions τ_i in $\mathfrak{S}_d$ satisfying $\prod_i \tau_i = 1$. Since the cover is connected, the subgroup generated by the τ_i must be transitive. Conversely, any choice, up to simultaneous $\mathfrak{S}_d$ conjugacy, of τ_i's meeting these conditions determines a unique connected cover simply branched over B. The rest follows easily. The Hurwitz variety has one other property that lies somewhat deeper and that we have therefore set off.

THEOREM (1.54) $\tilde{\mathcal{H}}_{d,g}$ *is connected. Equivalently, in view of (1.53).1,* $\tilde{\mathcal{H}}_{d,g}$ *is irreducible.*

The connectedness depends on an analysis (first carried out by Klein, Clebsch, Lüroth and Hurwitz) of the braid monodromy of $\tilde{\mathcal{H}}_{d,g}$ over $\tilde{\mathcal{B}}_{d,g}$. Essentially, this involves calculating the action of certain loops in $\tilde{\mathcal{B}}_{d,g}$ on the combinatorial description of the monodromy of a cover $[\pi]$ in $\tilde{\mathcal{H}}_{d,g}$ and then building a loop that takes a given combinatorial description to a standard one. The classic reference is [26]; a good modern one is Moishezon's paper [115].

Clearly, $\tilde{\mathcal{H}}_{d,g}$ is too small for many purposes. When we try to enlarge it naively, however, we run into trouble. When the map $C \to \mathbb{P}^1$ has nonstandard ramification (i.e., the branch divisor B has multiple points), then the number of possible combinatorial forms for the monodromy drops. Hence the most we can hope for is to extend $\tilde{\mathcal{H}}_{d,g}$ to a ramified cover $\mathcal{H}_{d,g}$ of some compactification $\mathcal{B}_{d,g}$ of $\tilde{\mathcal{B}}_{d,g}$.[11] Whatever $\mathcal{B}$ we choose, the existence of a universal family of curves C and maps π becomes a much more subtle question. All that is clear is that $\tilde{\mathcal{H}}_{d,g}$ will be a dense open subset in the space $\mathcal{H}_{d,g}$ no matter how we define the latter.

When we return to this subject in more depth in Section 3.G, we'll study a very pretty and useful resolution of this difficulty, due to Knudsen and Mumford. The key idea is to find a compactification $\mathcal{B}_{d,g}$ of $\tilde{\mathcal{B}}_{d,g}$ in which branch points *always* remain distinct: this definition then leads one naturally to the compactification of $\tilde{\mathcal{H}}_{d,g}$ by the space $\mathcal{H}_{d,g}$ of *admissible covers*; this has virtually all of the properties we might desire.

---

[11]The obvious compactification $\mathbb{P}^b$ is ruled out by degree considerations as noted above

Chapter 2

Basic facts about moduli spaces of curves

This chapter is an essentially expository one which summarizes the major approaches to the construction of moduli spaces of curves and states some of the most important results and open problems about their local and global geometry.

We have two principal reasons for inserting this summary. The first is to introduce the topics that will occupy the remainder of the book. The second is to state a number of important results that do *not* reappear. Indeed, a careful treatment of all the results stated in this chapter would be impossible in a single volume. Rather than simply passing such results by, we've chosen to record their statements and provide references for them here.

Even with this proviso, this chapter is far from complete. Our choice of results reflects our tastes and interests and we ask your indulgence if your own preferences differ from ours.

A Why do fine moduli spaces of curves not exist?

Most of the moduli spaces of curves that we'll be studying are only coarse moduli spaces. The obstructions to representing the corresponding moduli functors (equivalently, to constructing fine moduli spaces) come from automorphisms of the data of the problems.

In this section, we wish first to give some elementary examples which illustrate the phenomena involved and then to take a look at the various approaches which have been developed to work around them. We begin by looking again at moduli of curves of genus 1.

Recall that in Exercise (1.6), we constructed an algebraic family

$$\mathcal{X} = \{y^2 = x(x-1)(x-\lambda)\} \subset \mathbb{P}^2_{x,y} \times \mathbb{A}^1_\lambda$$

$$\varphi \Big\downarrow$$

$$\mathbb{A}^1_\lambda - \{0,1\}$$

of smooth curves of genus 1 in which every such curve appears. Moreover,

$$X_\lambda \cong X_{\lambda'} \iff \lambda' \in \left\{ \lambda,\, 1-\lambda,\, \frac{1}{\lambda},\, \frac{1}{1-\lambda},\, \frac{\lambda-1}{\lambda},\, \frac{\lambda}{\lambda-1} \right\}$$

and there is a j-map $j : \mathbb{A}^1_\lambda \to \mathbb{A}^1_j$ whose fibers are these $\mathfrak{S}^3$-orbits. In other words, the j-line is a coarse moduli space for curves of genus 1. Exercise (1.6) gave two ways of seeing that the j-line was not a fine moduli space. Here, we want to see this from yet a third point of view.

If the j-line were a fine moduli space, there would be a universal curve $\mathbf{1} : \mathcal{C} \to \mathbb{A}^1_j$ and a fiber-product diagram

$$
\begin{array}{ccc}
\mathcal{X} & \xrightarrow{\;\mathcal{J}\;} & \mathcal{C} \\
\varphi \downarrow & & \downarrow \mathbf{1} \\
\mathbb{A}^1_\lambda & \xrightarrow{\;j\;} & \mathbb{A}^1_j
\end{array}
$$

Thus the $\mathfrak{S}^3$-action on $\mathbb{A}^1_\lambda$ would have to lift to $\mathcal{X}$. What could the lifting of the involution $\lambda \mapsto 1-\lambda$ be? Since X_λ is the curve ramified over $0, 1, \lambda$ and ∞, this lift would have to look like

$$(x, y, \lambda) \mapsto (1-x, \pm iy, 1-\lambda).$$

Either of these choices acts nontrivially on the fiber $X_{1/2}$ and the quotient of $\mathcal{X}$ by this involution would have a rational fiber over $1/2$. (In fact, $X_{1/2}$ is the curve corresponding to the lattice $\mathbb{C}^2/(\mathbb{Z}1 + \mathbb{Z}\sqrt{-1})$ which has j-invariant 1728 and the involution above is the extra automorphism of order 2 of this lattice.)

There is also a more global obstruction to lifting λ due to the fact that multiplication by (-1) is an automorphism of any elliptic curve. Either of the potential choices for the lifting above has order 4 on $\mathcal{X}$:

$$(x, y, \lambda) \mapsto (1-x, \pm iy, 1-\lambda) \mapsto (x, -y, \lambda).$$

In other words, while the square of such a candidate lifting would give an automorphism of X_λ, this automorphism would have to be nontrivial.

There are a number of approaches to dealing with the obstructions to the existence of fine moduli spaces due to automorphisms. To simplify, we'll restrict our discussion to moduli problems of curves but all these techniques are more generally applicable.

The simplest is to eliminate the locus of varieties with automorphisms.[1] If $\mathcal{M}$ is the coarse moduli space for a moduli problem $\mathbf{F}$ in which we're interested, we'll denote by $\mathcal{M}^0$ the locus of curves C in $\mathcal{M}$ such that $\mathrm{Aut}(C) = \{id_C\}$. This will, in general, be a fine moduli space for the open subfunctor of F of "curves without automorphisms". This solution is often extremely unnatural since our interest in the objects of $\mathcal{M}$ may be completely unrelated to their automorphisms. Moreover, we can almost never hope to find complete moduli spaces without allowing some varieties with automorphisms. On the other hand, the complement of $\mathcal{M}^0$ is often of high codimension in $\mathcal{M}$ — for example, in the moduli space $\mathcal{M}_g$ of smooth curves of genus g, $\mathcal{M}_g^0$ has codimension $g - 2$ for $g \geq 2$ — so this approach does allow us to use a fine moduli space to deal with many low-codimension questions.

The second approach is to find some extra structure that can be added to the moduli problem that is sufficiently fine that no automorphisms of an underlying curve can fix the extra structure. This approach is called *rigidifying* the problem. For curves, the most common extra structures to use are sets of *marked* or *distinguished* smooth points on the curve and *level structures*.

The existence of the bound $84(g - 1)$ for the order of the automorphism group of a smooth curve C of genus $g \geq 2$ in terms of g *alone* ensures that no nontrivial automorphism of such a curve can fix n distinct points of C for any sufficiently large n: thus we get a fine moduli space, denoted $\mathcal{M}_{g,n}$, for such marked curves. The defect of this approach is that each *marked point* increases the dimension of the moduli space by 1. This makes it unclear how much of the geometry of the original moduli space of unmarked curves can be captured from that of the marked curves. On the other hand, there are a number of interesting geometric questions that deal directly with marked curves so these spaces often arise naturally.

Level structures are a second method of rigidifying moduli problems that avoid changing the dimension of the moduli space. A full level n structure on a curve C of genus g is a symplectic basis $\{\alpha_1, \ldots, \alpha_g, \beta_1, \ldots, \beta_g\}$ for $H_1(C, \mathbb{Z}/n\mathbb{Z})$: here *symplectic* means that, in terms of the basis, the intersection pairing on $H_1(C, \mathbb{Z}/n\mathbb{Z})$ has ma-

---

[1]Here, and in the rest of this chapter, we've allowed ourselves to use the phrases "with/without automorphisms" as an admittedly slightly abusive shorthand for "with/without automorphisms other than the identity".

trix of the form

$$\begin{pmatrix} 0 & I_g \\ -I_g & 0 \end{pmatrix}.$$

This data is equivalent to the choice of a basis $(L_1, \ldots, L_{2g})$ of the space $\mathrm{Jac}(C)_n$ of n-torsion points in the Jacobian of C that is symplectic with respect to the Weil pairing on this space. The moduli space of curves of genus g with full level n structure is denoted $\mathcal{M}_g^{(n)}$. Since the spaces $H^1(C, \mathbb{Z}/n\mathbb{Z})$ (or $\mathrm{Jac}(C)_n$) are isomorphic as symplectic spaces for every curve C, we have a finite Galois covering map $\mathcal{M}_g^{(n)} \to \mathcal{M}_g$ with Galois group $\mathrm{Sp}(2g, \mathbb{Z}/n\mathbb{Z})$ that is unramified exactly over $\mathcal{M}_g^0$. Monodromy arguments — see [77] — show that these covers are all connected.

Another way to rigidify moduli spaces of curves so as to obtain finite coverings is to use ordered sets of Weierstrass points on the curve. This isn't enough to get a fine moduli space over the locus of hyperelliptic curves (why?) so sets of higher-order Weierstrass points are also sometimes used. The covers obtained by these methods have somewhat larger ramification loci: using ordinary Weierstrass points yields a cover that is ramified over curves without automorphisms but having Weierstrass points of weight greater than 1. When such coverings are connected is very much an open question that has only been settled by ad hoc methods in a few cases.

EXERCISE (2.1) 1) Show that, for $n \geq 3$, no curve of genus $g \geq 1$ with level n structure has any automorphisms — i.e., there does not exist a curve C and an automorphism φ of C fixing all points of order n on $\mathrm{Jac}(C)$.

2) Show that this is false for $n = 2$.

3) Show that the map $j : \mathbb{A}^1_\lambda \to \mathbb{A}^1_j$ discussed above is the covering map $\mathcal{M}_1^{(2)} \to \mathcal{M}_1$ associated to level two structures of curves of genus 1.

4) Show that, although the fiber of family $\varphi : \mathcal{X} \to \mathbb{A}^1_\lambda$ over λ is the curve with level two structure whose moduli point is λ, the space $\mathbb{A}^1_\lambda$ is *not* a fine moduli space for curves of genus 1 with level 2 structure.

This example shows the need to use the notion of a "universal family" with care: the existence of a family over a coarse moduli space with the correct fibers need not imply the universal functorial property that characterizes the universal family over a fine moduli space.

EXERCISE (2.2) Consider the family of curves $x^3 + y^3 + z^3 + m \cdot xyz$, parameterized by the affine line $\mathbb{A}^1$ with coordinate m.

1) Find the open set U in $\mathbb{A}^1$ over which the fibers in this family are smooth, and compute the j-function $j = j(m)$ on U.

2) Show that j expresses U as a Galois cover of the j-line, with Galois group $SL_2(\mathbb{Z}/3\mathbb{Z})$.

3) Show that U is a fine moduli space for curves of genus 1 with full level 3 structure.

Hint: The curves in this family are the plane cubics whose flexes are located at $[0, 1, -\omega]$, $[-\omega, 0, 1]$ and $[1, -\omega, 0]$ where ω is a primitive cube root of unity.

EXERCISE (2.3) Show that there does not exist a universal family of curves of genus 2 over any open subset $U \subset \mathcal{M}_2$. In general, if $H_g \subset \mathcal{M}_g$ is the locus of hyperelliptic curves, for which g does there exist a universal family over some open subset $U \subset H_g$? *Answer*: For g odd.

EXERCISE (2.4) Construct examples of:

1) A nontrivial family of smooth curves of genus 3 over a smooth, one-dimensional base B, all of whose fibers over closed points are isomorphic.

2) A map $\varphi : B \rightarrow \mathcal{M}_3$ from a smooth curve B to $\mathcal{M}_3$ that doesn't come from any family of curves of genus 3 over B.

3) A map $\varphi : B \rightarrow \mathcal{M}_3$ from a smooth surface B to $\mathcal{M}_3$ that doesn't come from any family of curves of genus 3, but whose restriction to each open set U_α of a cover of B does.

To say that a moduli functor doesn't admit a fine moduli space simply means that it cannot be represented in the category of schemes. The third approach to this failure is to look for a larger category in which the functor *can* be represented. In order to make such an approach worthwhile, we must understand the larger category well enough to be able to carry out geometric investigations in it. If this can be achieved, it becomes a matter of taste whether the advantages of having a moduli space with good universal properties are sufficient compensation for the additional technical difficulties of working with these more general objects. The mildest generalization of schemes that has proven useful is Artin's category of algebraic spaces [10]. An *algebraic space* looks locally like the quotient of a scheme by an étale equivalence relation. Unfortunately, this category is still too small to provide representing spaces for most moduli problems.

A larger category is that of functors from schemes to sets. This category has the advantage that moduli functors are, by definition, objects in it. What isn't so clear is how we are to interpret geometric notions in this category. Here we shall simply state that this can be done fairly satisfactorily for moduli functors and refer you to Mumford's seminal article [119] if you want to get a sense of the flavor of the arguments needed. In that paper, Mumford shows how to extend the notions of invertible sheaf and Picard group to such functors and, as evidence that it's possible to work with such notions, calculates the Picard group of the moduli space of elliptic curves as $\mathbb{Z}/12\mathbb{Z}$!

Categories of *algebraic stacks* are other enlargements of the category of schemes that have been widely used to study moduli problems. Very roughly, a stack — say, the moduli stack of curves of genus g for concreteness — is itself a category. In our example, typical objects would be families $\mathcal{C} \to A$ and $\mathcal{D} \to B$ of such curves and a morphism between two such objects would be a morphism of schemes from B to A plus an isomorphism of $\mathcal{D}$ with the fiber product

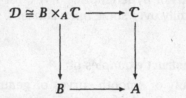

of B and $\mathcal{C}$ over A. In essence, twisting by automorphisms is prohibited by definition. We leave it to you to formulate or find the correct definition of a morphism between stacks (a task that starts to bring out the flavor of the subject).

The stack approach has the advantage of being somewhat closer to geometric intuition: for example, a line bundle on a stack is simply a system of line bundles on the base of each family, together with, for each morphism of families as above, an isomorphism of the line bundle associated to the family $\mathcal{D} \to B$ with the pullback to B of the line bundle associated to $\mathcal{C} \to A$. We won't work with stacks here, but we'll be working with related notions (see the discussion in Section 3.D). If you're interested, you can look at [29] for a first discussion of stacks in the present context; or, if you're prepared for a considerable effort, go to [106] for a full treatment. There is also a forthcoming book [14] that may finally clear up what has traditionally been a murky area.

The approaches to extending the category of schemes via "categories of functors" and via "algebraic stack" are not comparable: that is, neither category faithfully contains the other. There is a common extension, "fibered categories" due (naturally, as it were) to Grothendieck [68] which we shall pass by in complete silence.

B   Moduli spaces we'll be concerned with

We've already mentioned the moduli space $\mathcal{M}_g$ (though we have yet to prove its existence). It is the coarse moduli space for smooth, complete, connected curves C of genus g over $\mathbb{C}$. For the rest of this section, we use "curve" to abbreviate this package. The space $\mathcal{C}_g$ is simply the coarse moduli space of pairs (C, p) where C is a curve and p a point of C. Note that $\mathcal{C}_g$ naturally maps to $\mathcal{M}_g$ by forgetting the point p. In fact, $\mathcal{C}_g$ may look at first glance like a universal curve over $\mathcal{M}_g$, but on closer examination we see that this is true only over the open set $\mathcal{M}_g^0$: the set-theoretic fiber of $\mathcal{C}_g$ over a point $[C] \in \mathcal{M}_g$ is the quotient $C/\mathrm{Aut}(C)$. Thus, for example, over an open subset of $\mathcal{M}_2$, $\mathcal{C}_2$ is a $\mathbb{P}^1$-bundle (in the analytic topology; in the Zariski topology it's a conic bundle). This is even true scheme-theoretically in this example. You may wish to consider the question: what, in general, are the scheme-theoretic fibers of the map from $\mathcal{C}_g$ to $\mathcal{M}_g$? Despite this, we'll occasionally abuse language in order to honor custom by calling $\mathcal{C}_g$ the *universal curve* over moduli.

The space $\mathcal{M}_{g,n}$ is a direct generalization of $\mathcal{C}_g$: it is the coarse moduli space for $(n+1)$-tuples $(C, p_1, \ldots, p_n)$ where C is a curve and $p_1, \ldots, p_n \in C$ are *distinct* points. Thus $\mathcal{C}_g$ equals $\mathcal{M}_{g,1}$. (Because the justification for requiring the points to be distinct comes from the way in which the compactifications of these spaces are constructed, we'll postpone discussion of it until we come to consider these spaces. One can also construct moduli spaces involving marked sets of unordered points (distinct or otherwise); in practice, working with these involves the additional aggravation of keeping track of the $\mathfrak{S}_d$-action without any compensating advantages so $\mathcal{M}_{g,n}$ is the space most commonly dealt with.) Once again, it's tempting to view $\mathcal{M}_{g,n}$ as the open subset of the fiber product $\mathcal{C}_g \times_{\mathcal{M}_g} \mathcal{C}_g \times_{\mathcal{M}_g} \cdots \times_{\mathcal{M}_g} \mathcal{C}_g$ obtained by removing all diagonals; but automorphisms, as for $\mathcal{C}_g$, make this correct only over a sublocus.

The next space we wish to mention is $\mathcal{P}_{d,g}$[2], the coarse moduli space of pairs (C, L) where C is a curve and L a line bundle of degree d on C. Again, the fiber of $\mathcal{P}_{d,g}$ over a point $[C] \in \mathcal{M}_g$ corresponding to a curve C *without* automorphisms is the connected component $\mathrm{Pic}^d(C)$ of the Picard variety of C; in particular, $\mathcal{P}_{0,g}$ is sometimes called the *Jacobian bundle* over moduli. Despite the fact that all the fibers $\mathrm{Pic}^d(C)$ of the varieties $\mathcal{P}_{d,g}$ for various values d are isomorphic over $\mathcal{M}_g^0$, $\mathcal{P}_{d,g}$ will not in general be isomorphic to $\mathcal{P}_{d',g}$ even over $\mathcal{M}_g^0$: see the exercises below. For example, it follows from the Harer-Mestrano proof of the Franchetta conjecture (discussed later in this chapter; or see [112]) that $\mathcal{P}_{0,g} \cong \mathcal{P}_{d,g}$ if and only if $(2g - 2)|d$. We may, however, note that

$$\mathcal{P}_{d,g} \cong \mathcal{P}_{d+2g-2,g}$$

and

$$\mathcal{P}_{d,g} \cong \mathcal{P}_{-d,g},$$

---

[2]Also denoted Jac_g^d in some references.

the isomorphisms being provided by the maps $(C, L) \mapsto (C, L \otimes K_C)$ and $(C, L) \mapsto (C, L^{-1})$ respectively. Thus, for each g there can be *at most g* of these objects — $\mathcal{P}_{0,g}, \dots, \mathcal{P}_{g-1,g}$ — that are distinct up to isomorphism. Note also that in the special case $d = g - 1$ there is a natural theta-divisor θ in $\mathcal{P}_{g-1,g}$, restricting on each fiber to the corresponding class: it's the locus of pairs (C, L) in $\mathcal{P}_{g-1,g}$ with $H^0(C, L) \neq 0$. Beware, however, that we cannot, as for individual curves, define such a class in every degree.

EXERCISE (2.5) Show that for $d = 0, \dots, g - 2$ there does not exist any line bundle on $\mathcal{P}_{d,g}$ whose restriction to the fiber $\mathrm{Pic}^d(C)$ of $\mathcal{P}_{d,g}$ over a general point $[C] \in \mathcal{M}_g$ is the line bundle associated to some translate of the Θ-divisor on $\mathrm{Pic}^d(C) \cong J(C)$.

EXERCISE (2.6) Show that no two of the moduli spaces $\mathcal{P}_{0,g}, \dots, \mathcal{P}_{g-1,g}$ are isomorphic.

For a general curve C of genus $g \geq 1$, the Jacobian $J(C)$ has Picard number (i.e., rank of Neron-Severi group) equal to 1 and the Neron-Severi group is generated by a translate of the Θ divisor. It follows that for each d and g, the Picard group of $\mathcal{P}_{d,g}$ has rank 1 over the Picard group $\mathrm{Pic}(\mathcal{M}_g)$, with the generator restricting to some multiple $m(d, g) \cdot \Theta$ of the general fiber of $\mathcal{P}_{d,g}$ over $\mathcal{M}_g$, and we may ask what the coefficient $m(d, g)$ is for each d and g. For example, the existence of the natural theta-divisor $\theta \subset \mathcal{P}_{g-1,g}$ shows that $m(g - 1, g) = 1$ for all g. But, $m(d, g) = 1$ only rarely. The following exercise suggests some of the naive ways of approaching the problem; following it, we'll give the general formula for $m(d, g)$ found by Kouvidakis.

EXERCISE (2.7) 1) Show that $m(0, g) \neq 1$.

2) Show that the locus of pairs (C, L) where L is a line bundle on C of the form $\mathcal{O}_C(p_1 + \cdots + p_{g-1} - (g - 1)p)$ with p a Weierstrass point on C forms a divisor in $\mathcal{P}_{0,g}$. Use this to deduce that $m(d, g) | (g^3 - g)$ for any d and g.

3) Show that there exists a divisor on $\mathcal{C}_g$ whose fiber over a general point $[C] \in \mathcal{M}_g$ is a canonical divisor on C, and deduce that $m(d, g) | (2g - 2)$ for any d and g.

4) Find an example where $m(d, g) = 5$.

THEOREM (2.8) (KOUVIDAKIS [105]) *For all d and g,*

$$m(d, g) = \frac{2g - 2}{\gcd(2g - 2, g + d - 1)}.$$

C   Constructions of $\mathcal{M}_g$

As we indicated earlier, every construction of the moduli space $\mathcal{M}_g$ amounts to looking a priori at curves with some additional structure, so that a parameter space can be described, and then taking the quotient of this space by the relation that identifies these additional structures. In this section, we look at the three most common approaches.

The Teichmüller approach

Here we consider the space of pairs $(C; \gamma_1, \ldots, \gamma_{2g})$ where C is a curve and $\{\gamma_1, \ldots, \gamma_{2g}\}$ is a normalized set of generators for $\pi_1(C)$ — that is, one that may be drawn (in genus 2) as shown in Figure (2.9). This

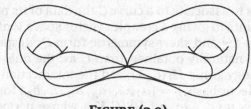

FIGURE (2.9)

is equivalent to choosing a homeomorphism of the underlying topological manifold of C with the National Bureau of Standards' compact orientable surface X_0 of genus g, up to isotopy. The basic theorem here, due to Bers [15], then says that: *the space of such data is naturally an open subset T_g in $\mathbb{C}^{3g-3}$ homeomorphic to a ball.* This open set is called *Teichmüller space.* The group Γ_g of diffeomorphisms of X_0 modulo isotopy then acts on Teichmüller space, and we may realize the moduli space $\mathcal{M}_g$ as the quotient of this action. Note that since the stabilizer of any point is finite (it's simply the group of automorphisms of the underlying curve C), this quotient exists as an analytic variety.

Probably the most important thing about this approach is that it gives us a handle on the topology of $\mathcal{M}_g$: since $\mathcal{M}_g$ is a quotient of a contractible space by the group Γ_g, we see that for small k, the cohomology groups $H^k(\mathcal{M}_g, \mathbb{Q})$ are just the cohomology groups of the group Γ_g tensored with $\mathbb{Q}$. We may then try to calculate these by examining an action of Γ_g on another contractible space $\mathcal{N}$.[3] By using a more tractable $\mathcal{N}$ that is combinatorially defined, this approach has

---

[3]We only get information about the rational cohomology because Γ_g has finite isotopy subgroups at points of T_g corresponding to curves with automorphisms. There is, however, an analogue $T_{g,n}$ of T_g parameterizing marked surfaces of genus g with n marked points (or to use the indigenous terminology, with n punctures) from which

been fruitfully exploited by Harer (whose results we'll describe later) and, more recently, by Kontsevich.

We should also mention that this approach provides $\mathcal{M}_g$ with a natural metric, called the Weil-Petersson metric whose positivity properties have been used by Wolpert ([153], [154]) to construct an embedding of $\mathcal{M}_g$ in a projective variety with many of the nice properties of the Deligne-Mumford stable compactification which we will introduce later in this section. An excellent survey of what is known along these lines can be found in the paper of Hain and Looijenga [70].

The Hodge theory approach

The idea here is to associate to a curve C the data of its polarized Jacobian: this amounts to giving a complex vector space V of dimension g with lattice $\Lambda \cong \mathbb{Z}^{2g}$ and skew-symmetric form Q. Respectively, these ingredients are naturally obtained from C as: the dual of $H^0(C, K_C)$; the first homology group $H_1(C, \mathbb{Z})$; and the intersection pairing. If we choose a symplectic basis $\beta = \{a_1, \ldots, a_g, b_1, \ldots, b_g\}$ for $H_1(C, \mathbb{Z})$ and a complex basis $\omega_1, \ldots, \omega_g$ of $H^0(C, K_C)$ whose period matrix with respect to the a-cycles is I_g, we may in turn associate to these data the *period matrix* $P \in \mathbb{C}^{g^2}$ given by integrating the ω's around the b-cycles. The Riemann bilinear relations then say that P is symmetric with positive definite imaginary part. These last two conditions define, respectively, a subspace and an open subset of the space of $g \times g$ complex matrices whose intersection is called the *Siegel upper halfspace* of dimension g and is denoted $\mathfrak{h}_g$. The group $\mathrm{Sp}(2g, \mathbb{Z})$ of symplectic changes of basis acts on $\mathfrak{h}_g$ and this action corresponds exactly to the choice of symplectic basis made above.

Here the main facts are that: *period matrices of curves form a locally closed subset c_g of $\mathfrak{h}_g$; the quotient $\mathcal{A}_g$ of $\mathfrak{h}_g$ by $\mathrm{Sp}(2g, \mathbb{Z})$ is a coarse moduli space for abelian varieties of dimension g; and, $\mathcal{M}_g$ can be constructed by restricting this quotient map to the locus c_g.* Again, this construction yields $\mathcal{M}_g$ only as an analytic space but it has the important advantage over the Teichmüller approach that the group $\mathrm{Sp}(2g, \mathbb{Z})$ by which we're quotienting is more approachable than Γ_g. We pay for this, however, because we can say much less about the space c_g that we're quotienting. Describing the locus $\mathcal{M}_g$ in $\mathcal{A}_g$ (or c_g in $\mathfrak{h}_g$) is the Schottky problem. Formally, a number of solutions have recently been obtained ([8], [117], [125], [142]) but for practical

---

the moduli space $\mathcal{M}_{g,n}$ can be constructed by forming a quotient by a suitable group $\Gamma_{g,n}$. For n large enough that such marked curves have no automorphisms, we can obtain information about $H^k(\mathcal{M}_{g,n}, \mathbb{Z})$ by this method.

purposes — such as determining whether a given period matrix P in $\mathfrak{h}_g$ lies in $\mathfrak{c}_g$ — they are little help.

This construction has one other important consequence. Since $\mathcal{A}_g$ is a hermitian symmetric domain, it has by [12] a natural Baily-Borel compactification $\widehat{\mathcal{A}}_g$. (We will discuss compactifications at length below: for the present, when we say that $\overline{\mathcal{M}}$ is a compactification of $\mathcal{M}$, we'll mean that $\overline{\mathcal{M}}$ is a compact analytic variety that contains $\mathcal{M}$ as an analytic open subset.) The compactification $\widehat{\mathcal{A}}_g$ was historically the first such compactification to be constructed [136] and it remains known as the Satake compactification in honor of its discoverer.

Taking the closure of $\mathcal{M}_g$ in $\widehat{\mathcal{A}}_g$ yields a compactification of $\mathcal{M}_g$ which we'll denote by $\widetilde{\mathcal{M}}_g$ and also refer to as the Satake compactification. Unfortunately, the Satake compactification isn't modular. Recall that this means that $\widetilde{\mathcal{M}}_g$ is not a moduli space for any moduli functor of curves that contains the moduli functor of smooth curves as an open subfunctor. (In fact, the points of $\widetilde{\mathcal{M}}_g \setminus \mathcal{M}_g$ do correspond to isomorphism classes of smooth curves of lower genus, but these don't naturally fit into families with curves of genus g. Thus while we can associate to families of curves with some singular fibers a moduli map to $\widetilde{\mathcal{M}}_g$, we can't go back and interpret subvarieties of $\widetilde{\mathcal{M}}_g$ not contained in $\mathcal{M}_g$ in terms of families of curves.) This greatly lessens the usefulness of $\widetilde{\mathcal{M}}_g$ for the study of most questions about families of curves or about $\mathcal{M}_g$ itself.

There is one important exception to this last statement. It depends on the following two properties of $\widetilde{\mathcal{M}}_g$: first, $\widetilde{\mathcal{M}}_g$ *is projective*; and second, *the codimension of the complement $\widetilde{\mathcal{M}}_g \setminus \mathcal{M}_g$ in $\widetilde{\mathcal{M}}_g$ is equal to 2 for $g \geq 3$*. By intersecting $\mathcal{M}_g$ with generic divisors in some large multiple $\mathcal{O}(n)$ of a very ample invertible sheaf on $\widetilde{\mathcal{M}}_g$ through any point, we see that *through any point of $\mathcal{M}_g$ there passes a complete curve lying entirely in $\mathcal{M}_g$*. In fact, there is a complete curve through any finite collection of points of $\mathcal{M}_g$: see the exercise below. Using a curve through two points, on which any holomorphic function must be constant, we see that *there are no nonconstant functions on $\mathcal{M}_g$*.

EXERCISE (2.10) Assuming the facts cited above about the Satake compactification $\widetilde{\mathcal{M}}_g$, show that through any finite collection of points of $\mathcal{M}_g$ there passes a complete curve lying in $\mathcal{M}_g$. *Hint*: blowup the points and use the fact that the pullback to this blowup of a sufficiently large multiple of an ample linear series on $\widetilde{\mathcal{M}}_g$ minus the sum of the exceptional divisors of the blowup is very ample.

Together these facts show that $\mathcal{M}_g$ is neither projective nor affine. In the next section, we will look at some more refined results about complete subvarieties of $\mathcal{M}_g$ which shed light on where in the range between these extremes $\mathcal{M}_g$ lies.

The geometric invariant theory (G.I.T.) approach

This attack is quite distinct in flavor from the previous two. Simply put, in each of the last two cases, the extra data attached to a curve C was essentially analytic. Correspondingly, the parameter space of curves with this extra data was not an algebraic but a complex analytic variety, and the group acting on this space with quotient $\mathcal{M}_g$ was not an algebraic group. In the G.I.T. approach, however, everything is algebraic.

The idea is straightforward: for any integer $n \geq 3$, any curve C may be embedded as a curve of degree $2(g-1)n$ in projective space $\mathbb{P}^N = \mathbb{P}^{(2n-1)(g-1)-1}$ by the complete linear series $|nK_C|$. We may accordingly attach to a curve C the data of such an embedding — i.e., we consider pairs consisting of a curve C and an n-canonical embedding $\varphi : C \to \mathbb{P}^N$. Now, we've already seen how to parameterize such pairs: the family of all such corresponds to a *locally closed* subset $\mathcal{K}$ of the Hilbert scheme $\mathcal{H} = \mathcal{H}_{2(g-1)n,g,(2n-1)(g-1)-1}$ of smooth curves of degree $2(g-1)n$ and genus g in $\mathbb{P}^N$. (Over the open set of *smooth* curves in $\mathcal{H}$, the universal curve $\mathcal{C}$ carries two natural invertible sheaves, the hyperplane sheaf, $\mathcal{O}_{\mathcal{C}}(1)$, and the relative dualizing sheaf, $\omega_{\mathcal{C}/\mathcal{H}}$; $\mathcal{K}$ is simply the locus where $\mathcal{O}_{\mathcal{C}}(1)$ and $(\omega_{\mathcal{C}/\mathcal{H}})^{\otimes n}$ are isomorphic.) Moreover, the ambiguity in choosing the map φ is simply a matter of choosing a basis for the space $H^0(C, K_C^{\otimes n})$ of n-canonical differentials on C — in other words, the group $\mathrm{PGL}(N+1, \mathbb{C})$ acts on $\mathcal{K}$, and the quotient (if one exists) should be $\mathcal{M}_g$.

One problem with this approach is that, since the group $\mathrm{PGL}(N+1, \mathbb{C})$ is continuous rather than discrete, the existence of a nice quotient is by no means assured. This is shown by using the techniques of *geometric invariant theory*, which we'll discuss later. Assuming, for the moment, that we've constructed this quotient, however, the approach has two signal advantages:

- It exhibits the $\mathcal{M}_g$ as a quasiprojective algebraic variety.

- It leads to an explicit, modular projective compactification of $\mathcal{M}_g$.

Briefly, we've indicated that it requires some nontrivial work to show that the quotient $\mathcal{K}$ by $\mathrm{PGL}(N+1, \mathbb{C})$ exists. Having undertaken this work, however, it's tempting to try to compactify $\mathcal{M}_g$ by taking a quotient of the closure $\overline{\mathcal{K}}$ of $\mathcal{K}$ in $\mathcal{H}$. However, this is only possible for an open subset $\widetilde{\mathcal{K}}$ of $\overline{\mathcal{K}}$ containing $\mathcal{K}$. To get an idea of why some such restriction is necessary, consider the family $\mathcal{C}_{(a,b)}$ of smooth cubics over the affine t-line whose fibers C_t are given by

$$(2.11) \qquad yz^2 = x^3 - t^2axz^2 - t^3bz^3.$$

The curves C_t for $t \neq 0$ are all isomorphic to the smooth curve C_1 of genus 1 but the curve C_0 is a rational cuspidal curve. Clearly, this sort of jump discontinuity rules out the existence of any kind of good moduli space containing both C_1 and C_0. By varying the choice of a and b, we can arrange for C_1 to have any desired j-invariant so the blame for this pathology clearly belongs with C_0. We're thus faced with the problem of determining what abstract curves to admit into the enlarged parameter space $\widetilde{\mathcal{K}}$. The right curves, which emerge naturally from studying this quotienting problem (and, as we'll see later, from several other points of view), are stable curves.

DEFINITION (2.12) A stable curve *is a complete connected curve that has only nodes as singularities and has only finitely many automorphisms.*

In view of the connectedness of C, its automorphism group can fail to be finite only if C contains rational components. Thus, the finiteness condition can be equivalently reformulated as:

- every smooth rational component C meets the other components of C in at least 3 points; or,
- every rational component of the normalization of C has at least 3 points lying over singular points of C.

If we weaken either of these conditions by replacing the number 3 by 2, the resulting curves are called *semistable*. Geometrically, this amounts to allowing chains $C_1, \ldots, C_k$ of smooth rational curves as subcurves of C. More precisely, saying that we have a chain means that: C_1 and C_k each meet the complement of the chain in C in a single node; the other C_i are disjoint from this complement; and, each C_i for i between 2 and $(k-1)$ meets each of C_{i-1} and C_{i+1} in a single node and meets no other components of the chain. Later on, we'll introduce other notions of stability for curves connected with the quotienting process and, to distinguish the curves described above, will call them *moduli stable curves* or *Deligne-Mumford stable curves*.

Stable curves with marked points are defined analogously:

DEFINITION (2.13) A stable n-pointed curve *is a complete connected curve C that has only nodes as singularities, together with an ordered collection $p_1, \ldots, p_n \in C$ of distinct smooth points of C, such that the $(n+1)$-tuple $(C; p_1, \ldots, p_n)$ has only finitely many automorphisms.*

As in the definition of stable curve, the finiteness condition can be equivalently reformulated as saying that every rational component of the normalization of C has at least 3 points lying over singular and/or marked points of C. Also as before, if we weaken either of these conditions by replacing the number 3 by 2, the resulting pointed curves are called *semistable*.

As for smooth curves, the arithmetic genus $g = h^1(C, \mathcal{O}_C)$ of a stable curve C is a primary invariant. As will be verified in Exercise (3.2), we can reexpress the genus more geometrically as follows: if C has δ nodes and ν irreducible components $C_1, \ldots, C_\nu$ of geometric genera $g_1, \ldots, g_\nu$, then

$$g = \sum_{i=1}^{\nu} (g_i - 1) + \delta + 1$$

(2.14)

$$= \left(\sum_{i=1}^{\nu} g_i \right) + \delta - \nu + 1.$$

The fact that stable curves of genus g are the right class of curves to consider is expressed in:

THEOREM (2.15) (DELIGNE-MUMFORD-KNUDSEN) *There exist coarse moduli spaces* $\overline{\mathcal{M}}_g$ *and* $\overline{\mathcal{M}}_{g,n}$ *of stable curves and n-pointed stable curves; and these spaces are projective varieties.*

The spaces $\overline{\mathcal{M}}_g$ and $\overline{\mathcal{M}}_{g,n}$ are called the *stable compactifications* of $\mathcal{M}_g$ and $\mathcal{M}_{g,n}$.

It's hard to overestimate the importance of having such a modular compactification: i.e., one that is actually a moduli space for a well-behaved class of (possibly singular) curves. Clearly, being able to deal with a projective variety like $\overline{\mathcal{M}}_g$ rather than just a quasiprojective one like $\mathcal{M}_g$ allows us to bring to bear many of the tools of projective algebraic geometry in the study of these spaces; this is what will allow us, for example, to answer in Section 6.F the classical question about the unirationality of $\mathcal{M}_g$.

Beyond that, and perhaps even more significantly, the existence of a compact moduli space for curves has changed the way we view them. Now, anytime we have a one-parameter family of curves $\{C_t\}$ in projective space, or simply mapping to projective space — a family of plane curves acquiring a nasty singularity, or becoming reducible or nonreduced, or a family of branched covers of $\mathbb{P}^1$ in which a large number of branch points coalesce at once — we know that however wild the singularities of the flat limit C_0 of these curves, there is also a well-defined limit of the arc $\{[C_t]\} \subset \overline{\mathcal{M}}_g$; in other words, a canonical limit Y_0 of the abstract curves C_t that has only nodes as singularities and whose geometry will illuminate that of the curve C_0. This notion, expressed formally in Proposition (3.47), underlies almost all of the constructions and applications in this book. It would not be an exaggeration to say that Theorem (2.15) has played as fundamental a role in the theory of algebraic curves in the last thirty years as the notion of abstract curve did in the preceding sixty.

As suggested above, $\overline{\mathcal{M}}_g$ may be realized by geometric invariant theory: if we define $\widetilde{\mathcal{K}}$ to be the locus of stable curves C embedded by the nth power of their dualizing sheaves ω_C and define $\mathcal{M}_g$ to be the quotient $\widetilde{\mathcal{K}}/\text{PGL}(N+1,\mathbb{C})$, then $\overline{\mathcal{M}}_g$ is the coarse moduli space of stable curves. While this will be carried out in detail in Chapter 4, we'd like now to introduce a few problems related to this construction. If you haven't seen the basics of the geometric invariant theory of Hilbert points, you may want to skip the next few paragraphs until after you've read Chapter 4.

The first problem is to show that the orbits that the G.I.T. quotient of $\widetilde{\mathcal{K}}$ "throws away" are exactly those that are *not* pluricanonically embedded stable curves. The analysis of this and related questions that arise in the G.I.T. construction of $\overline{\mathcal{M}}_g$ is quite intricate. Having carried out this analysis, it's natural to ask what sort of compact quotient we can build by considering not just pluricanonically embedded curves but all embedded curves with semistable Hilbert points. This amounts to trying to find moduli for pairs (C, Λ) where C is a curve of genus g and Λ is a linear system of degree d and dimension r on C. For smooth C, the answer is both easy to state and relatively straightforward to verify. If $d \gg g$, then the orbit of such a pair produces a point in the quotient whenever Λ is complete. The resulting quotient is a universal Picard bundle $\mathcal{P}_{d,g}$. The full quotient again yields a modular compactification $\overline{\mathcal{P}}_{d,g}$ of $\mathcal{P}_{d,g}$. Using the isomorphism $\mathcal{P}_{d,g} \cong \mathcal{P}_{d+2g-2,g}$ discussed in Section B, this gives a stable compactification of the Jacobian bundle. Determining which orbits, or, more generally, which finite sets of orbits, determine points of this quotient involves a lengthy and delicate combinatorial analysis that has only recently been carried out by Caporaso [16]: see the discussion following Theorem (4.45) for more details.

We will discuss $\overline{\mathcal{M}}_g$ in more detail later. For the time being, we want only to make some elementary observations. Fix a curve C with δ nodes and ν irreducible components $C_1, \ldots, C_\nu$ of geometric genera $g_1, \ldots, g_\nu$. Now, to specify such a stable curve we have to specify the normalizations $\tilde{C}_i$ of the C_i, and then specify the points on each that will be identified to form the nodes of C — there will be 2δ such points in all. The family of such curves thus has dimension

$$\sum_{i=1}^{\nu} (3g_i - 3) + 2\delta \, ,^4$$

---

[4]You may be worried about this parameter count when g_i equals 0 or 1. In the rational case, the correct contribution should be 0 not -3. Fortunately, this is exactly compensated for by the fact that the $\delta_i \geq 3$ marked points on such a component actually depend on only $\delta_i - 3$ parameters because of the automorphisms of the rational curve $\tilde{C}_i$. The number $(3g_i - 3)$ actually counts moduli of C_i minus moduli

which, in view of the genus formula (2.14) equals,

$$3g - 3 - \delta.$$

In other words, *the locus in $\overline{\mathcal{M}}_g$ of curves with exactly δ nodes has pure codimension δ in $\overline{\mathcal{M}}_g$*. Moreover, a local computation in deformation theory which we'll carry out in the next chapter shows that the locus of curves with more than δ nodes lies in the closure of the locus of those with exactly δ. In particular, the boundary $\Delta = \overline{\mathcal{M}}_g - \mathcal{M}_g$ is a divisor, with each component the closure of a locus of curves with 1 node. In this case, the combinatorics are easy to work out: a stable curve with one node is either irreducible, or the union of smooth curves of genera i and $g - i$ meeting at one point. These give rise to divisors Δ_0 and Δ_i, $i = 1, \ldots, \lfloor g/2 \rfloor$.

Parenthetically, we should say that it's at this point that one in general stops drawing curves as two-dimensional objects and starts using the less suggestive but more efficient one-dimensional representation. Thus, instead of drawing general curves in the boundary components Δ_0 and Δ_1 in $\overline{\mathcal{M}}_3$ as in Figure (2.16), we would draw them simply as

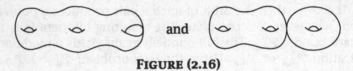

and

FIGURE (2.16)

in Figure (2.17). The surface pictures are actually rather misleading:

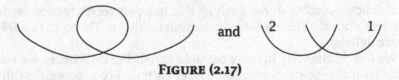

and

FIGURE (2.17)

locally, a node looks like a pair of (real) two-manifolds meeting transversely in a point; this can and does occur in real fourspace, but not in threespace. In order to fit our pictures in a two-dimensional representation of threespace, we're obliged to either pinch these planes, as on the left, or show them as tangent as on the right, either of which is incorrect.

The boundary components of $\overline{\mathcal{M}}_{g,n}$ may be listed analogously; here is the statement for $n = 1$: the space $\overline{\mathcal{M}}_{g,1}$ is often denoted $\overline{\mathcal{C}}_g$, and called (misleadingly; see the discussion in Section B) the *universal curve* over $\overline{\mathcal{M}}_g$.

---

of its automorphisms and this makes all parameter counts like that above come out right. We leave you to check that the count is also correct in the genus 1 case.

EXERCISE (2.18) Show that the boundary $\overline{\mathcal{C}}_g \setminus \mathcal{C}_g$ consists of exactly g divisors: the closure Σ_0 of the locus of pairs (C, p) where C is an irreducible curve with a single node; and the closures Σ_i of the locus of pairs (C, p) where C is the union of smooth curves of genera i and $g - i$ meeting at a single point, and p lies on the component of genus i.

EXERCISE (2.19) Even rational curves can have moduli when marked points are added. Show that $\mathcal{M}_{0,3}$ and $\overline{\mathcal{M}}_{0,3}$ are both simply a point by using an automorphism to fix the 3 marked points. Likewise, show that $\mathcal{M}_{0,4} \cong \mathbb{P}^1 \setminus \{0, 1, \infty\}$. Show that any singular stable curve in $\overline{\mathcal{M}}_{0,4}$ must consist of a pair of smooth rational curves meeting in a point with each carrying two of the four marked points and that two such are isomorphic if and only if the induced decompositions of $\{1, 2, 3, 4\}$ into two pairs agree. For more moduli spaces of rational curves, see [91].

Here are some exercises on the stable compactification of $\mathcal{M}_g$.

EXERCISE (2.20) As a consequence of the dimension computation above, we may deduce that *no stable curve can have more than* $3g - 3$ *nodes*. Prove this directly.

EXERCISE (2.21) How many stable curves of genera 2, 3 and 4 are there up to homeomorphism (in the analytic topology)? For each homeomorphism type, find the dimension of the locus of the corresponding curves in $\overline{\mathcal{M}}_g$ and say which of these loci are in the closure of which others.

EXERCISE (2.22) Show that the normalization $\tilde{C}$ of a stable curve C with $3g - 3$ nodes is a union of rational curves, each having 3 marked points. Up to isomorphism, how many such curves C are there for $g = 2, 3, 4$ and 5? Harder: for general g?

EXERCISE (2.23) How many components are there in the locus of stable curves with 2 or more nodes? Which lie in the closure of each boundary component Δ_i?

Let B be a smooth curve, $p \in B$ any point, and let $\tilde{\mathcal{X}} \to B - \{p\}$ be any family of stable curves. Let $\tilde{\varphi} : B - \{p\} \to \overline{\mathcal{M}}_g$ be the corresponding map to moduli. Since $\overline{\mathcal{M}}_g$ is projective, the valuative criterion for properness implies that there is a unique extension of $\tilde{\varphi}$ to a map $\varphi : B \to \overline{\mathcal{M}}_g$. In this circumstance, the curve corresponding to $\varphi(p)$ is called the *stable limit* of the curves $\{X_q\}_{q \in B - \{p\}}$ as q approaches p. The determination of such limits by the process of *semistable reduction* will be discussed in considerable detail in the next chapter. Here is a warm-up exercise for those of you already familiar with this process.

EXERCISE (2.24) Let B be a smooth curve of genus $g - 1$, $p \in B$ any point, and for any $q \in B \setminus \{p\}$ let X_q be the stable curve obtained by identifying p and q on B. What is the stable limit of the family $\{X_q\}$ as q approaches p?

EXERCISE (2.25) [posed by Jean-Francois Burnol] Let $\Delta^{(\alpha)} \subset \overline{\mathcal{M}}_g$ be the locus of stable curves with α or more nodes. For which α is $\Delta^{(\alpha)}$ connected?

EXERCISE (2.26) It's a classical fact that the automorphism group of a smooth curve of genus g can have order at most $84(g - 1)$. Does the same statement hold for stable curves?

D   Geometric and topological properties

Basic properties

We've already said that $\mathcal{M}_g$ is irreducible of dimension $3g - 3$. Any of the standard ways of establishing the dimension amounts to making the computation of the Hilbert number in reverse: we used the dimension of $\mathcal{M}_g$ in computing the dimension $(r + 1)d - (r - 3)(g - 1)$ of any component of the Hilbert scheme $\mathcal{H}_{d,g,r}$ whose general member was nonspecial. Conversely, if we exhibit such a component having dimension exactly $h_{d,g,r}$ and dominating $\mathcal{M}_g$ we will have verified its dimension. This is straightforward either using Hurwitz schemes (this is the more usual, since the dimension of any component of $\mathcal{H}_{d,g}$ is visibly $b = 2d + 2g - 2$) or Severi varieties.

Irreducibility comes a little harder. Again, the standard approaches invoke the parameter spaces $\mathcal{H}_{d,g}$ or $V_{d,g}$. Thus, for example, Clebsch analyzed the Hurwitz scheme $\mathcal{H}_{d,g}$ as a covering space of an open subset of $\mathbb{P}^b$, and showed that the monodromy acted transitively on the sheets; he deduced that $\mathcal{H}_{d,g}$ was irreducible for any d and g and hence that $\mathcal{M}_g$ was. Likewise, the fact that the Severi variety $V_{d,g}$ is irreducible for any d and g implies that $\mathcal{M}_g$ is (although historically, the irreducibility of $\mathcal{M}_g$ was known long before the irreducibility of $V_{d,g}$). Although we'll only prove the irreducibility of $\mathcal{M}_g$ in Section 6.A, we'll make free use of it in the interim. You can easily check that we introduce no circular dependencies in doing so.

Local properties

The local structure of the moduli space $\mathcal{M}_g$ is very well understood. The basic facts, which we'll state here, are all consequences of the deformation theory we'll describe in detail in Chapter 3.

To begin with, the moduli space $\mathcal{M}_g$ is smooth at a point $[C]$ corresponding to a curve without automorphisms. For genus $g \geq 4$, the singular locus of $\mathcal{M}_g$ is exactly the locus of curves with automorphisms, and the singularities of $\mathcal{M}_g$ are finite quotient singularities — more precisely, in an analytic neighborhood of any point $[C] \in \mathcal{M}_g$, $\mathcal{M}_g$ looks like a quotient of an open subset of $\mathbb{C}^{3g-3}$ by a linear action of $\text{Aut}(C)$, where the fixed point sets of elements $\varphi \in \text{Aut}(C)$ are exactly the curves nearby to which the automorphism φ deforms. We will see in the following chapter how to describe this linear action more explicitly. In particular, whenever the locus of curves with automorphisms has codimension two or more, a curve with an automorphism must be a singular point of $\mathcal{M}_g$. Of course, for $g = 2$, every curve has a hyperelliptic involution, but $\mathcal{M}_2$ is smooth except at one point (corresponding to the curve given by $y^2 = x^5 - 1$, which has additional automorphisms). For $g = 3$, the hyperelliptic curves form a divisor whose generic point corresponds to a curve with a single nontrivial automorphism, the hyperelliptic involution. At such points, the space $\mathcal{M}_3$ is smooth. This explains the restriction $g \geq 4$ in the second sentence of this paragraph. The exercise below checks that these are the only divisorial components in the locus of curves with automorphisms.

EXERCISE (2.27) Consider a smooth curve C of genus g with a nontrivial automorphism σ of prime order p. Let $f : C \to D$ be the quotient of C by the group generated by σ, let h be the genus of D, and let b be the number of points of D over which f ramifies.

1) Show that all ramification points of f have ramification index $(p-1)$ — i.e., exactly p sheets meet at each. Use the Riemann-Hurwitz formula to derive the relation

$$2g - 2 = p(2h - 2) + (p - 1)b.$$

2) Show that the curve D together with the branch points of f is a stable b-pointed curve of genus h. Then, use the fact that such a D depends on $3h - 3 + b$ moduli to deduce that the pairs (C, σ) depend on at most $2g - 1$ moduli with equality only if $h = 0$ and $p = 2$.

3) Show that the only component in the locus of smooth curves of genus g with automorphisms which is a divisor in $\mathcal{M}_g$ is the hyperelliptic locus in genus 3.

Analogous statements hold for moduli spaces of curves with marked points (with the additional exceptional case of $\mathcal{M}_{1,1}$ in which every curve has an automorphism given by inversion with respect to the marked point).

A similarly explicit description may be given of the local structure of $\overline{\mathcal{M}}_g$. Precisely, near a point $[C] \in \overline{\mathcal{M}}_g$, $\overline{\mathcal{M}}_g$ looks like a quotient of an open subset of $\mathbb{C}^{3g-3}$ by a linear action of $\text{Aut}(C)$; thus $\overline{\mathcal{M}}_g$ is smooth at points $[C]$ with $\text{Aut}(C)$ trivial, and at worst has finite quotient singularities. Note that any curve in $\overline{\mathcal{M}}_g$ with an *elliptic tail* — that is, an elliptic component joined to the rest of the curve at a single point — has a nontrivial automorphism, namely, the involution on the tail fixing the join point. Thus, for all g, the boundary component Δ_1 is a locus of codimension 1 in $\overline{\mathcal{M}}_g$ consisting entirely of curves with automorphisms. Since a general point of Δ_1 has automorphism group $\mathbb{Z}/2\mathbb{Z}$, such a point will be a smooth point of $\overline{\mathcal{M}}_g$. With this exception, $\overline{\mathcal{M}}_g$ ($g \geq 4$) is again singular at moduli points of curves with automorphisms.

EXERCISE (2.28) 1) Show that, for $g \geq 4$, Δ_1 is the *only* component of Δ whose generic element has a nontrivial automorphism.

2) Find all divisors in $\overline{\mathcal{M}}_g$, $g \geq 3$, whose generic element has a non-trivial automorphism.

We can likewise describe the structure of the boundary $\Delta = \bigcup \Delta_i$ by appealing to results about deformations of stable curves that will be discussed in Section 3.C. For simplicity, assume C is a stable curve with δ nodes $p_1, \ldots, p_\delta$ and without automorphisms. Then in a neighborhood of $[C]$, the boundary Δ is a union of smooth hypersurfaces S_i intersecting transversely. The hypersurface S_i is the locus of deformations C' of the curve C not smoothing the node p_i (i.e., such that C' has a node near p_i). Thus, for example, if C is a stable curve that looks schematically like the curve in Figure (2.29) then in a neighborhood

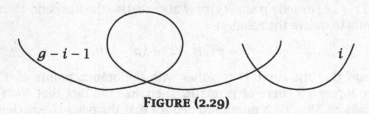

$$g - i - 1 \qquad\qquad\qquad\qquad\qquad\qquad i$$

FIGURE (2.29)

of $[C]$ the boundary will, schematically, look like Figure (2.30).

Finally, we note that the loci Δ_i are, as our language has been implicitly assuming, the irreducible components of Δ. Assuming the irreducibility of $\overline{\mathcal{M}}_g$ itself, and hence of all moduli spaces of pointed curves, the irreducibility of each Δ_i is easily checked by exhibiting it as the closure of the image of an irreducible scheme X under a map $X \to \overline{\mathcal{M}}_g$. A general point of Δ corresponds to a curve C with a single node. If C lies in Δ_0, then its normalization may be viewed as a smooth curve of genus $(g-1)$ with the two preimages of the node as marked points. Identifying the two marked points thus defines a dominating

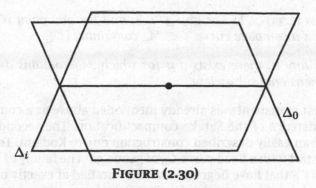

FIGURE (2.30)

map $\mathcal{M}_{g-1,2} \to \Delta_0$. When C lies on Δ_i for some $i > 0$, its normalization has two components of genera i and $g - 1$ each with one marked point: this determines a dominating map $\mathcal{M}_{i,1} \times \mathcal{M}_{g-i,1} \to \Delta_i$.

Complete subvarieties of $\mathcal{M}_g$

Intuitively, $\mathcal{M}_g$ isn't a projective variety, since smooth curves do degenerate to singular stable ones. (The implication is only a naive one because we'll see later that the moduli map may send the base of a family of curves with singular fibers into $\mathcal{M}_g$.) On the other hand, we've also said that it isn't affine: for one thing, it has no nonconstant holomorphic functions. The question arises, then, of how close to being either affine or projective $\mathcal{M}_g$ is. One way to quantify this is to observe that an affine variety contains no projective subvarieties of positive dimension, while a projective variety X of dimension n tautologically contains an n-dimensional one. We may thus ask the question:

QUESTION (2.31) What is the largest dimension r_g of a complete (i.e., projective) subvariety contained in $\mathcal{M}_g$?

However, the formulation of this question permits some misleading responses. For example, if we took an affine variety X of dimension n and blew up a point, the resulting variety $\tilde{X}$ would contain a projective subvariety of dimension $n - 1$. Accordingly, a better posed question is:

QUESTION (2.32) If $[C] \in \mathcal{M}_g$ is a general point, what is the largest dimension $\tilde{r}_g$ of a complete (i.e., projective) subvariety contained in $\mathcal{M}_g$ and passing through $[C]$?

Here is the current state of our knowledge on these issues, in one direction:

THEOREM (2.33)　　1) *For any $g \geq 3$, and for any point $[C] \in \mathcal{M}_g$, there is a complete curve $X \subset \mathcal{M}_g$ containing $[C]$.*

2) *For any n, there exists a g for which $\mathcal{M}_g$ contains a complete, n-dimensional subvariety.*

The first statement was already mentioned above as a consequence of the existence of the Satake compactification. The second result is based on an easily described construction due to Kodaira. To give one variant, start with a fixed curve C_0 of genus g_0. The family of branched covers of C_0 that have degree 3 and are ramified at exactly one point[5] is a complete one-parameter family $\{C_\lambda\}$ of smooth curves of genus $g_1 = 3g_0 - 2$. The key point in verifying this is that we have just one branch point (which by Riemann-Hurwitz forces us to use a covering of odd degree). As we'll soon see, the minimum genus in which the Kodaira construction based on covers of degree d yields a complete n-dimensional subvariety is roughly d^n; hence the choice of degree 3. If we iterate this construction — that is, consider all covers of degree 3 of curves in the family $\{C_\lambda\}$ ramified at one point — we get a complete, 2-dimensional family of curves of genus $g_2 = 3g_1 - 1 = 9g_0 - 4$. In general, we obtain in this way a complete n-dimensional family of curves of genus $g = 3^n g_0 - (3^n - 1)/2$. The dimension of the family must increase with each iteration because any smooth curve covers only finitely many curves of positive genus (see Exercise (1.21)). Note that we don't claim that these families are connected or that they map birationally to moduli.

For g large, this yields the complete subvariety of $\mathcal{M}_g$ of largest dimension known as of this writing. At the same time, it's clear that the dimension of the families produced is only logarithmic in the genus g. One idea for improving the bound is to use more branch points gaining more than a single dimension at each stage and perhaps allowing the ratio of g_{i+1}/g_i to be smaller as well. If we do this, however, we must somehow ensure that when these branch points meet, as they very much tend to do in any complete family, the corresponding covers do not acquire singularities. One condition that would ensure this, but that seems to be hard to arrange, is that as branch points meet, their ramification cycles have disjoint supports in the corresponding fiber of the covering.

We cannot resist mentioning one trick for forcing the branch points to remain distinct. If C_0 is a curve with a fixed point free involution i, we consider the family of all double covers of C_0 branched at a pair of points of the form $(P, i(P))$ getting a complete 1-parameter family of curves of genus $2g_0$. The set of all *unramified* double covers of curves

---

[5]Note that these cannot be cyclic covers.

in this family will be a complete family of curves of genus $4g_0 - 1$ *with fixed point free involutions*. Iterating this pair of steps, we get complete n-dimensional families in genus $4^n g_0 - (4^n - 1)/3$.

We should remark that all the curves in any of these families for reasonably large n are very special, so the examples are relevant only to Question (2.31).

In the negative direction, the outstanding result is Diaz' theorem [30], whose proof we'll give in Section 6.B after we've introduced the notion of admissible covers.

THEOREM (2.34) (DIAZ' THEOREM) *There does not exist a complete, $(g - 1)$-dimensional subvariety of $\mathcal{M}_g$ for any g.*

The results above leave a large gap that cries out to be filled. Specifically, we don't know whether there exist complete subvarieties of $\mathcal{M}_g$ of any dimension between that produced by the Kodaira construction (roughly $\log_3(g)$) and the bound $g - 2$ given by Diaz. Thus, we know $r_2 = 0$ and $r_3 = 1$, but already have only the inequalities $1 \leq r_4 \leq 2$, $1 \leq r_5 \leq 3$, and $2 \leq r_6 \leq 4$, in which the gap is growing roughly like g. Even this pales before our almost complete ignorance about $\tilde{r}_g$. Here we know that $\tilde{r}_g \geq 1$ for $g \geq 3$ with equality for $g = 3$. But, we don't know whether there is any g for which there exists a complete surface $S \subset \mathcal{M}_g$ passing through a general point $[C] \in \mathcal{M}_g$.

To close, let's pose the:

PROBLEM (2.35) Give an explicit complete one-parameter family $\mathcal{X} \to B$ of plane quartics whose generic element is smooth and whose associated map $B \to \overline{\mathcal{M}}_3$ is nonconstant but lies entirely in $\mathcal{M}_3$. (Since the discriminant locus is a hypersurface in the projective space of all quartic curves, some of the fibers of $\mathcal{X}$ *must* be singular, but we ask that their semistable models — see Section 3.C — be smooth. As we'll see, this could happen, for example, if they are all double conics.)

We might mention parenthetically here that there is an analogous question for Hilbert schemes that is likewise open. The problem is: if $\mathcal{R} \subset \mathcal{H} = \mathcal{H}_{d,g,r}$ is the open subset of smooth irreducible nondegenerate curves in the Hilbert scheme $\mathcal{H}$ of curves of degree d and genus g in $\mathbb{P}^r$, how large a complete subvariety may $\mathcal{R}$ contain? The best result along these lines is a beautiful theorem of Mei-Chu Chang and Ziv Ran ([24], [23]), which states that $\mathcal{R}$ *cannot contain a complete variety of dimension $r - 1$*. At the same time, it is possible to construct families of smooth irreducible nondegenerate curves $C \subset \mathbb{P}^r$ of dimension $r - 3$ for any r — see the exercises below — and $r - 2$ for some special r. Thus, the answer is nearly known, but a gap remains. In particular, *it isn't known whether there exist complete, positive-dimensional families of smooth, irreducible and nondegenerate curves in $\mathbb{P}^3$*.

EXERCISE (2.36) Show that there does not exist a complete, positive-dimensional family of twisted cubics in $\mathbb{P}^3$.

EXERCISE (2.37) Let $A \subset \mathbb{P}^N$ be an abelian variety of dimension n, embedded in projective space of large dimension, let $\pi : A \rightarrow \mathbb{P}^{n+3}$ be a general projection, and let $C \subset A$ be a curve. Show that the map π restricted to any translate of C is an embedding, and deduce that there exists a complete, $(r - 3)$-dimensional family of smooth irreducible nondegenerate curves in $\mathbb{P}^r$.

EXERCISE (2.38) Let $\sigma : \mathbb{P}^r \times \mathbb{P}^{r-3} \rightarrow \mathbb{P}^N$ be the Segre embedding, and let φ be the map

$$\varphi : \mathbb{P}^{r-3} \rightarrow \mathbb{G}(n - r - 1, N)$$

sending a point p to the subspace $\text{Ann}(\sigma(\mathbb{P}^r \times \{p\}))$.

1) Show that the pullback $\varphi^*(Q)$ of the universal quotient bundle Q on $\mathbb{G}(n - r - 1, N)$ is projectively trivial, i.e., is a line bundle tensored with a trivial bundle of rank $r + 1$.

2) Now let $C \subset (\mathbb{P}^N)^\vee$ be the general translate (under the action of $\text{PGL}(N + 1, \mathbb{C})$) of a smooth curve. Show that the projections $\pi_{\varphi(p)}$ of C from the subspaces $\varphi(p)$ give a family of smooth curves in $\mathbb{P}^r$ of dimension $r - 3$.

EXERCISE (2.39) More generally, let $X \rightarrow B$ be any family of smooth abstract curves with n-dimensional complete base B, and let L be a very ample line bundle on X. Suppose that $h^0(X_b, L|_{X_b}) \geq n + 4$ and that, for $b \in B$, the restriction map $H^0(X, L) \rightarrow h^0(X_b, L|_{X_b})$ is surjective. Show that if $\sigma_0, \ldots, \sigma_{n+3}$ are $n + 4$ general sections of L, then the map $\varphi_\sigma : X \rightarrow \mathbb{P}^{n+3}$ embeds each fiber X_b of X as a smooth nondegenerate curve in $\mathbb{P}^{n+3}$, again giving us an $(r - 3)$-dimensional family of smooth curves in $\mathbb{P}^r$.

Cohomology of $\mathcal{M}_g$: Harer's theorems

We come now to the fascinating question of the cohomology and/or cycle structure of $\mathcal{M}_g$. Most of what we know about the first of these questions is due to work of John Harer ([72], [74]), who uses the description of $\mathcal{M}_g$ as the quotient of the contractible Teichmüller space by the Teichmüller modular group Γ_g to derive results on $H^*(\mathcal{M}_g, \mathbb{Q})$. (As remarked above, these methods lead only to results about rational cohomology so we shall suppress the coefficients in what follows.)

The first result of Harer's that we'll give is the one that in some sense frames all the others. This is the *stability theorem*, and it says that the low-dimensional cohomology ring of $\mathcal{M}_g$ is independent of g. Specifically, Harer [73] shows that we have isomorphisms

(2.40) $\qquad H^k(\mathcal{M}_g) \cong H^k(\mathcal{M}_{g+1}) \qquad$ for $3k - 1 \leq g$.

Moreover, when all the relevant isomorphisms are defined they commute with the cup product. The bound for k in this result is probably far from sharp: Harer conjectures that the isomorphism should continue to hold until k is roughly equal to g. The most important possibility opened up by the stability theorem is that of defining what is called the *stable cohomology ring* $H^*(\mathcal{M})$ by setting $H^k(\mathcal{M}) = H^k(\mathcal{M}_g)$ for any $g \geq 3k - 1$; it is this ring that is the focus of most of the results that follow. (We emphasize that $H^*(\mathcal{M})$ is a purely algebraic object: there is no actual moduli space $\mathcal{M}$. Although it's possible to construct objects that are topologically like this imaginary $\mathcal{M}$, it's necessary to take a limit over g of spaces parameterizing the universal curve $\mathcal{C}_g$ plus additional analytic data in the form of a local coordinate at the marked point. The resulting spaces are infinite-dimensional for all g [9].)

There is a conjecture about the stable cohomology of moduli, called the *standard conjecture* which expresses $H^*(\mathcal{M})$ in terms of certain *standard cohomology classes* κ_i in $H^{2i}(\mathcal{M}_g)$. (These classes are also known as *tautological classes*.) To motivate the definition of these classes, think of a cohomology class on a space X as a functional that attaches to each cycle of the appropriate dimension a number, measuring the nontriviality of that cycle in some respect. A cycle in $\mathcal{M}_g$ corresponds to a family of curves, and the cycle is trivial if the family is, so we may accordingly think of a cohomology class in $\mathcal{M}_g$ as something that attaches to a family $X \to B$ of curves (of the appropriate dimension) a number measuring the variation of the curves in that family.

For example, suppose $\pi : X \to B$ is a family of curves with one-dimensional base. To say that the family is trivial — i.e., that $X = B \times C$, with π the projection on the first factor — implies that the relative dualizing sheaf $\omega_{X/B}$ (to be defined in the following chapter; for families of smooth curves it is simply the relative cotangent bundle) is a pullback from C, and hence in particular that its first Chern class has square $c_1(\omega_{X/B})^2 = 0$. (Conversely, it's not hard to see that if $c_1(\omega_{X/B})^2 = 0$, then the family is isotrivial.) We may thus think of the degree of $c_1(\omega_{X/B})^2$ as a measure of the nontriviality of the family.

With a bit more care, we can use this procedure to describe a cohomology class of codimension 2 on $\mathcal{M}_g$. The extra care is needed because $\mathcal{M}_g$ is neither smooth nor a fine moduli space. The construction that follows is a first example of the kind of persistent, but fundamentally minor, irritation that these facts cause. We will give an informal sketch here assuming $g \geq 4$. We first throw away the locus of curves with automorphisms: for $g \geq 4$ this has complex codimension ≥ 2, and so won't affect $H^2(\mathcal{M}_g)$. (Alternatively, we may make definitions analogous to those that follow using a rigidified moduli problem $\pi^{(n)} : \mathcal{C}_g^{(n)} \to \mathcal{M}_g^{(n)}$ and then push the corresponding classes down to $\mathcal{M}_g$ by the corresponding finite covering map. This method works for all $g \geq 2$.) We then have a universal curve $\pi : \mathcal{C}_g^0 \to \mathcal{M}_g^0$ with a smooth base. On $\mathcal{C}_g^0$, we have a relative dualizing sheaf $\omega = \omega_{\mathcal{C}_g^0/\mathcal{M}_g^0}$. Therefore, we can define the class

$$\eta = c_1(\omega_{\mathcal{C}_g^0/\mathcal{M}_g^0})$$

on $\mathcal{C}_g^0$ as the first Chern class of ω, and, on $\mathcal{M}_g^0$ the class

$$\kappa_1 = \pi_*(\eta^2) = \pi_*(c_1(\omega_{\mathcal{C}_g^0/\mathcal{M}_g^0})^2)$$

as the Gysin image of π of η^2.

In similar fashion, we can define classes $\kappa_i \in H^{2i}(\mathcal{M}_g)$ by setting $\kappa_i = \pi_*(\eta^{i+1})$. The standard conjecture [122] then states that the stable cohomology ring is freely generated by these classes, that is:

CONJECTURE (2.41) (MUMFORD'S STANDARD CONJECTURE)

$$H^*(\mathcal{M}) = \mathbb{Q}[\kappa_1, \kappa_2, \ldots].$$

Ed Miller [114] has shown that $\mathbb{Q}[\kappa_1, \kappa_2, \ldots]$ injects into $H^*(\mathcal{M})$ by constructing, for any finite set of classes in the standard ring, cycles on a suitable $\mathcal{M}_g$ on which these classes take on independent values. In the other direction, the evidence we have for this conjecture is all due to Harer.[6] Specifically, Harer has shown in [72] and [75] that

$$H^1(\mathcal{M}_g) = H^3(\mathcal{M}_g) = 0,$$

(2.42) $$H^2(\mathcal{M}_g) = \mathbb{Q} \cdot \kappa_1,$$

$$H^4(\mathcal{M}_g) = \mathbb{Q} \cdot \kappa_2 \oplus \mathbb{Q} \cdot (\kappa_1)^2.$$

The first two results show that $\mathrm{Pic}(\mathcal{M}_g) \otimes \mathbb{Q}$, the rational Picard group of line bundles on $\mathcal{M}_g$, is of rank 1, and that it is generated by

---

[6]Except for examples of natural geometric cohomology classes defined in other terms that do lie in the standard subring. For example, loci of curves that carry a divisor with Brill-Noether number $\rho < 0$ lie in this subring *when they are of codimension ρ*. This, however, is inductive evidence of the nonblack noncrow variety. (Since the statement "all crows are black" is the logical equivalent of "all things that are not black are not crows", a white piece of chalk, for example, could be seen as positive evidence for this assertion.) The evidence above simply shows that certain classes we know about lie in the space of classes we're sure we know about. Even the conjectured range of codimensions in which the underlined hypothesis holds is roughly the same — those less than roughly g — as the stable range.

the line bundle with first Chern class κ_1. Recently, Arbarello and Cornalba [6] have discovered a beautiful algebro-combinatorial approach which allows them to calculate the first, second, third and fifth cohomology groups of $\mathcal{M}_g$ and also provides some information about $\overline{\mathcal{M}}_g$.

EXERCISE (2.43) Show that if B is any complete curve that maps finitely to $\mathcal{M}_g^0$ and $\pi : \mathcal{C} \to B$ is the corresponding family then κ_1 is nonzero on B. Conclude that κ_1 isn't a torsion class on $\mathcal{M}_g$ and hence that $H^2(\mathcal{M}_g)$ does indeed have rank at least 1.

Harer's results also show that $\mathrm{Pic}(\overline{\mathcal{M}}_g) \otimes \mathbb{Q}$ is freely generated by κ_1 and the classes δ_i of the boundary components Δ_i, and, as discussed in the next section, allow us to determine $\mathrm{Pic}(\mathcal{C}_g) \otimes \mathbb{Q}$.

There is another approach to generating cohomology classes in $\mathcal{M}_g$ that should be mentioned. Another way of measuring the nontriviality of a family $\pi : X \to B$ is by its *Hodge bundle* Λ , which can be viewed informally as the vector bundle of rank g whose fiber over a point $b \in B$ is the space of holomorphic forms $H^0(X_b, K)$ on the fiber X_b. (More precisely, the Λ is the direct image $\pi_*(\omega_{X/B})$ of the relative dualizing sheaf.) In particular, we can associate to any family $\pi : X \to B$ the Chern classes $c_i(\Lambda)$. This suggests looking at the Hodge bundle Λ on $\mathcal{M}_g$ associated to the universal curve[7] $\mathcal{C}_g \to \mathcal{M}_g$ and taking its Chern classes

$$\lambda_i = c_i(\Lambda).$$

These also give cohomology classes on $\mathcal{M}_g$. As it turns out, these are polynomials in the classes κ_i, although the converse is not true.

One other beautiful result in this line is the calculation of the orbifold Euler characteristics of the moduli spaces $\mathcal{M}_{g,n}$ by Harer and Zagier [76]. The answers are striking: for example, they show that the orbifold Euler characteristic of the universal curve is

$$\chi(\mathcal{M}_{g,1}) = \zeta(1 - 2g) = -\frac{B_{2g}}{2g},$$

where ζ is the Riemann ζ-function and B denotes the Bernoulli number.

One consequence of their results is that the standard classes do not generate the full cohomology ring for large g. This can be seen by bounding the total number of standard classes and comparing to the absolute value of the Euler characteristic. Already for $g \geq 15$ the Euler characteristic is clearly larger, but it may well be that there are

---

[7] We leave you to supply the incantations analogous to those above needed to make formal sense of this.

nonstandard classes for all $g \geq 3$. For instance, Looijenga has shown that $H^6(\mathcal{M}_3)$ is nonzero and not tautological.

Cohomology of the universal curve

Harer has also produced analogous conjectures and results for the moduli spaces of pointed curves. For example, in the case of $\mathcal{C}_g$, Harer has shown that the stable cohomology of $\mathcal{C}_g$ is generated over $H^*(\mathcal{M})$ by the class ω of the relative dualizing sheaf.

One immediate consequence of this is that the Picard group of $\mathcal{C}_g$ is generated by the classes λ and ω. In other words, *the line bundle ω generates the Picard group of $\mathcal{C}_g$ over the Picard group of $\mathcal{M}_g$.* By the same token, this implies that $\mathrm{Pic}(\overline{\mathcal{C}}_g) \otimes \mathbb{Q}$ is freely generated by the class ω, together with λ and the classes σ_i of the boundary components Σ_i, as described in Exercise (2.18). (For a more complete statement over $\mathbb{Z}$, see Arbarello and Cornalba, [5].)

An equivalent restatement of the fact that the Picard group of $\mathcal{C}_g$ is generated by the classes λ and ω is that any line bundle on $\mathcal{C}_g$ must restrict to a multiple of K_C on a general fiber $C \subset \mathcal{C}_g$. This is almost, but not quite, the statement of what is called the *Enriques-Franchetta conjecture.*

This conjecture states that any assignment to a general curve C of a line bundle on C — precisely, a section of the universal Picard variety $\mathcal{P}_{d,g}$ over an open subset of $\mathcal{M}_g$, or equivalently a rational section of the map $\mathcal{P}_{d,g} \to \mathcal{M}_g$ — must be a power of the canonical line bundle. This was classically stated in the form, "the only rationally determined line bundles over moduli are the powers of the canonical bundle".

This problem was first considered by Enriques (cf. [43]) who found a defective proof; noting that his proof was faulty, he formulated the conjecture. (The statement of Enriques' conjecture, which is actually a stronger assertion implying the Enriques-Franchetta conjecture, is discussed in the following subsection.) Later, Franchetta [51] gave another false proof; unaware of the defect in his proof, he claimed the result, with the consequence that the theorem, finally proved by Harer and Mestrano, is often referred to as "Franchetta's conjecture", or even "Franchetta's Theorem".

The Enriques-Franchetta conjecture doesn't follow directly from Harer's results because giving a line bundle on every curve C in a family $X \to B$, as in the conjecture, doesn't necessarily give a line bundle on the total space — it does so locally over B, but the pieces don't necessarily patch together. In the end, however, Mestrano [112] overcame this obstacle and was able to deduce the conjecture from Harer's result.

One hypothesis that would guarantee the existence of the desired patchings would be the existence of a *Poincaré line bundle* over $\mathcal{P}_{d,g}$ — that is, a line bundle on the fiber product $\mathcal{P}_{d,g} \times_{\mathcal{M}_g} \mathcal{C}_g$ whose restriction to a fiber of the product over a point $(C, L) \in \mathcal{P}_{d,g}$ is L. A theorem due to Mestrano and Ramanan [113] asserts that there is a Poincaré line bundle on $\mathcal{P}_{d,g}$ if and only if $d - g + 1$ is relatively prime to $2g - 2$. A spectral sequence argument does show that, given any rationally determined line bundle with "fibers" L_x at the points of a variety X, there is a number n such that all the "fibers" $(L_x)^{\otimes n}$ do come from a line bundle on X. This, combined with Harer's results, implies that any rationally determined line bundle over moduli is a rational multiple of the canonical bundle.

To finish off the Enriques-Franchetta conjecture, then, it remains to analyze the finite covers of $\mathcal{M}_g$ obtained by taking roots of powers of K: that is, for any d, n and g such that $(2g - 2)|nd$, the spaces

$$\mathcal{I}_{n,d,g} = \left\{ (C, L) : L^{\otimes n} \cong K_C^{\otimes \left(\frac{nd}{2g-2} \right)} \right\} \subset \mathcal{P}_{d,g}.$$

We would like to be able to say that the space $\mathcal{I}_{n,d,g}$ has no section over $\mathcal{M}_g$, except for the obvious ones when $2g - 2$ divides d. It would be independently interesting to know more about the monodromy of this covering space as well in order to understand its component structure. When d is divisible by $2g - 2$, for example, the $(\frac{d}{2g-2})$-canonical bundles form an isolated sheet. When $d = 0$ so that we're considering n-torsion line bundles, the monodromy group is known to be $\mathrm{Sp}(2g, \mathbb{Z}/n\mathbb{Z})$ and the cover turns out to be irreducible if and only if n is prime; of course, this is then also true when d is a multiple of $2g - 2$. The other classically understood case is when $d = g - 1$ and $n = 2$ when there are exactly two components corresponding to even and odd theta characteristics. A first question might be: Is the cover irreducible when d is relatively prime to $2g - 2$?

Cohomology of Hilbert schemes

Two considerations prompt us to look for algebraic approaches to the standard conjecture. The first is simply our pride as algebraic geometers. The second is that Harer's approach to the calculation of $H^i(\mathcal{M}_g)$ becomes much harder to carry out with each increase in i; already with $i = 4$ we appear to be reaching the limits of human patience and perseverance (although Harer and some of his students have done work on the next cases). It seems unlikely that his methods can be pushed much further.

The most promising strategy is to try to solve the problem in two steps: first, understand the cohomology of some parameter space or spaces; second, by studying the maps from these spaces to moduli, deduce the cohomology of $\mathcal{M}_g$. For example, the irreducibility of $\mathcal{M}_g$ (which amounts to the calculation of $H^0(\mathcal{M}_g)$) is proved by showing that the Hurwitz schemes are irreducible and then showing that every $\mathcal{M}_g$ is dominated by such a scheme.

Here we simply want to mention some conjectures on the cohomology of the Hilbert scheme that are in close analogy to the standard conjecture above. We may summarize the approach taken above to generating the stable cohomology of the moduli space of curves as follows. First, look for a canonically defined line bundle on the universal curve $\mathcal{C}_g$, or, roughly equivalently at least up to torsion, for a consistently defined line bundle on each fiber of $\mathcal{C}_g$. Then, take the Chern class of this bundle, raise it to various powers and take their Gysin images in the cohomology of $\mathcal{M}_g$.

We may do exactly the same thing in the case of Hilbert schemes $\mathcal{H} = \mathcal{H}_{d,g,r}$. The main difference is that there are now *two* canonically defined bundles on the universal curve $\pi : X \to \mathcal{H}$: the relative dualizing sheaf $\omega = \omega_{X/\mathcal{H}}$ as before, and the line bundle $\mathcal{O}_X(1)$ pulled back from $\mathcal{O}_{\mathbb{P}^r}(1)$ by the inclusion of X in $\mathbb{P}^r \times \mathcal{H}$. We may thus take all monomials in the Chern classes η and ξ of these two line bundles, and push them forward. We ask: do these Gysin images generate the Chow or homology ring of the restricted Hilbert scheme $\mathcal{R}$ in low codimension? For example, we might expect that the group $A_1(\mathcal{R})$ of codimension 1 cycle classes in $\mathcal{R}$ will have rank 3, being generated over $\mathbb{Q}$ by the classes

$$A = \pi_*(\xi^2), \quad B = \pi_*(\xi \cdot \eta) \quad \text{and} \quad C = \pi_*(\eta^2).$$

We may call this statement the *standard conjecture* for the low-dimensional cohomology of the Hilbert scheme.

We may make an analogous conjecture on the low-dimensional cohomology of the universal curve $X_{\mathcal{R}} \to \mathcal{R}$: that it's generated by the classes η and ξ over the ring $H^*(\mathcal{R})$. In particular, this would say that the relative Picard group of $X_{\mathcal{R}}/\mathcal{R}$ (the group of line bundles on $X_{\mathcal{R}}$ modulo those pulled back from $\mathcal{R}$) is generated by η and ξ. In the formulation "if L_C is a rationally determined line bundle on every curve C in an open set of the Hilbert scheme, then for some n and m, $L_C \cong \omega_C^{\otimes n} \otimes \mathcal{O}_C(m)$", this conjecture is already in Enriques. Just as for $\mathcal{M}_g$ and $\overline{\mathcal{M}}_g$ these conjectures lead to statements that the low-codimensional cohomology of $\mathcal{R}$ is generated by standard classes and boundary components which we won't formulate here.

Unfortunately, these conjectures are *false* in general. There are examples of components of restricted Hilbert schemes $\mathcal{R}$ such that the universal curve $X_{\mathcal{R}} \to \mathcal{R}$ admits many line bundles (modulo those pulled back from $\mathcal{R}$) other than $\omega_{X/B}$ and $\mathcal{O}_X(1)$. The simplest of these is given in the two exercises below.

EXERCISE (2.44) Let S_0 be a smooth cubic surface and $L_1, \ldots, L_6$ be disjoint lines on S_0. Let $C_0 \subset S_0$ be a general curve in the linear system

$$C_0 \in |nH - L_1 - 2L_2 - \cdots - 6L_6|.$$

Show that if n is sufficiently large, then
1) C_0 is smooth and irreducible;
2) a general curve C in a component $\mathcal{H}$ of the Hilbert scheme containing $[C_0]$ also lies on a smooth cubic surface S; and,
3) the class of the curve $C \subset S$ is expressible in the form $nH - \sum i \cdot L_i$ for a *unique* choice of 6 skew lines $L_1, \ldots, L_6 \subset S$.

How large does n have to be for each of these assertions to hold?

EXERCISE (2.45) Let $\mathcal{X} \to \mathcal{H}$ be the universal curve over the component $\mathcal{H}$ of the Hilbert scheme described in the preceding exercise and $\mathcal{X}_\mathcal{R} \to \mathcal{R}$ its restriction to the open set of smooth curves. Show that the classes H and $L_1, \ldots, L_6$ give rise to seven independent line bundles on $\mathcal{X}_\mathcal{R}$, whose restrictions to a general fiber $C \subset \mathcal{X}_\mathcal{R}$ are independent. In other words, show that the group of rationally determined line bundles on $\mathcal{X}_\mathcal{R} \to \mathcal{R}$ has rank at least 7. For extra credit, show that the rank is exactly 7.

EXERCISE (2.46) Continuing our analysis of the Hilbert scheme described in the preceding exercise, consider the Gysin images of the pairwise products of the classes H and $L_1, \ldots, L_6$. Show that these give rise to at least four independent divisor classes on $\mathcal{H}$, thus violating the standard conjectures on the Picard group of the Hilbert scheme.

It's therefore somewhat remarkable that for $r = 1$ and 2 (that is, for the Hurwitz scheme and Severi variety), *the standard conjectures do seem to hold.* Why this should be is unclear. The situation is completely analogous with those considered in Chapter 1: the Hurwitz scheme and Severi variety are always irreducible of the correct dimension, while the Hilbert scheme is in general neither. The basic references for the Severi variety case of these conjectures are [33], [32] and [36]: in the last it's shown that a verification of the standard conjecture on the Picard group of Severi varieties would imply Harer's theorem on the Picard group of $\mathcal{M}_g$. Diaz and Edidin [31] have some results in the Hurwitz scheme case.

There is a further point to be made about the standard conjectures for Hilbert schemes of curves in higher-dimensional space. This is that, empirically, the components of the Hilbert scheme that violate the conjectures all lie over relatively small subvarieties of $\mathcal{M}_g$ — ones of codimension on the order of g or more. This phenomenon has been sufficiently often observed that we may include it in the statement of the conjectures. We include these here (for the Picard groups of the Hilbert schemes and their universal curves) for reference. We start with the Enriques conjecture:

Conjecture (2.47) (Enriques conjecture) *1) Let $\tilde{\mathcal{H}}_{d,g}$ be the space of branched covers $\pi : C \to \mathbb{P}^1$ of $\mathbb{P}^1$ of degree d and genus g. If L_C is a rationally determined line bundle on every curve C in an open set of $\tilde{\mathcal{H}}_{d,g}$, then for some n and m, $L_C \cong \omega_C^{\otimes n} \otimes \pi^* \mathcal{O}_{\mathbb{P}^1}(m)$.*
2) Similarly, let $V_{d,g}$ be the locus of reduced and irreducible plane curves of degree d and genus g. If L_C is a rationally determined line bundle on every curve C in an open set of $V_{d,g}$, then for some n and m, $L_C \cong \omega_C^{\otimes n} \otimes \mathcal{O}_C(m)$.
3) There exist real numbers $\alpha = \alpha(g) > 0$ and $\beta = \beta(g)$ such that the following statement holds: if $\mathcal{H}$ is any component of the restricted Hilbert scheme of curves of degree d and genus g in $\mathbb{P}^r$ such that the induced rational map $\varphi : \mathcal{H} \to \mathcal{M}_g$ has image of codimension less than or equal to $\alpha \cdot g + \beta$, and if L_C is a rationally determined line bundle on every curve C in an open set of $V_{d,g}$, then for some n and m, $L_C \cong \omega_C^{\otimes n} \otimes \mathcal{O}_C(m)$.

Exercise (2.48) Show that any one of these three statements implies the Enriques-Franchetta conjecture. (The converse, in case d is sufficiently large with respect to g, was established by Ciliberto [25].)

The statement of the standard conjecture for the low-dimensional cohomology of the Hilbert scheme (again just for the Picard group) is slightly more delicate, since it matters just what open subset of the Hurwitz/Severi/Hilbert scheme we choose. Probably the cleanest statement in the case of the Hurwitz and Severi schemes involves the smallest open subset: in the case of the Hurwitz scheme, the variety $\tilde{\mathcal{H}}_{d,g}$ of branched covers with simple branching, and, in the case of the Severi variety, the open set $U_{d,g} \subset \overline{V}_{d,g}$ of irreducible plane curves of degree d and geometric genus g with only nodes as singularities. In both cases, we expect first that the classes A, B and C will be torsion and second that the rank of the Picard group is 0. For the Severi variety, the first expectation is proved in [33]. Of course, this would give the second if we knew that these classes also generated the Picard group. Diaz and Harris actually show that the converse also holds: if the rank is 0, then these classes must actually generate $\mathrm{Pic}(V_{d,g}) \otimes \mathbb{Q}$.

Conjecture (2.49) (Standard conjecture for Picard groups of parameter spaces)
1) $\mathrm{Pic}(\tilde{\mathcal{H}}_{d,g}) \otimes \mathbb{Q} = 0$;
2) $\mathrm{Pic}(U_{d,g}) \otimes \mathbb{Q} = 0$; and

3) There exist real numbers $\alpha = \alpha(g) > 0$ and $\beta = \beta(g)$ such that the following statement holds: if $\mathcal{H}$ is any component of the restricted Hilbert scheme of curves of degree d and genus g in $\mathbb{P}^r$ and $\widetilde{\mathcal{H}} \subset \mathcal{H}$ the open subset parameterizing smooth curves, such that the induced map $\varphi : \widetilde{\mathcal{H}} \to \mathcal{M}_g$ has image of codimension less than or equal to $\alpha \cdot g + \beta$, then the Picard group $\mathrm{Pic}(\widetilde{\mathcal{H}}) \otimes \mathbb{Q}$ of $\mathcal{H}$ is generated by the classes A, B and C.

You may feel that the formulation of the third part of each of these conjectures in terms of unspecified constants $\alpha(g)$ and $\beta(g)$ is a cheat: the statement as a result is so vague as to be virtually immune to counterexample. We agree. The problem is, the evidence available doesn't give a clear indication of what the correct values of these constants should be. For the examples of which we know, $\alpha = 1$ and $\beta = 0$ should work; but that may not be the strongest possible statement. We leave it instead as a challenge:

PROBLEM (2.50) Can you find a component $\widetilde{\mathcal{H}}$ of the Hilbert scheme whose image in $\mathcal{M}_g$ has codimension less than g that violates the statement of either conjecture above?

EXERCISE (2.51) Calculate the codimension in $\mathcal{M}_g$ of the image of the component of the Hilbert scheme introduced in Exercise (2.44); in particular, observe that it's greater than g.

Finally, we should say that there are analogous conjectures about the dimension and irreducibility of the Hilbert scheme. As we remarked, while the Hurwitz and Severi varieties are always irreducible of the expected dimension, neither is true of the Hilbert scheme $\mathcal{H}_{d,g,r}$ in general. But, it may be conjectured that, as in the two conjectures above, the corresponding statements *do* hold for components of $\mathcal{H}_{d,g,r}$ whose images in $\mathcal{M}_g$ have relatively small codimension. We will discuss this briefly following the proof of the Brill-Noether theorem in Chapter 5.

Structure of the tautological ring

Having produced, at least conjecturally, classes which generate the stable cohomology ring of $\mathcal{M}_g$, a natural problem is to understand the relations amongst various products of these classes. The results and conjectures about this question are most easily discussed in the setting of the Chow ring $A^*(\mathcal{M}_g)$ rather than that of the cohomology ring $H^*(\mathcal{M}_g)$. To set this up, we define the *tautological subring* $R^*(\mathcal{M}_g)$ of the Chow ring $A^*(\mathcal{M}_g)$ to be the subring generated by the tautological classes κ_i and λ_i.

The first result about $R^*(\mathcal{M}_g)$ is due to Mumford [122] who showed that it's generated by the $g-2$ classes $\kappa_1, \ldots, \kappa_{g-2}$. The first element of the proof is a Grothendieck-Riemann-Roch calculation which shows, as was already noted in the discussion of Harer's theorem following Exercise (2.43), that the λ's are all expressible in terms of the κ's. (The first two such expressions are calculated in Section 3.E: see equations (3.106) and (3.107)). To express κ_i for $i \geq g + 1$ in terms of the lower κ's, Mumford uses the natural surjection between the pullback $\pi^*(\Lambda)$ of the Hodge bundle under the map $\pi : \mathcal{C}_g \to \mathcal{M}_g$ and the relative dualizing sheaf $\omega_{\mathcal{C}_g/\mathcal{M}_g}$. The kernel of this map is a locally free sheaf of rank $g - 1$ on $\mathcal{C}_g$ and so its Chern classes vanish in degrees above $g - 1$. Pushing down to $\mathcal{M}_g$ gives a relation in each degree i greater than $g - 2$ between κ_i and lower κ's. He applies the same technique to κ_{g-1} and κ_g except that now the pushed-down relation also involves the λ's. Since he has already shown how to express these in terms of the κ's, he is able to handle this by showing that the first two such relations are independent. Both steps yield a slew of relations not used in the proof but, at least initially, it was not clear that these could be summarized in any concise form.

Looijenga recently showed that Mumford's result is a shadow of a much stronger vanishing result:

THEOREM (2.52) (LOOIJENGA) *In any degree* $i > g - 2$, $R^i(\mathcal{M}_g) = \{0\}$, *and* $R^{g-2}(\mathcal{M}_g)$ *is generated by either the class of the hyperelliptic locus or the class* κ_{g-2}. *These classes are nonzero so* $R^{g-2}(\mathcal{M}_g)$ *is one-dimensional.*

We won't discuss the proof here — it's given in [110] (except for the nonvanishing of $R^{g-2}(\mathcal{M}_g)$ which is shown in [48]). However, you may get an idea of the force of this result by noting that Diaz' theorem [Theorem (2.34)] is an immediate corollary.

Even before Looijenga's result was established, Faber [48] had included it as part of still more precise conjectures about the structure of the tautological ring. The first part can be stated immediately.

CONJECTURE (2.53) (FABER'S CONJECTURE, FIRST PART)

1) *The tautological ring* $R^*(\mathcal{M}_g)$ *"looks like" the algebraic cohomology ring of a nonsingular projective variety of dimension* $g - 2$. *More precisely, it's Gorenstein with socle in degree* $g - 2$ — *that is, it vanishes above degree* $g - 2$, *it's one-dimensional in degree* $g - 2$ *and the pairing* $R^i(\mathcal{M}_g) \times R^{g-2-i}(\mathcal{M}_g) \to R^{g-2}(\mathcal{M}_g)$ *is perfect* — *and satisfies the conclusions of the Hard Lefschetz theorem and the Hodge index theorem with respect to the class* κ_1.

2) *The* $\lfloor \frac{g}{3} \rfloor$ *classes* $\kappa_1, \ldots, \kappa_{\lfloor g/3 \rfloor}$ *generate the ring with no relations in degrees less than or equal to* $\lfloor \frac{g}{3} \rfloor$.

The other parts of Faber's conjecture deal with the nature of the relations in the tautological ring. The first question to ask is: how can we produce such relations? To answer this, we introduce the space $\mathcal{C}_g^d$, the d-fold fiber product of $\mathcal{C}_g$ over $\mathcal{M}_g$, which parameterizes a curve C of genus g plus a d-tuple of points $p_1, p_2, \ldots, p_d$, *not* necessarily distinct. This comes equipped with diagonal divisor classes $D_{d,ij}$ (the locus where $p_i = p_j$) and with bundles which we denote $\omega_{d,i}$ obtained by pulling back the bundle $\omega_{\mathcal{C}_g/\mathcal{M}_g}$ via the projection onto the ith factor. There are lots of other projection maps: if $I \subset \{1, 2, \ldots, d\}$ is a subset of order e, we have a map $\pi_{d,I} : \mathcal{C}_g^d \to \mathcal{C}_g^{d-e}$ by forgetting the points p_i for $i \in I$. It also carries a particularly interesting bundle $\mathbb{F}_d$, of rank d, which can be described informally as the bundle whose fiber over $[(C, D)]$ is $H^0(C, K_C/K_C(-D))$ or, more precisely, as

$$(\pi_{d+1,\{d+1\}})_* \left(\left(\sum_{i=1}^d \mathcal{O}_{D_{d+1,i}} \right) \otimes \omega_{d+1,d+1} \right).$$

Faber studies the evaluation map $\varphi : \Lambda \to \mathbb{F}_d$ — we use Λ to denote the pullback of the Hodge bundle to $\mathcal{C}_g^d$ — which is fiberwise the map $H^0(C, K_C) \to H^0(C, K_C/K_C(-D))$ and shows, by a straightforward application of Porteous' formula (which we'll study in Chapter 3 starting on page 161) that:

PROPOSITION (2.54) *In* $A^*(\mathcal{C}_g^{(2g-1)})$, $c_j(\mathbb{F}_{2g-1} - \Lambda) = 0$ *for* $j \geq g$.

On the other hand, pushing down to $\mathcal{M}_g$ any monomial in the classes $D_{d,ij}$ and $\omega_{d,i}$ involved in the definition of the $\mathbb{F}_d$'s turns out to give an element of the tautological ring on $\mathcal{M}_g$ so this relation pushes down to one in $R^*(\mathcal{M}_g)$. The rules for carrying this out were already written down in [82] (in 1982!). Unfortunately, the codimension in which this relation lives is negative so all terms are 0. We can cure this by multiplying this relation with suitable classes before pushing down. The relations thus obtained are then highly nonobvious. In fact, Faber conjectures that:

CONJECTURE (2.55) (FABER'S CONJECTURE, SECOND PART) *The ideal of relations in the tautological ring* $R^*(\mathcal{M}_g)$ *is generated by those of the form*

$$\pi_*(M \cdot c_j(\mathbb{F}_{2g-1} - \Lambda))$$

for all monomials M in the classes $D_{d,ij}$ and $\omega_{d,i}$, and all $j \geq g$. (Here $\pi = \pi_{2g-1,\{1,2,\ldots,2g-1\}}$ is the projection which forgets all points).

We won't go into the mechanics of how individual relations like those in the second conjecture are unwound nor into the details of how the conclusions of the conjecture may be deduced from sets of such relations, except to say that the difficulties are due mainly to the rapidly growing combinatorial complexity of the calculations (cf. [47]). Faber works things out for $g = 2, 3$ and 4 on pages 10-12 of [48]. In practice, a computer is required to carry out all but the very simplest cases. Even using a symbolic calculation program, a naive approach bogs down around genus 6. However, by cleverly factoring the push-down operations involved using some auxiliary rules, Faber, assisted by a computer, has been able to verify the conjecture for $g \leq 15$, thus providing very strong evidence for it in general.

Having produced relations, it's natural to ask whether they can be explicitly expressed in terms of the κ's themselves. Here again Faber has a beautiful proposal for which there is also a computer assisted verification for $g \leq 15$.

CONJECTURE (2.56) (FABER'S CONJECTURE, THIRD PART) *Fix   any partition P of the integer $g - 2$ as a sum $d_1 + d_2 + \cdots + d_{k(P)}$ of positive integers. For each subset, $S \subset \{1, 2, \ldots, k(P)\}$, define $d_S = \sum_{j \in S} d_j$. For each permutation $\sigma \in \mathfrak{S}_{k(P)}$, let*

$$\sigma = \alpha_1 \cdot \alpha_2 \cdot \ldots \cdot \alpha_{\nu(\sigma)}$$

be its expression as a product of disjoint cycles, let S_l be the support of α_l (so that the sets S_i partition the set $\{1, 2, \ldots, k(P)\}$), let $d_i = d_{S_i}$, and let

$$\kappa_{\sigma, P} = \prod_{i=1}^{\nu(\sigma)} \kappa_{d_i} \in R^{g-2}(\mathcal{M}_g).$$

If we let

$$\tau_P = \sum_{\sigma \in \mathfrak{S}_{k(P)}} \kappa_{\sigma, P},$$

then, in $R^{g-2}(\mathcal{M}_g)$, we have the relation

$$\tau_P = \frac{(2g - 3 + k(P))!(2g - 1)!!}{(2g - 1)! \prod_{j=1}^{k(P)}(2d_j + 1)!!} \kappa_{g-2}$$

in which $(2n - 1)!!$ denotes, as usual, the product $1 \cdot 3 \cdot 5 \cdot \ldots \cdot (2n - 1)$.

Although we won't enter into this, these relations can be unwound to yield relations for individual products of κ's of total degree $g - 2$. For example, Zagier showed that the third conjecture implies that

$$\kappa_1^{g-2} = \frac{1}{g-1} 2^{2g-5}((g-2)!)^2 \kappa_{g-2}.$$

Witten's conjectures and Kontsevich's theorem

Faber's motivation for looking at the class τ_P in the preceding subsection comes from work of Witten [152]. There he defined a much larger family of such τ-classes — in Witten's notation, $\tau_P = \langle \tau_{d_1+1}, \tau_{d_2+1}, \ldots, \tau_{d_{k(P)}+1} \rangle$ — and conjectured that a generating function F encoding all the intersection numbers associated to these classes satisfies two distinct systems of differential equations (one of which is the Korteweg-deVries system and one of which is associated to the Virasoro algebra).

At the time it was stated, this seemed, to us at least, plausible — there was numerical evidence for it coming from earlier work of Mumford on $R^*(\mathcal{M}_2)$ [122] and Faber on $R^*(\mathcal{M}_3)$ ([45], [46]) — but out of reach. However, a tour-de-force proof was soon provided by Kontsevich [102]. He uses a combinatorial version of the moduli space based on ribbon graphs to express the generating function F in terms of matrix integrals. This allows him to show that F satisfies both systems of differential equations by applying properties of these matrix models (cf. [89] and [34]). Probably the best place to begin if you want to understand the proof is Looijenga's Bourbaki Seminar talk [109].

In the rest of this subsection, we will simply state Witten's conjectures, suppressing, as usual, details of how various definitions which follow are made precise.[8] We begin by defining the line bundle $\mathcal{L}_i$ on $\overline{\mathcal{M}}_{g,n}$ to be the unique line bundle whose fiber over each pointed stable curve $(C; p_1, \ldots, p_n)$ is the cotangent space of C at p_i and letting $\psi_i \in A^1(\overline{\mathcal{M}}_{g,n})$ be the first Chern class of $\mathcal{L}_i$. It is convenient, and as we will see, nearly always harmless, to suppress the dependence of the classes on n and g: when we need to make this dependence explicit, we will write $\mathcal{L}_{g,n,i}$ and $\psi_{g,n,i}$.

Witten's conjectures concern the intersection products of the classes ψ_i. A concise notation for these products which exploits the symmetry in the markings is given by

$$(2.57) \qquad \langle \tau_{k_1} \tau_{k_2} \cdots \tau_{k_n} \rangle_g = \langle \prod_{i=1}^{n} \tau_{k_i} \rangle_g = \int_{\overline{\mathcal{M}}_{g,n}} \psi_1^{k_1} \psi_2^{k_2} \cdots \psi_n^{k_n}.$$

Such products are well-defined when all the k_i are nonnegative integers and the dimension condition $3g - 3 + n = \sum_{i=1}^{n} k_i$ holds. In all other cases, $\langle \tau_{k_1} \tau_{k_2} \cdots \tau_{k_n} \rangle_g$ is defined to be zero. The empty product $\langle 1 \rangle_1$ is also set to zero. The simplest nonzero integral (on $\overline{\mathcal{M}}_{0,3}$) is $\langle \tau_0^3 \rangle_0 = 1$. This evaluation and the first evaluation in the following

---

[8]We are grateful to Rahul Pandharipande for notes on which this subsection is based.

exercise act as set of initial conditions for the recursions which follow (cf. Exercise (2.63)).

EXERCISE (2.58) 1) Verify the evaluation $\langle \tau_1 \rangle_1 = \frac{1}{24}$ (on $\overline{\mathcal{M}}_{1,1}$).
2) Verify the evaluation $\langle \tau_0^3 \tau_1 \rangle_0 = 1$ (on $\overline{\mathcal{M}}_{0,4}$).

Let $\mathbf{t} = (t_0, t_1, \ldots, t_i, \ldots)$ be an infinite vector of variables indexed by $i \geq 0$ and let y denote the formal sum $y = \sum_{i=0}^{\infty} t_i \tau_i$. Witten next considers the formal generating function for the products (2.57):

$$F_g(\mathbf{t}) = \sum_{n=0}^{\infty} \frac{\langle y^n \rangle_g}{n!}$$

in which the expression $\langle y^n \rangle_g$ is defined by monomial expansion and multilinearity in the variables t_i. Thus, more concretely,

$$F_g(\mathbf{t}) = \sum_{\{n_i\}} \Big(\prod_{i=1}^{\infty} \frac{t_i^{n_i}}{n_i!} \Big) \langle \tau_0^{n_0} \tau_1^{n_1} \tau_2^{n_2} \cdots \rangle_g,$$

where the sum is over all sequences of nonnegative integers $\{n_i\}$ with finitely many nonzero terms. The generating function F defined by

$$F = \sum_{g=0}^{\infty} \lambda^{2g-2} F_g$$

arises as a partition function in two-dimensional quantum gravity. Based on a different physical realization of this function in terms of matrix integrals, Witten [152] conjectured that F satisfies two distinct systems of differential equations. Each system determines F uniquely and provides explicit recursions which compute all the products (2.57).

Before describing the full systems, two basic properties are needed for products with $2g - 2 + n > 0$. Under this hypothesis, the *string equation* says that

(2.59) $$\langle \tau_0 \prod_{i=1}^{n} \tau_{k_i} \rangle_g = \sum_{j=1}^{n} \langle \tau_{k_j-1} \prod_{i \neq j} \tau_{k_i} \rangle_g,$$

and the *dilaton equation* says that

(2.60) $$\langle \tau_1 \prod_{i=1}^{n} \tau_{k_i} \rangle_g = (2g - 2 + n) \langle \prod_{i=1}^{n} \tau_{k_i} \rangle_g.$$

Both the string and dilaton equations are derived from a comparison result describing the behavior of the ψ classes under pullback via the map $\pi : \overline{\mathcal{M}}_{g,n+1} \to \overline{\mathcal{M}}_{g,n}$ which forgets the $(n+1)^{\text{st}}$ point. If $i \in \{1, \ldots, n\}$, the basic formula is

(2.61) $$\psi_{g,n+1,i} = \pi^*(\psi_{g,n,i}) + [D]$$

where $D \cong \overline{\mathcal{M}}_{0,2} \times \overline{\mathcal{M}}_{g,n-1}$ is the boundary divisor on $\overline{\mathcal{M}}_{g,n+1}$ whose generic point parameterizes the join of a curve of genus 0 containing the marked points p_i and p_{n+1} with a curve of genus g containing the other $n-1$ marked points.

EXERCISE (2.62) Prove equation (2.61) and use it to deduce the string and dilation equations (2.59) and (2.60).

The next exercise gives a first glimpse of the force of these two equations.

EXERCISE (2.63) 1) Show the string equation and the initial condition $\langle \tau_0^3 \rangle_0 = 1$ determine all the genus 0 products in (2.57). More precisely, show that

$$\langle \prod_{i=1}^n \tau_{k_i} \rangle_0 = \frac{(n-3)!}{\prod_{i=1}^n k_i!}.$$

2) Show that the string equation, the dilaton equation, and the initial condition $\langle \tau_1 \rangle_1 = \frac{1}{24}$ determine all the genus 1 products.

Associated to each of the string and dilaton equations is a differential operator which annihilates $\exp(F)$. The operator associated to the string equation is

(2.64) $$L_{-1} = -\frac{\partial}{\partial t_0} + \frac{\lambda^{-2}}{2} t_0^2 + \sum_{i=0}^{\infty} t_{i+1} \frac{\partial}{\partial t_i},$$

and that associated to the dilaton equation is

(2.65) $$L_0 = -\frac{3}{2} \frac{\partial}{\partial t_1} + \sum_{i=0}^{\infty} \frac{2i+1}{2} t_i \frac{\partial}{\partial t_i} + \frac{1}{16}.$$

EXERCISE (2.66) 1) Show that the string equation and the initial condition $\langle \tau_0^3 \rangle_0 = 1$ imply the equation $L_{-1}(\exp(F)) = 0$.
2) Show that the dilaton equation and the initial condition $\langle \tau_1 \rangle_1 = \frac{1}{24}$ imply the equation $L_0(\exp(F)) = 0$.

The first system of differential equations which Witten conjectured the τ-products must satisfy are the Korteweg-deVries or KdV equations. To give his very compact formulation, let us set

(2.67) $$\langle\!\langle \tau_{k_1} \tau_{k_2} \cdots \tau_{k_n} \rangle\!\rangle = \frac{\partial}{\partial t_{k_1}} \frac{\partial}{\partial t_{k_1}} \cdots \frac{\partial}{\partial t_{k_1}} F,$$

so that, in particular, $\langle\!\langle \tau_{k_1} \tau_{k_2} \cdots \tau_{k_n} \rangle\!\rangle|_{t=0} = \langle \tau_{k_1} \tau_{k_2} \cdots \tau_{k_n} \rangle$. Then, Witten shows that the KdV equations for F take the form:

THEOREM (2.68) (WITTEN-KONTSEVICH FORMULAS, KdV FORM)
For all $n \geq 1$, the generating function F satisfies the equations

$$(2n+1)\lambda^{-2}\langle\!\langle \tau_n\tau_0^2 \rangle\!\rangle = \langle\!\langle \tau_{n-1}\tau_0 \rangle\!\rangle\langle\!\langle \tau_0^3 \rangle\!\rangle + 2\langle\!\langle \tau_{n-1}\tau_0^2 \rangle\!\rangle\langle\!\langle \tau_0^2 \rangle\!\rangle + \frac{1}{4}\langle\!\langle \tau_{n-1}\tau_0^4 \rangle\!\rangle .$$

Witten further showed that these equations and equation (2.64) for L_{-1} *together* determine all the products (2.67) and thus uniquely determine F.

As an example of how this works in practice, try taking $n = 3$ and evaluating equation (2.68) at **t = 0**. We obtain

$$7\langle \tau_3\tau_0^2 \rangle_1 = \langle \tau_2\tau_0 \rangle_1 \langle \tau_0^3 \rangle_0 + \frac{1}{4}\langle \tau_2\tau_0^4 \rangle_0.$$

Applying the string equation (2.59) yields

$$7\langle \tau_1 \rangle_1 = \langle \tau_1 \rangle_1 + \frac{1}{4}\langle \tau_0^3 \rangle_0.$$

Hence, we have rederived the equation $\langle \tau_1 \rangle = \frac{1}{24}$.

To describe the second system of differential equations for F, we introduce a Lie algebra L of holomorphic differential operators[9]: L is the algebra spanned by the operators L_n, where for $n \geq -1$

$$L_n = -z^{n+1}\frac{\partial}{\partial z},$$

and in which the bracket is given by $[L_n, L_m] = (n-m)L_{n+m}$. That this notation is consistent with the definitions for L_{-1} and L_0 introduced above follows from:

EXERCISE (2.69) Show that the differential operators defined in (2.64) and (2.65) satisfy

$$[L_{-1}, L_0] = -L_{-1}.$$

This suggests that equations (2.64) and (2.65) may be viewed as the beginning of a representation of L in a Lie algebra of differential operators. In fact, it turns out that, with certain homogeneity restrictions, there is a *unique* way to extend this assignment of L_{-1} and L_0 to such a representation of L. For $n \geq 1$, the expression for L_n takes the form

---

[9]The physical motivation for considering the Lie algebra L in this context comes from the fact that it is a sub-algebra of the Virasoro algebra.

$$L_n = -\left(\frac{(2n+3)!!}{2^{n+1}}\right)\frac{\partial}{\partial t_{n+1}} + \sum_{i=0}^{\infty}\left(\frac{(2i+2n+1)!!}{(2i-1)!!\,2^{n+1}}\right)t_i\frac{\partial}{\partial t_{i+n}}$$

(2.70)

$$+ \frac{\lambda^2}{2}\sum_{i=0}^{n-1}\left(\frac{(2i+1)!!\,(2n-(2i+1))!!}{2^{n+1}}\right)\frac{\partial^2}{\partial t_i\,\partial t_{n-1-i}}.$$

Recall that $(2n-1)!! = 1\cdot 3\cdot 5\cdot\ldots\cdot(2n-1)$ with the convention that $(-1)!! = 1$.

EXERCISE (2.71) Prove that the formula (2.70) defines a representation of L. *Hint*: Use the identity

$$(2i+1)!!(2n-(2i+1))!! = (-1)^{i+1}\left((-2i-1)(-2i+1)\cdots(-2i+2n-1)\right).$$

THEOREM (2.72) (WITTEN-KONTSEVICH FORMULAS, LIE FORM) *For all $n \geq -1$, $L_n(\exp(F)) = 0$.*

As with the KdV form, it's straightforward to see that the system of equations (2.72) also uniquely determines F. As a practical example of how this may be used, consider the equation determined by the operator L_3:

$$-\frac{945}{16}\frac{\partial F}{\partial t_4} + \sum_{i=0}^{\infty} t_i\frac{\partial F}{\partial t_{i+3}}$$

$$+ \lambda^2\left(\frac{15}{16}\left(\frac{\partial^2 F}{\partial t_0\partial t_2} + \frac{\partial F}{\partial t_0}\frac{\partial F}{\partial t_2}\right) + \frac{9}{32}\left(\frac{\partial^2 F}{\partial t_1\partial t_1} + \frac{\partial F}{\partial t_1}\frac{\partial F}{\partial t_1}\right)\right) = 0.$$

The constant term of the above relation reads

$$-\frac{945}{16}\langle\tau_4\rangle_2 + \frac{15}{16}\langle\tau_0\tau_2\rangle_1 + \frac{9}{32}\left(\langle\tau_1\tau_1\rangle_1 + \langle\tau_1\rangle_1^2\right) = 0.$$

EXERCISE (2.73) Use part 2 of Exercise (2.63) to compute the genus 1 numbers above and show that

$$\langle\tau_4\rangle_2 = \frac{1}{1152}.$$

Kontsevich's proof of both sets of formulas is fundamentally analytic. It would be very nice to have direct algebraic arguments, but as yet few cases have been treated; for example, for the Lie version, only L_{-1} and L_0.

Finally, we note that both Witten and Kontsevich had in mind, more than the *values* of these invariants, their *applications*! More precisely, Witten conceived a generalization of moduli spaces of stable curves to moduli spaces of stable maps to a fixed target variety (moduli spaces of curves being those in which the target of the map was a point) and saw the resulting intersection numbers as a way of producing invariants of the target. These moduli spaces of maps are the topic of our last section.

E Moduli spaces of stable maps

The ideas of Witten and Kontsevich in [152] and [102] have inspired a theory of moduli spaces of stable maps.[10] This theory is still undergoing very active development and many of the conjectured results are currently only known in very special cases. However, it has already yielded solutions to a wide range of enumerative problems dealing with rational curves and seems certain to have much wider applications.

We won't use moduli spaces of stable maps elsewhere in this book. In this section, we just want to introduce these spaces and state the main properties known and conjectured about them. For further details, we refer the reader to the excellent set of expository notes of Fulton and Pandharipande [54] which we've relied on heavily for this sketch.

First of all, what is a stable map $(C, (p_1, \ldots, p_n), \mu)$ of genus g with n marked points? As the name suggests, it's a map μ from a connected nodal curve C of arithmetic genus g with a collection $(p_1, \ldots, p_n)$ of n distinct smooth marked points to a projective scheme X satisfying the stability condition that the number of automorphisms of the map μ — that is, maps $\varphi : C \to C$ fixing the marked points and satisfying $\mu \circ \varphi = \mu$ — is finite.

EXERCISE (2.74) Show that this stability condition is equivalent to the condition that, if a smooth rational component D of C is mapped by μ to a point of X, then the number of marked points on D plus the number of nodes in which D meets the rest of C be at least 3, plus the condition that, if $g = 1$ and $n = 0$, then μ is nonconstant.

If γ is an element of $H_2(X, \mathbb{Z})$ (possibly 0), then we let $\overline{\mathcal{M}}_{g,n}(X, \gamma)$ denote the set of isomorphism classes of stable maps with target X for which the pushforward $\mu_*(\{C\})$ of the fundamental class $\{C\}$ of C equals γ. Although this set may be empty — for example, if $\gamma \neq 0$ and X contains no curves of genus g or less — it can always be made into a projective coarse moduli space. If l is the class of a line in $\mathbb{P}^n$, we write $\overline{\mathcal{M}}_{g,n}(\mathbb{P}^n, d)$ for $\overline{\mathcal{M}}_{g,n}(\mathbb{P}^n, dl)$.

EXERCISE (2.75) Show that if X is a point and hence the class γ is 0, then we recover the usual moduli space of stable curves, i.e., $\overline{\mathcal{M}}_{g,n}(pt, 0) \cong \overline{\mathcal{M}}_{g,n}$.

To get a tractable space, it is, at least at present, necessary to set $g = 0$ and to make strong positivity assumptions about X. We call a

---

[10]Gromov in [66] had introduced many of the ideas involved from a symplectic point of view.

nonsingular variety X *convex* if, for every map $\mu : \mathbb{P}^1 \to X$, we have $H^1(\mathbb{P}^1, \mu^*(T_X)) = \{0\}$. This follows if T_X is generated by global sections; hence, any variety admitting a transitive group action is convex. For convex X, the space $\overline{\mathcal{M}}_{0,n}(X, \gamma)$ has properties very much like those of moduli spaces of stable curves. In particular,

- $\overline{\mathcal{M}}_{0,n}(X, \gamma)$ is locally normal of pure dimension

$$\dim(X) + \int_{\gamma} c_1(T_X) + n - 3;$$

- $\overline{\mathcal{M}}_{0,n}(X, \gamma)$ is locally a quotient of a smooth variety by a finite group;

- The locus of maps without automorphisms in $\overline{\mathcal{M}}_{0,n}(X, \gamma)$ is a fine moduli space for such maps; and,

- The boundary of $\overline{\mathcal{M}}_{0,n}(X, \gamma)$ — that is, the locus of maps with reducible domain C — is a normal crossing divisor.

The next exercise makes this more concrete by analyzing the simple examples.

EXERCISE (2.76) 1) Show that an open set of $\overline{\mathcal{M}}_{0,0}(\mathbb{P}^2, 2)$ with domain a smooth rational curve parameterizes nonsingular conics. Maps with this domain also give all double covers of a line (which are determined, up to isomorphism, by the line and the two branch points on it). Next, maps with domain the union of two rational curves meeting at a point parameterize those singular conics that are the union of two distinct lines as well as double lines with a distinguished point (the image of the point where the two rational curves meet). Conclude that $\overline{\mathcal{M}}_{0,0}(\mathbb{P}^2, 2)$ is isomorphic to the classical space of *complete conics* (see for example Vainsencher [146]).

2) Show that the same classification of stable maps extends to $\overline{\mathcal{M}}_{0,0}(\mathbb{P}^n, 2)$ when $n \geq 3$. However, what we obtain is *not* the classical space of complete conics (classifying a plane plus a complete conic in that plane): when the map has image a line, the locus of planes containing the line is blown down to a point in $\overline{\mathcal{M}}_{0,0}(\mathbb{P}^n, 2)$.

The boundary of $\overline{\mathcal{M}}_{0,n}(X, \gamma)$ can, once again, be broken up into subloci indexed by the ways in which a stable map can have a reducible domain. Now, however, it's necessary to keep track not only of the decomposition of the curve, but also of the set of marked points and the class γ. For each partition of $\{1, \ldots, n\}$ into disjoint subsets A and B and each decomposition $\gamma = \alpha + \beta$, we let $\Delta(A, B; \alpha, \beta)$ be the closure of the locus of stable maps for which $C = C_A \cup C_B$, the points indexed by A lie on C_A and those indexed by B on C_B, and the restrictions of μ to C_A and C_B represent α and β, respectively. (Note that, if

$\alpha = 0$, then the stability of μ forces $\#A \geq 2$ and that $\Delta(A, B; \alpha, \beta)$ is empty unless the classes α and β are represented by stable maps of genus 0. Also, the irreducibility of $\Delta(A, B; \alpha, \beta)$ is only known when X is a projective space.)

If we define

$$\Delta(\{i, j\} | \{k, l\}) = \sum_{\substack{\{i,j\} \subset A \\ \{k,l\} \subset B \\ \alpha+\beta=\gamma}} \Delta(A, B; \alpha, \beta),$$

then we have the fundamental linear equivalence

(2.77) $\Delta(\{i, j\} | \{k, l\}) \sim \Delta(\{i, l\} | \{j, k\})$.

In the case of the moduli space $\overline{\mathcal{M}}_{0,4} \cong \mathbb{P}^1$, there are three such divisors corresponding to the three points 0, 1 and ∞ in this space parameterizing reducible curves — see Exercise (2.19) — and the linear equivalence is simply that of points on $\mathbb{P}^1$. In general, the equivalence follows by pulling back this special case under the map $\overline{\mathcal{M}}_{0,n}(X, \gamma) \to \overline{\mathcal{M}}_{0,4}$ which forgets the map γ and the points not indexed by $\{i, j, k, l\}$.

When we take X to be a projective space (certain other homogeneous spaces can be used as well), equation (2.77) can be used to obtain recursions for solutions to a wide range of enumerative questions about rational curves. The most direct approach is to write down a suitable curve Y in $\mathcal{M}_{0,n}(\mathbb{P}^n, d)$ and to interpret its intersection numbers with various boundary divisors as enumerative quantities. Applying these interpretations to the two sides of (2.77) then produces relations between the enumerative quantities.

Rather than even attempt to write down general results of this type, we sketch the now classic use of $\mathcal{M}_{0,3d}(\mathbb{P}^2, d)$ to calculate the number N_d of rational plane curves of degree d through $3d-1$ general points. In this case, the "right" curve Y consists of those stable maps which send the first two marked points to points lying on two fixed but general lines and the other $3d - 2$ marked points to fixed general points. The only maps

$$\mu : C = C_{\{1,2\}} \cup C_{\{3,\ldots,3d\}} \to \mathbb{P}^2$$

at which Y can meet $\Delta(\{1, 2\}, \{3, \ldots, 3d\}; 0, d)$ are those which collapse $C_{\{1,2\}}$ to the point of intersection of the two fixed lines. Since the map must also take the points of $C_{\{3,\ldots,3d\}}$ to the $3d - 2$ fixed points, $\#(Y \cap \Delta(\{1, 2\}, \{3, \ldots, 3d\}; 0, d)) = N_d$.

Since the images of the marked points are general, Y is disjoint from every other $\Delta(A, B; 0, d)$ for which A contains $\{1, 2\}$ and, for $0 < e < d$, it can meet $\Delta(A, B; e, d-e)$ only when $\#A = 3e + 1$. If so, we can count as follows: there are $\binom{3e-4}{3e-1}$ partitions for which $\{1, 2\} \in A$

and $\{3, 4\} \in B$; N_e choices for the image of C_A and N_{d-e} for the image of C_B; e choices for the image of each the points p_1 and p_2 which must map to points of intersection of $\mu(C_A)$ with the corresponding fixed line; and $e(d - e)$ choices for the image of the intersection $C_A \cap C_B$ which must lie in the intersection $\mu(C_A) \cap \mu(C_B)$. Thus,

$$\#\left(Y \cap \Delta(\{1,2\}|\{3,4\})\right) = N_d + \sum_{0<e<d} N_e N_{d-e} e^3 (d - e) \binom{3d - 4}{3e - 1}$$

EXERCISE (2.78) Suppose now that $\{1, 4\} \in A$ and $\{2, 3\} \in B$. Show that $\#(Y \cap \Delta(A, B; e, d - e)) = 0$ if $e = 0$ or $e = d$. Otherwise, show that Y meets $\Delta(A, B; e, d - e))$ only if $\#A = 3e$ and that, in each of these $\binom{3d - 4}{3e - 2}$ cases, we have

$$\#(Y \cap \Delta(A, B; e, d - e)) = N_e N_{d-e} e^2 (d - e)^2 .$$

Finally, use $\Delta(\{1, 2\}|\{3, 4\}) = \Delta(\{1, 4\}|\{2, 3\})$ (cf. (2.77)) to deduce the recursion

$$N_d = \sum_{0<e<d}^{\cdot} N_e N_{d-e} \left(e^3 (d - e) \binom{3d - 4}{3e - 1} - e^2 (d - e)^2 \binom{3d - 4}{3e - 2} \right) .$$

Of course, this argument depends coming up with the right curve Y. This can be avoided by rephrasing (2.77) in terms of the formalism of Gromov-Witten invariants and quantum cohomology. Even defining these terms carefully would take us too far afield, so we give only the barest sketch and refer to [54] for all details.

Gromov-Witten invariants are numerical invariants of suitable collections of cohomology classes on X obtained by: pulling the classes back to $\overline{\mathcal{M}}_{0,n}(X, \gamma)$ using the evaluation maps which send a stable map to its value at a marked point; cupping together the pullbacks; and finally integrating them over the fundamental class of $\overline{\mathcal{M}}_{0,n}(X, \gamma)$. When X is a homogeneous space, they have enumerative interpretations. If, in addition, all classes on X which represent a stable map are expressible as nonnegative linear combinations of a finite number $\gamma_1, \ldots, \gamma_m$, the Gromov-Witten invariants can be used to define an extension of the ring structure on the Chow ring of $A^*(X)$ to its tensor product with the formal power series ring $\mathbb{Q}[[\gamma_1, \ldots, \gamma_m]]$: the extended ring is called a *quantum cohomology ring* . Equation (2.77) amounts to the associativity of this extended product. Once this is set up, enumerative results can be obtained by simply writing down the associativity equations and applying the enumerative interpretations of the Gromov-Witten invariants.

These ideas are currently being pursued in a number of different directions. The basic program is laid out in [103] and [104]. On the one hand, there is a symplectic approach to quantum cohomology (cf. [135]) which has been used to prove associativity in some cases not yet handled by the kinds of methods discussed here. On the other, the genus 0 results have motivated work by Ran, Caporaso-Harris, Pandharipande, Vakil, Getzler and others on enumerative geometry of curves of higher genus (cf. [133], [17], [18], [19], [129], [128], [147], [55] and [56]) to which the quantum cohomological formalism doesn't seem to extend directly.

Chapter 3

Techniques

A Basic facts about nodal and stable curves

In this short section, we collect various elementary facts about stable (or, more generally, nodal) curves that we'll need later, as well as a few which have already been used in Chapter 2. Most of the verifications are straightforward and will be left to you as exercises. Almost everything we discuss here is treated in the original paper of Deligne and Mumford [29].

We begin with the *genus formula* already stated in (2.14): if C is a connected nodal curve with δ nodes $p_1, \ldots, p_\delta$ and ν irreducible components $C_1, \ldots, C_\nu$ of geometric genera $g_1, \ldots, g_\nu$, then in Section 2.C

$$(3.1) \qquad g = \sum_{i=1}^{\nu}(g_i - 1) + \delta + 1 = \left(\sum_{i=1}^{\nu} g_i\right) + \delta - \nu + 1.$$

EXERCISE (3.2) Let us, as we'll also do in the sequel, abuse notation by identifying the sheaf

$$\mathcal{O}_{\tilde{C}} = \sum_{i=1}^{\nu} \mathcal{O}_{\tilde{C}_i}$$

on the normalization $\tilde{C} = \bigcup \tilde{C}_i$ of C with its direct image on C. This is harmless, since the Leray spectral sequence identifies all cohomology groups of these two sheaves. Show that we have an exact sequence

$$0 \longrightarrow \mathcal{O}_C \longrightarrow \mathcal{O}_{\tilde{C}} \longrightarrow \sum_{j=1}^{\delta} \mathbb{C}_{p_j} \longrightarrow 0$$

and verify (3.1) by using the associated long exact sequence.

Dualizing sheaves

Next, we list some basic properties of the *dualizing sheaf* ω_C of a nodal curve C, which will play a role for such curves analogous to that of the canonical bundle of a smooth curve. We will start with the most concrete definition. Let C be a nodal curve, with normalization $\nu : \tilde{C} \to C$; let $p_1, \ldots, p_\delta$ be the nodes of C, and $\{q_i, r_i\} = \nu^{-1}(p_i)$ the pair of points q_i and r_i of $\tilde{C}$ lying over each node p_i. The dualizing sheaf ω_C may be defined as a subsheaf of the pushforward of the sheaf of rational differentials on $\tilde{C}$: it's the sheaf associating to each open $U \subset C$ the space of rational one-forms η on $\nu^{-1}(U) \subset \tilde{C}$ having at worst simple poles at the pairs of points q_i and r_i of $\tilde{C}$ lying over each node $p_i \in U$ of C, and such that for each such pair of points

$$(3.3) \qquad \qquad \mathrm{Res}_{q_i}(\eta) + \mathrm{Res}_{r_i}(\eta) = 0 .$$

The following exercise establishes some of the basic properties of the dualizing sheaf of a nodal curve that we'll be using:

EXERCISE (3.4) Let C be a nodal curve of arithmetic genus g.

1) Show that the dualizing sheaf of C is an invertible sheaf.

2) Show that the degree of ω_C is $2g - 2$.

3) Show that if C is connected, then the space $H^0(C, \omega_C)$ of global sections of the dualizing sheaf has dimension g; more generally, $h^0(C, \omega_C) = g + \mu - 1$, where $\mu = h^0(C, \mathcal{O}_C)$ is the number of connected components of C.

More generally, for any curve C with normalization $\nu : \tilde{C} \to C$, the dualizing sheaf ω_C associates to each $U \subset C$ the rational one-forms η on $\nu^{-1}(U) \subset \tilde{C}$ such that for each $p \in U$ and each $f \in \mathcal{O}_{C,p}$,

$$(3.5) \qquad \qquad \sum_{q \in \nu^{-1}(p)} \mathrm{Res}_q(\nu^* f \cdot \eta) = 0$$

All three parts of Exercise (3.4) are true more generally for any curve with *Gorenstein singularities*, a class that includes all local complete intersection curves; in fact, part 1 of Exercise (3.4) is equivalent to C being Gorenstein.

EXERCISE (3.6) 1) Show that in the case of a node the requirement (3.5) implies that η has at most simple poles.

2) What order of poles are allowed if C has a tacnode? A planar triple point? A spatial triple point?

3) Show that the dualizing sheaf ω_C is invertible at a planar triple point of C, but not at a spatial triple point.

The name "dualizing sheaf" derives from an identification

(3.7) $$H^1(C, \omega_C) = \mathbb{C}$$

such that for any coherent sheaf $\mathbf{F}$ on C the cup product map

$$H^1(C, \mathbf{F}) \otimes H^0(C, \underline{\mathrm{Hom}}_{\mathcal{O}_C}(\mathbf{F}, \omega_C)) \longrightarrow H^1(C, \omega_C) = \mathbb{C}$$

is a nondegenerate pairing, inducing a natural isomorphism

$$\mathrm{Hom}_{\mathbb{C}}(H^1(C, \mathbf{F}), \mathbb{C}) \cong \mathrm{Hom}_{\mathcal{O}_C}(\mathbf{F}, \omega_C).$$

(Recall that by $\mathrm{Hom}_{\mathcal{O}_C}(\mathcal{F}, \mathcal{G})$ we mean the space of global sections of the sheaf $\underline{\mathrm{Hom}}_{\mathcal{O}_C}(\mathcal{F}, \mathcal{G})$), that is, the space of sheaf morphisms $\mathcal{F} \to \mathcal{G}$.) For a proof, see [84]; alternatively, the following exercise sketches a proof in case $\mathbf{F}$ is invertible.

EXERCISE (3.8) Fix a nodal curve C together with an invertible sheaf which we write as $\mathcal{O}_C(\sum_{i=1}^k n_i s_i)$ with $s_1, \ldots, s_k$ are smooth points of C. Let $\Delta_i \subset C$ be an open disc around s_i. In terms of the covering of C by the open sets $U = \bigcup \Delta_i$ and $V = C \setminus \{s_1, \ldots, s_k\}$, any element of

$$H^1(C, \mathcal{O}_C(\sum_{i=1}^k n_i s_i))$$

may be represented by a collection of Laurent series $f_i \in \mathcal{O}_C(\Delta_i \setminus \{p_i\})$ on the punctured discs $\Delta_i \setminus \{p_i\}$. Use the analogous statement (that is, Kodaira-Serre duality) on the normalization $\tilde{C}$ of C to show that the pairing

$$H^1\left(C, \mathcal{O}_C(\sum_{i=1}^k n_i s_i)\right) \times H^0\left(C, \omega_C(\sum_{i=1}^k -n_i s_i)\right) \longrightarrow \mathbb{C}$$

given by

$$\{f_1, \ldots, f_k\} \times \eta \mapsto \sum_{i=1}^k \mathrm{Res}_{p_i}(f_i \cdot \eta)$$

is well-defined, and that it's a perfect pairing.

Note that, as a consequence, the Riemann-Roch theorem likewise extends to nodal (or more generally Gorenstein) curves, in the form

$$h^0(C, L) = d - g + 1 + h^0(C, \omega_C \otimes L^\vee)$$

for a curve C of arithmetic genus g and invertible sheaf L of degree d on C.

The next two exercises show that the dualizing sheaf of a *stable* curve has ampleness properties only very mildly weaker than those of the canonical bundle on a smooth curve, and give in these terms a characterization of moduli stable curves amongst all nodal ones.

EXERCISE (3.9) Let C be a stable curve of genus g.

1) Show that $H^0(C, \omega_C^{\otimes n}) = (2n - 1)(g - 1)$, and that for $n \geq 2$, $H^1(C, \omega_C^{\otimes n}) = 0$.

2) Show that for $n \geq 3$, $\omega_C^{\otimes n}$ is very ample on C.

EXERCISE (3.10) Let C be a complete connected nodal curve.

1) Show that, for $n \geq 3$, the sheaf $\omega_C^{\otimes n}$ is very ample if and only if C is moduli stable. Hence, ω_C is ample on a complete connected nodal curve C if and only if C is moduli stable.

2) Similarly, let $p_1, \ldots, p_n \in C$ be distinct smooth points of C. Show that $(C; p_1, \ldots, p_n)$ is a stable n-pointed curve if and only if the line bundle

$$\omega_C(\sum_{i=1}^{n} p_i)$$

is ample.

A fundamental and important fact about dualizing sheaves of nodal curves is that they fit together in families: if $\varphi : \mathcal{C} \to B$ is a flat family of nodal curves, we may define the *relative dualizing sheaf* $\omega_{\mathcal{C}/B}$ of the family to be the sheaf of rational relative differentials — that is, rational sections of the relative cotangent bundle $\mathrm{Coker}(d\varphi : \varphi^* \Omega_B \to \Omega_{\mathcal{C}})$ — satisfying the residue condition (3.3) on each fiber. That this is in fact an invertible sheaf on $\mathcal{C}$ may not be clear, but it may be readily verified by a local calculation. In fact, this virtually forces the definition of the dualizing sheaf on us:

EXERCISE (3.11) Let $\mathcal{C}$ be the locus $xy = t^k$ in $\mathbb{A}^3$, and consider the morphism $\varphi : \mathcal{C} \to \mathbb{A}^1$ given by $(x, y, t) \mapsto t$. Show that the relative cotangent bundle of φ on $\mathcal{C}^* = \mathcal{C} \setminus \{(0, 0, 0)\}$ extends (uniquely) to a line bundle on all of $\mathcal{C}$, and that the sections of this line bundle on the fiber $C_0 = \varphi^{-1}(0)$, viewed as rational differentials on C_0, satisfy the residue condition (3.3).

Given this, the relative dualizing sheaf of a family $\varphi : \mathcal{C} \to B$ of nodal curves whose general fiber is smooth may be characterized as the unique line bundle on $\mathcal{C}$ extending the relative cotangent bundle on the locus $\mathcal{C}^*$ of smooth points of φ. Moreover, if the total space $\mathcal{C}$ is smooth, we may the describe the relative dualizing sheaf as the canonical bundle of $\mathcal{C}$ tensored with the dual of the pullback of the canonical bundle of B:

$$\omega_{\mathcal{C}/B} = K_{\mathcal{C}} \otimes \varphi^* K_B^{\vee}.$$

Another feature of the relative dualizing sheaf is the naturality of the identification (3.7) as shown by the:

EXERCISE (3.12) Let $\varphi : \mathcal{C} \to B$ be a family of connected nodal curves and $\omega_{\mathcal{C}/B}$ the relative dualizing sheaf of the family. Using the concrete description of the relative dualizing sheaf in terms of rational differentials, show that

$$R^1\varphi_*\omega_{\mathcal{C}/B} = \mathcal{O}_B.$$

A consequence is that if L is a line bundle on $\mathcal{C}$ with $h^0(C_b, L)$ constant, then we have a very special case of *Grothendieck duality*, the relative version of Kodaira-Serre duality:

$$R^1\varphi_*(L^\vee \otimes \omega_{\mathcal{C}/B}) = (\varphi_*L)^\vee.$$

Observe that Exercise (3.12) characterizes yet again the relative dualizing sheaf of a family $\varphi : \mathcal{C} \to B$ of nodal curves among all line bundles whose restriction to each fiber is the dualizing sheaf of the fiber.

It follows from any of the descriptions above that the relative dualizing sheaf is functorial: if $B' \to B$ is any morphism and $\mathcal{C}' = \mathcal{C} \times_B B' \to B'$ the pullback family, then $\omega_{\mathcal{C}'/B'} = \pi^*(\omega_{\mathcal{C}/B})$, where $\pi : \mathcal{C}' \to \mathcal{C}$ is the projection.

EXERCISE (3.13) Let $\varphi : \mathcal{C} \to B$ be a flat family of stable curves of genus g. Show that for $n \geq 2$, the direct image $\varphi_*(\omega_{\mathcal{C}/B}^{\otimes n})$ is locally free of rank $(2n-1)(g-1)$.

Automorphisms

Our definition of a stable curve requires that it have only finitely many automorphisms. To verify the local description of $\overline{\mathcal{M}}_g$ given in the previous chapter (as smooth away from loci of curves with automorphisms at which it has quotient singularities), we'll need the slightly stronger assertion that the scheme-theoretic automorphism group of a stable curve is finite and *reduced*. This is our next goal.

To start with, we need to make precise the scheme structure on the automorphism group. This turns out to be a bit involved. We will just sketch the ideas here and refer you to Section 1 of [29] for more details. We also simplify by working over a point rather than a more general base. The first step is to define, for any two stable curves C and D, an isomorphism functor $\mathbf{Isom}(C, D)$ whose value on a scheme S' is the set of S'-isomorphisms between $C \times S'$ and $D \times S'$. Any such isomorphism must identify the relative dualizing sheaves of C and D and hence all powers of these sheaves. Using Exercise (3.9), this leads to a representation of the functor $\mathbf{Isom}(C, D)$ by a subscheme of a suitable projective linear group.

More precisely, fix $g \geq 2$ and an integer $n \geq 3$. Define integers r and d in terms of these by

(3.14)
$$r + 1 = (2 \cdot n - 1)(g - 1) \quad \text{and}$$
$$d = 2 \cdot n(g - 1).$$

Let $\mathcal{H} = \mathcal{H}_{d,g,r}$ and let $[C]$ and $[D]$ be the points of $\mathcal{H}$ determined by C and D. Define a map $\mu : \mathrm{PGL}(r+1) \to \mathcal{H} \times \mathcal{H}$ by $\mu(\alpha) = (\alpha \cdot [C], [D])$ and let $I(C,D) = \mu^{-1}(\Delta) \in \mathrm{PGL}(r+1)$, where Δ is the diagonal in $\mathcal{H} \times \mathcal{H}$. The scheme $\mathrm{Isom}(C,D)$ turns out to represent $\mathbf{Isom}(C,D)$ although we won't verify this here.

To define the automorphism group of a stable curve C, we just take $D = C$ in the foregoing, This amounts to identifying $\mathrm{Aut}(C)$ with the stabilizer in $\mathrm{PGL}(r+1)$ of $[C] \in \mathcal{H}$. The assertion we're after is then:

LEMMA (3.15) $\mathrm{Aut}(C) = \mathrm{stab}_{\mathrm{PGL}(r+1)}([C])$ *is reduced.*

If not, there would be a nonzero $\mathbb{I}$-valued point of $\mathrm{Aut}(C)$ lying over the identity, or equivalently, a nonzero regular vector field on C. The following exercise rules this out.

EXERCISE (3.16) Show that, on a stable curve C, there is no nonzero, everywhere regular vector field. Equivalently, $\mathrm{Ext}^0(\Omega_C, \mathcal{O}_C) = \{0\}$.
Hint: Such a vector field would correspond to a regular vector field on the normalization $\tilde{C}$ vanishing at all points lying over the nodes of C. Show that such a vector field must be identically 0 on every component of $\tilde{C}$.

We conclude by remarking that similar arguments give a relative version: given two stable curves $C \to S$ and $D \to S$, there is a scheme $\mathrm{Isom}_S(C,D)$ which is finite and unramified over S and represents the functor of S-isomorphisms between the curves. In particular, if S itself is a curve, 0 is a point of S and $S^* = S \setminus \{0\}$, then any S^*-isomorphism between the restrictions to S^* of $C \to S$ and $D \to S$ extends to an S-isomorphism.

B Deformation theory

Overview

In this short section, we want to quickly sketch the typical stages of an application of deformation theory[1] using the simplest example,

---

[1] A few words about other references are in order here. We know of no accessible reference that deals with all the variations we wish to discuss here. For a very read-

deformations of smooth varieties, as our model. A *deformation* of a smooth variety X with base a pointed scheme (Y, y_0) is a proper flat morphism $\varphi : X \to Y$, *together with* an isomorphism $\psi : X \to \varphi^{-1}(y_0)$ of X with the scheme-theoretic fiber of φ over the point $y_0 \in Y$. In other words, it's a fiber square

(3.17)

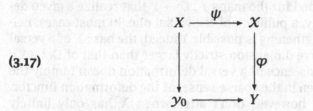

where the map $X \to y_0 \cong \text{Spec}(\mathbb{C})$ is the structure map and the map $y_0 \to Y$ is the inclusion. Two such squares are called *equivalent* if there is an isomorphism of fiber squares between them that equals the *identity* on X.

A *first-order deformation* of X is a deformation over the pointed space $(\mathbb{I}, 0)$ of dual numbers. The space $\text{Def}_1(X)$ of such deformations is extremely important for two reasons. First, it can almost always be identified with some cohomology group and hence is readily calculated. Second, if there is a moduli space $\mathcal{M}$ containing X then $\text{Def}_1(X)$ will usually equal the tangent space to $\mathcal{M}$ at the moduli point of X: this must, by definition, be the case if the moduli space is fine. An nth-order deformation of X is defined similarly as a deformation over $(\mathbb{I}^n, 0)$ but, except in the first-order case, it's generally very difficult to calculate these explicitly.

Having defined such infinitesimal deformations, it's natural to ask whether they can be integrated. More precisely, we ask whether there exists a deformation $\varphi : X \to Y$ with the *versality* property: any other deformation $\xi : X \to Z$ is analytically isomorphic in a neighborhood U of each point of Z to the pullback of $\varphi : X \to Y$ by a map $f : U \to Y$. Such families go by a number of other names. Analysts usually call them *Kuranishi families* (especially when deforming a complex manifold, the case originally studied by Kuranishi) and the term *complete deformation* is also seen.

---

able, detailed description of the deformation theory of compact, complex manifolds, see Kodaira's book [98]. This also contains references to the papers in which the basic theory was originally presented. The only drawback here is that the discussion is limited to smooth abstract varieties. Another good reference for this case is Palamodov's survey [126]; Vistoli's expository article [150] gives a very careful exposition of deformations of local complete intersection schemes, both embedded and abstract, from an algebraic viewpoint; and Kollár gives a thorough treatment of deformations of embedded varieties in [100]. However, none of these deals explicitly with such variations as deformations of varieties with additional structure (e.g., a line bundle) and all take distinctly different points of view.

We emphasize that versality is, in two important ways, formally weaker than the *uni*-versal properties we have dealt with in discussing moduli and parameter spaces. First, deformations are required to be pulled back from the versal deformation only locally on the base. Second, and in practice usually much more significant, no uniqueness properties are claimed for the maps $f : U \to Y$ that realize a given deformation as, locally, a pullback of the versal one. In most cases, neither potential strengthening is possible. Indeed, the base Y of a versal deformation can have dimension strictly larger than that of $\mathrm{Def}_1(X)$. In particular, the existence of a versal deformation doesn't imply the representability, even in the coarse sense, of the deformation functor.

These problems, however, don't arise when X has only finitely many automorphisms (or, more generally, when the group of automorphisms of X that extend to all small deformations of X has finite index in the full automorphism group of X). In this case, a minimal versal deformation Y (that is, one for which the induced map $T_Y \to \mathrm{Def}_1(X)$ is an isomorphism) will be universal so that the map that realizes a family as a pullback of the versal one will be unique. (This is one, but not the only, point at which making the identification of the central fiber part of the definition of a deformation is crucial.) As usual, such uniqueness properties mean that any two versal deformations of X are locally isomorphic. Since we'll be interested in versal deformations mainly for stable curves — which, by definition, have finite automorphism group — we'll almost always be able to make such uniqueness assumptions.

For general varieties, existence of versal deformations is usually a difficult question. Even the existence of liftings of first-order deformations to second order (much less to arbitrary order or to formal families) can be hard to decide. There is a general theory that describes groups in which the obstructions to such liftings lie. However, we'll only mention this theory briefly at a few points because, even in cases in which it's possible to calculate these obstruction groups explicitly, it's generally difficult to determine whether the obstruction defined by a given first-order deformation vanishes or not.

Fortunately, for curves, it is possible to give direct and explicit constructions of versal deformations. In the next section, we'll see how to do this in two ways: first, by integrating certain canonical first-order deformations called Schiffer variations; and second, by taking a suitable subscheme of a Hilbert scheme as the base of such a deformation and, in essence, inheriting the deformation and the desired (uni)versality property from the corresponding universal curve.

The basic model described above can be varied in many ways to adapt it to the study of a particular problem of interest. However, almost all applications of deformation theory involve three steps analogous to those outlined above:

1. Pose the appropriate deformation theoretic problem;

2. calculate the space of first-order deformations; and,

3. construct, if one exists, a versal deformation.

In the rest of this section, we first work out, in some detail, the three steps in the basic case where X is a smooth curve. Then, we carry out the first two steps in a variety of useful and representative examples, often omitting detailed proofs and leaving you to make the necessary modifications to the smooth curve model.

Deformations of smooth curves

Throughout this section, we fix a smooth curve C of genus $g \geq 1$. We begin by determining the first-order deformations of C. [2]
To begin with, let's fix an affine open cover U_α of C and a collection of linear maps $\varphi_{\alpha\beta} : \mathcal{O}_{U_\alpha \times 1}|_{U_{\alpha\beta}} \to \mathcal{O}_{U_\beta \times 1}|_{U_{\alpha\beta}}$ that restrict to the identity modulo ε — that is, we want the maps $\varphi_{\alpha\beta}$ to satisfy

$$\varphi_{\alpha\beta}(\varepsilon) = \varepsilon \quad \text{and}$$

$$\varphi_{\alpha\beta}(f) = f + \varepsilon D_{\alpha\beta} \quad \text{for } f \in \mathcal{O}_{U_{\alpha\beta}}$$

with each $D_{\alpha\beta}$ a $\mathbb{C}$-linear function of f. In order that such a collection of maps glue together to give a first-order deformation of C, two conditions are necessary and sufficient.
First, each $\varphi_{\alpha\beta}$ must be a ring homomorphism:

$$\varphi_{\alpha\beta}(fg) = \varphi_{\alpha\beta}(f)\varphi_{\alpha\beta}(g),$$

or, using the definitions above and $\varepsilon^2 = 0$,

$$fg + \varepsilon D_{\alpha\beta}(fg) = fg + \varepsilon(f D_{\alpha\beta}(g) + g D_{\alpha\beta}(f)).$$

In other words, the maps $D_{\alpha\beta}$ must be derivations and hence give a cochain, with respect to the cover of $U_{\alpha\beta}$'s, taking values in the tangent bundle to C.
Second, on the triple overlaps the $\varphi_{\alpha\beta}$'s must satisfy the multiplicative cocycle condition

$$\varphi_{\alpha\gamma} = \varphi_{\beta\gamma} \circ \varphi_{\alpha\beta}.$$

Plugging in again, this amounts to

$$f + \varepsilon D_{\alpha\gamma}(f) = D_{\beta\gamma}(f + \varepsilon D_{\alpha\beta}(f)) = f + \varepsilon(D_{\beta\gamma}(f) + D_{\alpha\beta}(f))$$

---

[2]The approach that follows is due to M. Artin and was shown to us by Angelo Vistoli.

or the assertion that the $D_{\alpha\beta}$'s are an additive cocycle.

Combining these observations, we see that

$$\mathrm{Def}_1(C) \cong H^1(C, T_C).$$

Notice also that, up to this point, what we've said applies equally well to any smooth variety.

Most references follow a somewhat more analytic path in calculating $\mathrm{Def}_1(C)$, which is outlined in the following exercise.

EXERCISE (3.18) Fix a cover as above and choose a local coordinate z_α on each U_α so that $\mathcal{O}_{U_\alpha \times \mathbb{I}}$ can be identified with the ring of convergent power series $\mathbb{C}\{z_\alpha, \varepsilon\}/(\varepsilon^2)$. Choose in addition a collection $\{\psi_{\alpha\beta}\}$ of coordinate transformations defined on the overlaps $U_{\alpha\beta}$ by

$$\psi_{\alpha\beta}(z_\alpha) = p_{\alpha\beta}(z_\beta) + \varepsilon q_{\alpha\beta}(z_\beta)$$

with p and q power series in one-variable convergent in some neighborhood of the origin.

1) Use Taylor expansions to show that the maps $\psi_{\alpha\beta}$ satisfy the cocycle condition $\psi_{\alpha\gamma} = \psi_{\beta\gamma} \circ \psi_{\alpha\beta}$ if and only if

$$p_{\alpha\gamma}(z_\gamma) = p_{\beta\gamma} \circ p_{\alpha\beta}(z_\gamma)$$

$$q_{\alpha\gamma}(z_\gamma) = \frac{\partial p_{\alpha\beta}}{\partial z} p_{\beta\gamma} + q_{\alpha\beta} p_{\beta\gamma}$$

and hence construct a map from $H^1(C, T_C)$ to $\mathrm{Def}_1(C)$.

2) Given a first-order deformation $\mathcal{X} \to \mathbb{I}$ of C, consider the normal bundle sequence

$$0 \to T_C \to T_{\mathcal{X}}|_C \to N_{C/\mathcal{X}} \to 0.$$

Show that $\mathcal{X}$ is trivial if and only if the derivation $\frac{\partial}{\partial \varepsilon}$ at 0 in $\mathbb{I}$ lifts to a derivation $D \in H^0(C, N_{C/\mathcal{X}})$ along C. Show that any two such lifts have the same image in $H^1(C, T_C)$ under the coboundary map $H^0(C, N_{C/\mathcal{X}}) \to H^1(C, T_C)$. Conclude that the map in 1) is an isomorphism.

Our calculation of $\mathrm{Def}_1(C)$ shows that it's isomorphic to $H^1(C, T_C) \cong H^0(C, K_C^{\otimes 2})^\vee \cong \mathbb{C}^{3g-3}$. We next want to integrate these, eventually obtaining a versal family. As a first step, we introduce an important family of one-parameter first order deformations that can be integrated explicitly.

Fix a point P on C and consider the cover of C by two open sets, a small disc U centered at P and the complement $V = C - \{P\}$ of $\{P\}$ in C. Since there is only a single overlap $W = U \cap V \cong \Delta^*$ the cocycle condition is vacuous and a first-order deformation is simply a holomorphic vector field on $U \cap V$. If z is a local coordinate on this overlap, we may choose the vector field to be, for example,

$$s_P = \frac{1}{z}\frac{\partial}{\partial z},$$

in which case the corresponding first-order deformation is called a *first-order Schiffer variation* at P.

EXERCISE (3.19) Using the Serre duality between $H^1(C, T_C)$ and $H^1(C, K_C^{\otimes 2})$, we may view elements of $H^0(C, K_C^{\otimes 2})$ as linear functionals on $H^1(C, T_C)$.

1) Show that the annihilator of the first-order Schiffer variation at P is $H^0(C, K_C^{\otimes 2}(-P))$. In particular, changing coordinates at P simply rescales the corresponding first-order Schiffer variation.

2) Show that the set of all first-order Schiffer variations spans $H^1(C, T_C)$.

These first-order deformations can be integrated to a deformation $\mathcal{X} \longrightarrow \Delta_t$ as follows. Let U_t be a constant unit disc Δ_z with coordinate z and W_t be the varying subannulus in which $|z| > t^{1/2}$. Next let w be a local coordinate on C centered at P and map W_t to V_t by $w = z + \frac{t}{z}$. This amounts to identifying the shaded region below (the ellipsoidal image of U_t in C minus the similarly shaped image of $U_{t^{1/2}}$) with the annulus W_t. Making these identifications for all t at once yields a

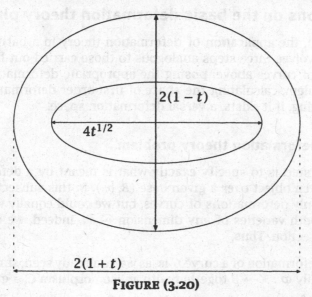

$2(1-t)$

$4t^{1/2}$

$2(1+t)$

FIGURE (3.20)

deformation $\mathcal{X} \longrightarrow \Delta_t$ that sends the vector field $\frac{\partial}{\partial t}$ to the first-order Schiffer variation of C at P given by s_P as desired.

This construction can now be carried out independently near each of several points $P_1, \ldots, P_k$ to give deformations of C over a product of k discs. Generalizing the exercise above, we see that the annihilator of the image in $\mathrm{Def}_1(C)$ of the tangent space to such a deformation is

$$H^0(C, K_C^{\otimes 2}(-P_1 - \cdots - P_k)).$$

Hence, if we choose any generic collection of $3g - 3$ points on C, we obtain a deformation over a $3g - 3$-dimensional polydisc whose tangent space maps isomorphically to $\mathrm{Def}_1(C)$.

EXERCISE (3.21) Use the change of coordinates

$$w = z + \frac{t_1}{z} + \frac{t_2}{z^2} + \cdots + \frac{t_{3g-3}}{z^{3g-3}}$$

to construct a kth order variation of complex structure over a polydisc of dimension $3g - 3$. Show that if p is generic then this variation also has a tangent space that maps isomorphically to $\mathrm{Def}_1(C)$.

The deformations constructed above are versal — even universal — but, unfortunately, there seems to be no direct method of verifying this. Instead, we give an alternate construction which has the advantage of working for all stable curves at the end of the next section.

Variations on the basic deformation theory plan

In general, the application of deformation theory in a particular instance involves three steps analogous to those carried out for deformations of curves above: posing the appropriate deformation theoretic problem, calculating the space of first-order deformations and constructing, if it exists, a versal deformation space.

Pose a deformation theory problem

The first step is to specify exactly what is meant by a deformation of the given object over a given base (B, b_0). In this subsection we'll discuss only deformations of curves, but we could equally well work with smooth varieties of any dimension — as, indeed, we do in the next subsection. Thus,

- A deformation of a curve C is, as we've already seen, simply a flat family $\varphi : X \to B$ together with an isomorphism $C \cong \varphi^{-1}(b_0)$.
- A deformation of a pointed curve $(C, p_1, \ldots, p_k)$ is a flat family $\varphi : X \to B$ with an isomorphism $\psi : C \cong \varphi^{-1}(b_0)$ and disjoint sections $\sigma_i : B \to X$ such that $\sigma_i(b_0) = \psi(p_i)$.

- A deformation of a curve C with line bundle L is a flat family $\varphi : X \to B$ with isomorphism $C \cong \varphi^{-1}(b_0)$, together with a line bundle $\mathcal{L}$ on X and isomorphism $\mathcal{L}|_{\varphi^{-1}(b_0)} \cong L$.

- A deformation of a map $f : C \to C'$ with C and C' fixed is a map $\overline{f} : C \times B \to C' \times B$ whose restriction to $C \times \{b_0\}$ is f. Note that if C' is a projective space $\mathbb{P}^n$ and f is an embedding then, in the absence of automorphisms of C, a deformation of f over B is just a map of B into the relevant Hilbert scheme.

- A deformation of a map $f : C \to C'$ with C' fixed is a deformation $X \to B$ of C (that is, a flat family $\varphi : X \to B$ with isomorphism $\psi : C \cong \varphi^{-1}(b_0)$), together with a map $\overline{f} : X \to C' \times B$ fitting into a commutative diagram:

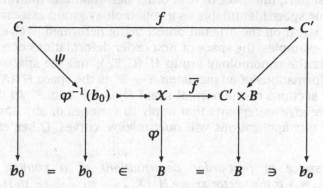

We could list many further types of deformation but these examples make it clear how to define such variations. In these examples, we haven't explicitly stated when two deformations are to be considered equivalent since, in each case, the desired relation is fairly unambiguous. But, differences in how we choose this relation can sometimes significantly affect what problem we're posing. For example, if we are interested in studying deformations of an isolated singular point of a variety X, we don't care about global deformations of X. We may therefore define a deformation of the singularity (X, p) to be a deformation $U \to B$ of a neighborhood U of p in X, with the equivalence relation generated by inclusions $U \subset V$.

Another issue that is sometimes delicate to address is that of imposing side conditions on deformations. For example, among deformations of a curve singularity (C, p) are a subclass called *equisingular deformations*. In a complex analytic setting, these are the deformations that are locally topologically trivial — i.e., deformations $U \to B$ of C such that some neighborhood of p is actually a topological fiber bundle over B. This definition seems perfectly reasonable but hides a subtle problem: what do we mean, for example, by an equisingular deformation over $\mathrm{Spec}(\mathbb{C}[\varepsilon]/\varepsilon^2)$? This question does have an answer, albeit a rather complicated one, due to Zariski. However, superficially very similar questions do not. For example, no satisfactory answer is known to the analogous question for *equigeneric deformation*. These are deformations that preserve the δ-invariant of a singularity (its contribution to the genus of the curve) or, equivalently, the length of the $\mathbb{C}$-vector space $\Gamma(\mathcal{O}_{\tilde{U}})/\Gamma(\mathcal{O}_U)$ where U is a neighborhood of P in C and $\tilde{U}$ its normalization.

Calculate the space of first-order deformations

The second step of the process is in general the most fun. Almost invariably, if the deformation theory problem has been set up correctly in the first part, the space of first-order deformations will turn out to be a vector space identifiable as a cohomology group associated to a coherent sheaf on the original object being deformed. We've already seen two examples: the space of first-order deformations of a smooth curve C is the cohomology group $H^1(C, T_C)$, and the space of first-order deformations of an inclusion $X \hookrightarrow \mathbb{P}^r$ is the space $H^0(X, N_{X/\mathbb{P}^r})$ of global sections of the normal sheaf $N_{X/\mathbb{P}^r}$ of X in $\mathbb{P}^r$. In this subsection we give statements that apply to varieties of any dimension, although our applications will only involve curves. Other examples are:

1) *The space of first-order deformations of a pointed variety* $(X, p_1, \ldots, p_k)$ *is the vector space* $H^1(X, T_X(-p_1 - \cdots - p_k))$.

Note that in this example there is a natural map to the space of deformations of X alone. The induced map on the space of first-order deformations is the obvious one: it's the map on H^1's associated to the inclusion $T_X(-\sum_i p_i) \hookrightarrow T_X$ in the exact sequence

$$0 \longrightarrow T_X\left(-\sum_i p_i\right) \longrightarrow T_X \longrightarrow T_X/T_X\left(-\sum_i p_i\right) \longrightarrow 0.$$

In particular, if X has no global vector fields (i.e., $H^0(X, T_X) = 0$), then the space of first-order deformations of the points $(p_1, \ldots, p_k)$ on the fixed variety X is just the space of global sections of $T_X/T_X(-\sum_i p_i)$, which is just to say the direct sum of the tangent spaces to X at the points p_i. In general, vector fields on X yield other such deformations.

2) *The space of first-order deformations of a pair* (X, D) *with X a fixed smooth variety and D an effective divisor on X is the space* $H^0(X, \mathcal{O}(D)/\mathcal{O})$.

EXERCISE (3.22) Let $X = C$ be a smooth curve of genus $g \geq 2$. The last statement may be rephrased as saying that the tangent space to the dth-symmetric product C_d of C at a point D is the space

$H^1(C, \mathcal{O}(D)/\mathcal{O})$, and hence that the cotangent space is

$$T^*(C_d)|_D = H^0(C, K/K(-D)).$$

Now let $\Sigma = \{(p, D) : D - p \geq 0\} \subset C \times C_d$ be the universal divisor and $\pi : \Sigma \to C_d$ and $\eta : \Sigma \to C$ the projections. Show that the cotangent bundle to C_d may be realized as

$$T^*(C_d) = \pi_*(\eta^* K_C \otimes \mathcal{O}_\Sigma).$$

3) *The space of first-order deformations of a line bundle L on a fixed variety X is the space $H^1(X, \mathcal{O})$.*

This is relatively easy: if L is given with respect to a suitable cover $\{U_\alpha\}$ of X by transition functions $g_{\alpha\beta}$ then a deformation of L will be given by transition functions of the form $\{g_{\alpha\beta} + \varepsilon \cdot h_{\alpha\beta}\}$ where the $h_{\alpha\beta}$ are holomorphic functions satisfying the cocycle rule but otherwise unrestricted (modulo checking, of course, that cohomologous cocycles give rise to equivalent deformations of L). Note that we could see this directly by observing that the connected components of $\mathrm{Pic}(X)$ are the tori $H^1(X, \mathcal{O})/H^1(X, \mathbb{Z})$.

In these terms, we can ask as well when a given section σ of a line bundle L extends to a deformation $\mathcal{L}$ of L. The answer is straightforward: if $\mathcal{L}$ is given by a cocycle $\{h_{\alpha\beta}\}$ as above and σ is given with respect to the corresponding trivializations of L by the sections $\sigma_\alpha \in \Gamma(\mathcal{O}_{U_\alpha})$, then an extension of σ to a section of $\mathcal{L}$ on $X \times \mathbb{I}$ will be given by sections $\sigma_\alpha + \varepsilon \cdot \tau_\alpha$ satisfying

$$(\sigma_\alpha + \varepsilon \cdot \tau_\alpha) = (g_{\alpha\beta} + \varepsilon \cdot h_{\alpha\beta}) \cdot (\sigma_\beta + \varepsilon \cdot \tau_\alpha),$$

i.e.,

$$\tau_\alpha = g_{\alpha\beta}\tau_\beta + h_{\alpha\beta}\sigma_\alpha.$$

To put this another way, the cup product of the classes $\sigma \in H^0(X, L)$ and $h \in H^1(X, \mathcal{O})$ is the class in $H^1(X, L)$ represented by the cocycle $\{h_{\alpha\beta}\sigma_\alpha\}$, and the section σ extends to $\mathcal{L}$ if and only if this cocycle is a coboundary.

We may in particular conclude from this that the tangent space at a point $[L]$ to the locally closed subscheme $W^r(X) \subset \mathrm{Pic}(X)$ of line bundles L having $h^0(X, L) = r + 1$ is the kernel of the map

$$H^1(X, \mathcal{O}) \to \mathrm{Hom}(H^0(X, L), H^1(X, L))$$

given by cup product. In the case of a curve C, we may dualize this to see that the cotangent space to the subscheme $W^r(C)$ is the annihilator of the image of the multiplication map

$$H^0(C, L) \otimes H^0(C, K \otimes L^{-1}) \to H^0(C, K).$$

4) *The space of deformations of a pair (X, L) with X a smooth variety and L a line bundle on X, X not taken to be fixed, is the space $H^1(X, \Sigma_L)$ where Σ_L is the sheaf of first-order differential operators in L.*

Note that Σ_L fits into a nice exact sequence

$$0 \longrightarrow \mathcal{O}_X \longrightarrow \Sigma_L \longrightarrow T_X \longrightarrow 0,$$

and that the induced maps of the H^1's representing various first-order deformations are the obvious ones.

EXERCISE-WARNING (3.23) Recall from Section 2.B the moduli space $\mathcal{P}_{d,g}$ of line bundles of degree d on smooth curves of genus g which is naturally a bundle over $\mathcal{M}_g$. Show that there is no splitting of the tangent bundle to $\mathcal{P}_{d,g}$ along the fiber of this map over a fixed curve C that realizes the splitting of H^1's in the exact sequence above.

5) *The space of first-order deformations of a map $f : X \to Y$ with X and Y both fixed is the space of sections $H^0(X, f^*T_Y)$ of the pullback to X of the tangent bundle of Y.*

Note in particular that the space of first-order deformations of the identity map $X \to X$ is just the space $H^0(X, T_X)$ of global vector fields on X.

6) *The space of first-order deformations of a map $f : X \to Y$ with only Y assumed fixed is the space of sections $H^0(X, N_f)$, where N_f is the normal sheaf of the map f, defined by*

$$N_f = \mathrm{Coker}(df : T_X \to f^*T_Y).$$

(We also, as usual, write $N_{X/Y}$ for N_f when the map f is understood.) For example, if f were the inclusion of X as a subvariety of Y then we get the usual definition of

$$N_f = T_Y|_X / T_X = (\mathcal{I}_{X/Y} / \mathcal{I}_{X/Y}^2)^{\vee}.$$

We can also give a deformation-theoretic interpretation to the long exact cohomology sequence associated to the short exact sequence

$$0 \longrightarrow T_X \longrightarrow f^*T_Y \longrightarrow N_f \longrightarrow 0.$$

The coboundary $\delta : H^0(N_f) \to H^1(T_X)$ takes a deformation of the map f to the corresponding deformation of X, forgetting the map; the kernel consists of the deformations of the map f fixing both X and Y, modulo automorphisms of X.

Note that if $X \subset Y \subset Z$ is a nested sequence of closed subvarieties, we can identify when a first-order deformation $\mathcal{X} \subset Z \times \mathbb{I}$ of X is contained in a first-order deformation $\mathcal{Y} \subset Z \times \mathbb{I}$ of Y: we look at the diagram

$$N_{X/Z} \xrightarrow{\;\alpha\;} N_{Y/Z}|_X$$

$$\Bigg\uparrow \beta$$

$$N_{Y/Z}$$

and if the image of the deformation $[\mathcal{X}] \in H^0(N_{X/Z})$ under the map α coincides with the image under β of the class $[\mathcal{Y}] \in H^0(N_{Y/Z})$, we'll have $\mathcal{X} \subset \mathcal{Y}$.

EXERCISE (3.24) The classical *Noether-Lefschetz theorem* says that a general surface $S \subset \mathbb{P}^3$ of degree $d \geq 4$ contains no curves other than complete intersections with S. There is in fact a stronger form of this, which says that if S is any smooth surface of degree $d \geq 4$ and $C \subset S$ any curve not a complete intersection with S, then a general first-order deformation of S contains no first-order deformation of C. Use this to prove the:

PROPOSITION (3.25) *If S is any smooth surface in $\mathbb{P}^3$ and $C \subset S$ is any smooth curve lying on it, then C is a complete intersection with S if and only if the normal bundle sequence*

$$0 \longrightarrow N_{C/S} \longrightarrow N_{C/\mathbb{P}^3} \longrightarrow N_{S/\mathbb{P}^3}|_C \longrightarrow 0$$

splits.

Warning. The locus of surfaces of degree d having "extra" curves is a countable collection of varieties, each of which is proper but whose union is everywhere locally dense. Note also that the proposition above is true even when S has degree 2 or 3.

7) *The space of first-order deformations of a singular point p of a plane curve $C \subset \mathbb{A}^2$ given by $f(x,y) = 0$ is the local ring of C at p modulo the* Jacobian ideal $\mathcal{J}$ *generated by the partial derivatives $\frac{\partial f}{\partial x}$ and $\frac{\partial f}{\partial y}$.* The elements of $\mathcal{J}$ amount essentially to trivial deformations obtained by translating the coordinate system at p: e.g., modulo ε^2,

$$f_\varepsilon(x,y) = f(x+\varepsilon, y) = f(x,y) + \varepsilon \frac{\partial f}{\partial x}(x,y).$$

Thus, first-order deformations are given simply by $(f + \varepsilon \cdot g)$ for $g \in \mathcal{O}_{C,p}/\mathcal{J}$. In this form, it's clear that the description is equally applicable to any *planar* curve singularity, i.e., one whose Zariski tangent space is two-dimensional. Note that the node $xy = 0$ has a one-dimensional space of first-order deformations, given by the family $xy - a$ as shown schematically in Figure (3.26) for real a centered at 0.

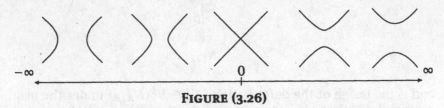

$$\text{FIGURE (3.26)}$$

The cusp, on the other hand, has a 2-dimensional deformation space with family
$$y^2 = x^3 + a \cdot x + b;$$
the tacnode $y^2 = x^4$ has a three-dimensional family
$$y^2 = x^4 + a \cdot x^2 + b \cdot x + c;$$
and the triple point $x^3 - y^3 = 0$ has a four-dimensional family
$$y^3 + x^3 + a \cdot xy + b \cdot x + c \cdot y + d = 0.$$

A fascinating question about these families is: what types of singularities occur in fibers near the central one and, over which loci in the base does each occur? For example, curves in the first family are either smooth (in general), nodal (for (a, b) lying on the cuspidal curve $4a^3 = 27b^2$) or cuspidal (only for $(a, b) = (0, 0)$).

EXERCISE (3.27) Show that in the deformation of a tacnode we may see curves with two nodes, one cusp or one node. For which values of (a, b, c) do each of these possibilities arise?

EXERCISE (3.28) What singularities or combinations of singularities occur in the fibers of the deformation of a triple point? Over what loci in (a, b, c, d)-space do they occur?

We should remark that while the analogous stratification of the deformation space of any given singularity can, like the simple cases above, be analyzed by hand, there is no systematic method for determining which "baskets" of singularities will appear on some fiber of the deformation. One important case is:

EXERCISE (3.29) Show that in the deformation space of an ordinary m-fold point there is a fiber containing every combination of singularities that appears on some plane curve of degree m.

There are, of course, some cases whose spaces of first-order deformations aren't immediately calculable. Examples are the spaces of equisingular deformations of a curve singularity (C, p) or of equigeneric deformations (those preserving the geometric genus). Such spaces are usually *not* smooth (as in the example of the cusp above where the equigeneric locus is itself a cuspidal curve). Examples suggest that they tend to be irreducible, but we know of no general results along this line. In such cases, it would be tremendously helpful for many reasons (some of which arise later on in Chapter 5) to have even an estimate on the dimension of the space of first-order deformations, such as might come from a cohomological interpretation; but none is known.

8) *A very important example is the space of first-order deformations of a singular variety.*

This is a subject that requires a fair bit of machinery even in simple cases and that can become arbitrarily elaborate. To give the flavor, we'll first state the basic result for local complete intersection varieties referring you to [150] for proofs and further details. In case X is a local complete intersection (say X is locally embedded in a smooth variety V, with ideal sheaf $\mathcal{I}$), we have the conormal sequence

$$0 \longrightarrow \mathcal{I}/\mathcal{I}^2 \longrightarrow \Omega_V|_X \longrightarrow \Omega_X \longrightarrow 0.$$

Since $\mathcal{I}/\mathcal{I}^2$ is locally free, when we dualize we get

$$0 \longrightarrow \Theta_X \longrightarrow T_V|_X \longrightarrow N_{X/V}$$

where Θ_X is the sheaf of derivations of $\mathcal{O}_X$ and $N = N_{X/V}$ is the dual of $\mathcal{I}/\mathcal{I}^2$. We then *define* the sheaf

$$T_X^1 = \mathrm{Coker}(T_V|_X) \to N_{X/V}.$$

An alternate description is

$$T_X^1 = \underline{\mathrm{Ext}}^1_{\mathcal{O}_X}(\Omega_X, \mathcal{O}_X),$$

so that T_X^1 is a sheaf supported on X_{sing}. If we then set

(3.30) $\mathbb{T}_X^1 = \mathrm{Ext}^1_{\mathcal{O}_X}(\Omega_X, \mathcal{O}_X),$

it turns out that $\mathbb{T}_X^1$ is the space of first-order deformations of X.

Note that the local-to-global spectral sequence for Ext gives us an exact sequence

$$0 \longrightarrow H^1(\Theta_X) \longrightarrow \mathbb{T}_X^1 \longrightarrow H^0(T_X^1) \longrightarrow H^2(\Theta_X)$$

which has a natural geometric interpretation: first-order deformations of X induce first-order deformations of the singularities of X; the space $H^1(\Theta_X)$ gives deformations of X preserving its singularities.

Of course, the case in which we want to apply this theory is that of nodal curves. Here we can be more concrete. For example, if C is a curve with one node at the point p, $\tilde{C}$ is its normalization and q_1 and $q_2 \in \tilde{C}$ are the points lying over the node of C, we can identify Θ_C as the pushforward of the sheaf of vector fields on $\tilde{C}$ vanishing at q_1 and q_2. We then have

$$H^1(C, \Theta_C) = H^1(\tilde{C}, \Theta_{\tilde{C}}(-q_1 - q_2));$$

that is, the deformations of C preserving the singularity correspond, as expected, to deformations of the 2-pointed curve $(\tilde{C}, q_1, q_2)$.

Still in the nodal curve case, we can identify the sheaf $\mathbb{T}^1$ (which is now a skyscraper sheaf supported at the nodes) as follows. Suppose the curve C is (locally) embedded in a smooth surface S with ideal sheaf $\mathcal{I}$ given in local analytic coordinates on S by $\mathcal{I} = (xy)$, with $p = (0,0)$ the node. Locally, the first two terms of the exact sequence

$$0 \longrightarrow \mathcal{I}/\mathcal{I}^2 \overset{\alpha}{\longrightarrow} (T_S^\vee)|_C \longrightarrow \Omega_C^1 \longrightarrow 0$$

look like

$$\mathcal{O}_{C,p}\langle xy \rangle \to \mathcal{O}_{C,p}\langle dx, dy \rangle$$

with the map α given by

$$xy \mapsto x\,dy + y\,dx.$$

Hence Ω_C^1 looks locally like

$$\frac{\mathcal{O}_{C,p}\langle dx, dy \rangle}{\langle x\,dy + y\,dx \rangle},$$

which is locally free of rank 1 *except* at the node p. Dualizing, we get the sequence

$$0 \longrightarrow \Theta \longrightarrow T_S|_C \overset{\alpha^\vee}{\longrightarrow} N_{C/S} \longrightarrow T^1 \longrightarrow 0$$

with $\alpha^\vee$ locally the map

$$\mathcal{O}_{C,p}\langle \tfrac{\partial}{\partial x}, \tfrac{\partial}{\partial y} \rangle \to \mathcal{O}_{C,p}\langle xy \rangle^\vee$$

defined by sending $\frac{\partial}{\partial x}$ and $\frac{\partial}{\partial y}$ to the linear functionals $xy \mapsto y$ and $xy \mapsto x$ respectively. Since $N_{C/S}$ is generated by the homomorphism $xy \mapsto 1$ the image of $\alpha^\vee$ is exactly $\mathcal{M}_{p,C} \cdot N_{C/S}$. Hence, at a node p the quotient T^1 is isomorphic to the stalk at p of $N_{C/S}$ and, in particular, has length 1. This description may seem to depend on the embedding, but in fact is intrinsic. If C_1 and C_2 are the branches of C at p, we have

$$T^1 = \mathcal{O}_S(-C)^\vee \otimes \mathcal{O}_p$$
$$= \mathcal{O}_S(C) \otimes \mathcal{O}_p$$
$$= \mathcal{O}_S(C_1) \otimes \mathcal{O}_S(C_2) \otimes \mathcal{O}_p$$
$$= T_p(C_1) \otimes T_p(C_2),$$

which is independent of the choice of S.[3]

Put another way, this chain of ideas says that:

PROPOSITION (3.31) *The normal space to the boundary $\Delta \subset \overline{\mathcal{M}}_g$ at a point corresponding to a curve C with one node is the tensor product of the tangent spaces to the branches of C at the node.*

This is an important fact, which will be essential to making enumerative calculations later on.

Finally, we can put this analysis together with what we've seen in the case of a smooth curve to give an infinitesimal description of the boundary $\Delta \subset \overline{\mathcal{M}}_g$ at an arbitrary point. First, because $H^2(\Theta_C) = 0$, we have an exact sequence

$$0 \longrightarrow H^1(\Theta_C) \longrightarrow \mathbb{T}^1_X \longrightarrow H^0(T^1_X) \longrightarrow 0.$$

A first consequence is that the space B of first-order deformations again has dimension $3g - 3$. Identifying a neighborhood of $[C]$ itself in B with a neighborhood of the origin in $\mathbb{C}^{3g-3}$ identifies the deformations of C preserving each node with a smooth divisor in $\mathbb{C}^{3g-3}$ and any two of these divisors meet transversely at the origin. Thus,

PROPOSITION (3.32) *Let C be a curve without automorphisms. In a neighborhood of the point $[C]$, the boundary Δ is a normal-crossings divisor, with branches corresponding one-to-one to the nodes of C and with the normal space to each branch isomorphic to the tensor product of the tangent spaces to the branches of C at the corresponding node.*

EXERCISE (3.33) Let B and C be smooth curves of genera g and h (both at least 2) respectively. Use first-order deformations to show that any deformation of $B \times C$ is again a product of two curves. Is it also true that a deformation of a symmetric product of curves is again a symmetric product?

---

[3]Further discussion and applications of this independence are given in [52]

Construct a versal deformation

As we remarked in the overview (page 86), this is usually very hard to do "from the inside" — that is, by building up from infinitesimal deformations. When it's possible at all, the machinery required goes well beyond our scope here. Instead, we'll refer you again to [150] and, in the next subsection, give a construction that uses the Hilbert scheme as a "deus-ex-machina".

Universal deformations of stable curves

In this subsection, we'll sketch how Hilbert schemes may be used to construct the universal deformation space of any stable curve. Since any two universal deformations are locally isomorphic near the corresponding base points, we'll henceforth be able to speak of *the* germ of the versal deformation of a stable curve.

We will need the following lemma, which follows immediately from the description given earlier of the versal deformation space of a node.

LEMMA (3.34) *Let* $\varphi : \mathcal{C} \to B$ *be a proper flat family of curves. Then the set* $U = \{b \in B | \mathcal{C}_b$ *is a nodal curve*$\}$ *is open in* B.

Next, fix $g \geq 2$ and an integer $n \geq 3$. Define integers r and d in terms of these by

$$r + 1 = (2 \cdot n - 1)(g - 1) \quad \text{and}$$

$$d = 2 \cdot n(g - 1).$$

Then, let $\mathcal{H} = \mathcal{H}_{d,g,r}$, let $P(m) = md - g + 1$, let $\varphi : \mathcal{C} \to \mathcal{H}$ be the corresponding universal curve, and let $\mathcal{L} = \mathcal{O}_{\mathcal{C}}(1)$ be the universal line bundle on $\mathcal{C}$. We will abuse language and also write $\mathcal{C}$ and $\mathcal{L}$ for the restrictions of the corresponding objects to subschemes of $\mathcal{H}$. The lemma shows that the subset $\mathcal{U}$ of $\mathcal{H}$ parameterizing connected nodal curves is *open* in $\mathcal{H}$. Since $\mathcal{C} \to \mathcal{U}$ is a family of nodal curves, it has a *relative dualizing sheaf* $\omega = \omega_{\mathcal{C}/\mathcal{U}}$. We define $\widetilde{\mathcal{K}}$ to be the closed subscheme of U over which the sheaves $\mathcal{L}$ and $\omega^{\otimes n}$ are equal. More formally, $\widetilde{\mathcal{K}}$ is the subscheme defined by the gth Fitting ideal of $R^1\varphi_*(\omega^{\otimes n} \otimes \mathcal{L}^{-1})$.

Naively, we refer to $\widetilde{\mathcal{K}}$ as the locus of n-canonically embedded stable curves. (By Exercise (3.10), we could replace "stable" by "semistable" or "connected, nodal" in this definition without altering $\widetilde{\mathcal{K}}$.) Our plan to produce a versal deformation of a stable curve C is simple: take a linear slice of $\widetilde{\mathcal{K}}$ passing through $[C]$ and transverse to the PGL$(r+1)$-orbit of $[C]$. Some extra care is needed in order to be able to check that the resulting slices are versal when C has nontrivial automorphisms. To simplify our presentation, we assume that C has no automorphisms and, after going through the construction, indicate what modifications are needed to handle curves with automorphisms.

The first key fact we need is:

LEMMA (3.35) $\widetilde{\mathcal{K}}$ *is smooth of dimension* $(3g - 3) + (r^2 + 2r)$.

PROOF. Fix an n-canonically embedded stable curve C with Hilbert point $[C]$ in $\widetilde{\mathcal{K}}$. The dimension count for $\widetilde{\mathcal{K}}$ is clear: the curve C depends on $3g - 3$ moduli and the choice of a basis of $H^0(C, \omega^{\otimes n})$ (modulo scalars) on $r^2 + 2r$.

Now, the restriction to C of the tangent bundle to $\mathbb{P}^r$ is a quotient of a direct sum of $r + 1$ copies of $\mathcal{O}_{\mathbb{P}^r}(1)$. Since $\mathcal{O}_{\mathbb{P}^r}(1)$ restricts to $\omega^{\otimes n}$ on C, it follows immediately from the standard normal sheaf sequence that $H^1(C, N_{C/\mathbb{P}^r}) = 0$ and hence that the tangent space $H^0(C, N_{C/\mathbb{P}^r})$ to $\mathcal{H}$ at $[C]$ has the expected dimension $(3g - 3) + g + (r^2 + 2r)$. (The extra g parameters are for the choice of a line bundle of degree d on C.) Thus, the smoothness of $\widetilde{\mathcal{K}}$ would follow if we could solve the:

PROBLEM (3.36) Describe the subspace of $H^0(C, N_{C/\mathbb{P}^r})$ corresponding to tangent vectors to $\widetilde{\mathcal{K}}$, and show that it has codimension g.

Lacking an answer to this problem, we take an indirect approach. Let $\mathcal{J}$ be the subscheme of $\mathcal{H}$ consisting of the Hilbert points of those subcurves $C' \subset \mathbb{P}^r$ that are abstractly isomorphic to C (i.e., that correspond to different choices of line bundle and basis of sections on C). The dimension of $\mathcal{J}$ is therefore $g + (r^2 + 2r)$. The intersection of $\mathcal{J}$ and $\widetilde{\mathcal{K}}$ is just the $\mathrm{PGL}(r + 1)$-orbit of $[C]$. We claim that the tangent space to $\mathcal{J} \cap \widetilde{\mathcal{K}}$ has dimension $(r^2 + 2r)$. In view of parameter counts above, this is only possible if the tangent spaces to both $\mathcal{J}$ and $\widetilde{\mathcal{K}}$ have the minimal possible dimension and hence will imply that $\widetilde{\mathcal{K}}$ (and incidentally $\mathcal{J}$) is smooth at $[C]$. Let $\mathcal{C} \to \mathcal{H}$ denote the universal curve over the Hilbert scheme $\mathcal{H}$. Consider the diagram determined by a general tangent vector $\chi : \mathbb{I} \to \mathcal{H}$ to $\mathcal{H}$:

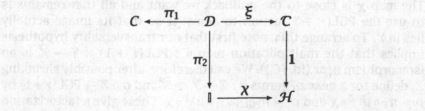

Such a vector is tangent to $\mathcal{J}$ if and only if $\mathcal{D}$ is a first order isotrivial deformation of C and thus corresponds to such a diagram for which the fiber product $\mathcal{D}$ is isomorphic to $C \times \mathbb{I}$. It's tangent to $\widetilde{\mathcal{K}}$ if and only if $\mathcal{D} \to \mathbb{I}$ is an n-canonical curve, i.e., $\xi^*(\mathcal{O}_C(1)) = \omega^{\otimes n}_{\mathcal{D}/\mathbb{I}}$. Therefore, it's tangent to $\mathcal{J} \cap \widetilde{\mathcal{K}}$ if and only if $\xi^*(\mathcal{O}_C(1)) = \omega^{\otimes n}_{C \times \mathbb{I}/\mathbb{I}} = \mathcal{O}_{\mathbb{I}} \otimes \omega^{\otimes n}_C$. In

other words, the choice of χ is equivalent to the choice of a basis of
the space

$$H^0(\mathbb{I}, (\pi_2)_*(\mathcal{O}_\mathbb{I} \otimes \omega_C^{\otimes n})) = \mathcal{O}_\mathbb{I} \otimes H^0(C, \omega_C^{\otimes n}).$$

Since $H^0(C, \omega_C^{\otimes n}) = r+1$, this choice depends on (r^2+2r) parameters
as claimed. ∎

Now we're ready to slice $\widetilde{\mathcal{K}}$ *assuming that C has no automorphisms*.
First, let $W = \Lambda^{P(m)}(H^0(\mathbb{P}^r, \mathcal{O}(m))^\vee)$ where m is taken large enough
that $\mathcal{H}$ embeds in $\mathbb{P}(W)$ as in Chapter 1. Next, choose a linear subspace V in $\mathbb{P}(W)$ containing $[C]$ that is complementary to the tangent
space to the orbit of $[C]$. Finally, choose an affine neighborhood Y of
$[C]$ in $\widetilde{\mathcal{K}} \cap V$ that is small enough that, for every point $[C'] \in Y$, the
curve C' has no automorphisms and its orbit meets Y transversely.

CLAIM (3.37) $\varphi : \mathcal{C} \to Y$ *is a universal deformation of any of its fibers.*

Let's first show this for C itself. Fix another deformation
$\psi : \mathcal{D} \to (Z, z_0)$ of C with identification of $\mathcal{D}_{z_0} \cong C$. The point $[C]$
determines a canonical basis of $H^0(C, \omega_C^{\otimes n})$ which we may view as a
basis of $H^0(\mathcal{D}_{z_0}, \omega^{\otimes n}|_{\mathcal{D}_{z_0}})$ and extend to a basis for $\psi_*(\omega_{\mathcal{D}/Z}^{\otimes n})$ near
z_0. This in turn embeds $\psi : \mathcal{D} \to Z$ as a family of subschemes of $\mathbb{P}^r$.
By the universal property of the Hilbert scheme, this is induced by a
unique map $\chi : Z \to \mathcal{H}$ which by construction has image in $\widetilde{\mathcal{K}}$ and fits
into a commutative square

(3.38)

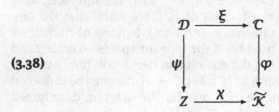

The map χ is close to the pullback we want and all that remains is
to use the PGL$(r + 1)$-action to adjust it so that its image actually
lies in Y. To arrange this, note first that our transversality hypothesis
implies that the multiplication map $\mu : \text{PGL}(r + 1) \times Y \to \widetilde{\mathcal{K}}$ is an
isomorphism near (id, $[C]$). We can therefore, after possibly shrinking
Z, define for z near z_0 germs $\rho : Z \to Y \subset \widetilde{\mathcal{K}}$ and $\sigma : Z \to \text{PGL}(r+1)$ by
$\rho = \pi_Y \circ \mu^{-1} \circ \chi$ and $\sigma = \pi_{\text{PGL}(r+1)} \circ \mu^{-1} \circ \chi$. These give a factorization
of χ in the sense that

$$\chi(z) = \mu(\sigma(z), \rho(z)) \text{ or } \chi = \mu \circ (\sigma, \rho).$$

On the other hand, σ is injective by construction, so we may also
define maps $\chi' : Z \to \widetilde{\mathcal{K}}$ and $\xi' : \mathcal{D} \to \mathcal{C}$ by setting $\chi' = \mu \circ (\sigma^{-1}, \chi)$

and $\xi' = \mu \circ ((\sigma \circ \psi)^{-1}, \chi)$, and, since $\mathrm{PGL}(r+1)$ acts equivariantly on $\varphi : \mathcal{C} \to \mathcal{K}$, the diagram (3.38) remains commutative if we replace the maps χ and ξ by χ' and ξ'. The factorization identity above, however, says that $\chi' = \rho$ and hence that χ' be viewed as a map $\chi' : Z \to Y$ and ξ' as a map to $\mathcal{C}|_Y$. We have therefore produced the desired pullback

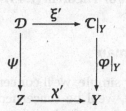

once we check that we have compatible identifications of the fiber of $\mathcal{C}|_Y$ over $\chi'(z_0)$ with $\mathcal{C}$ and $\mathcal{D}_{z_0}$. This follows immediately since, by construction, σ maps z_0 to the identity in $\mathrm{PGL}(r+1)$.

Uniqueness for χ' follows directly from the universality of the Hilbert scheme. Further, the claim follows for any fiber by observing that the only property of C itself that is used is transversality of the orbit of $[C]$ to Y and that this, by construction, holds for every fiber. We note, for future reference, that something almost as good — usually referred to as "openness of versality" is automatically true.

EXERCISE (3.39) Show that, if $\varphi : (\mathcal{X}, X) \to (Y, 0)$ is a versal deformation of X, then it's also a versal deformation of the fiber X_y for every y in some open neighborhood of 0.

Essentially the same ideas work when $[C]$ has a nontrivial stabilizer G in $\mathrm{PGL}(r+1)$ if we take account of these to maintain suitable equivariance at each step. The neighborhood Y must be chosen to be G-equivariant (by intersecting with any G-translates) and shrunk, if necessary, so that the stabilizer of *any* $y \in Y$ lies in G — this in turn requires showing that Y can be chosen so that if $g \in \mathrm{PGL}(r+1)$ and gY meets y, then $g \in G$.

Similar arguments also produce versal deformations for curves with marked points. We leave it to you to supply the necessary minor modifications in case you're interested.

Deformations of maps

We now consider a second example of deformation theory: the deformations of a map. We will describe here the space of first-order deformations of a map, and the *Kodaira-Spencer map* associated to a family of maps, which associates to a tangent vector to the base of a family of maps the corresponding first-order deformation. Versal deformation spaces for maps do exist (at least when the target and domain are reasonably well-behaved schemes, such as local complete intersections), but we won't prove this here; rather, we'll assert it and deduce as a consequence a dimension estimate for the space of maps of curves to the plane. In particular, we'll be able to conclude, as a corollary, the second part of Theorem (1.49)

The Kodaira-Spencer map

To keep things relatively simple, we'll concentrate on maps between smooth varieties; that is, we'll be concerned with families of maps from a possibly variable smooth domain to a fixed smooth target space. In other words, we'll consider a flat, smooth, proper family $f : X \to B$ over a smooth connected base B, a smooth variety Y and a morphism $\psi : X \to B \times Y$ of B-schemes. For each $b \in B$, we let $\psi_b : X_b \to Y$ be the restriction of ψ to the fiber X_b of X over b, and let

$$d\psi_b : TX_b \longrightarrow \psi_b^* TY$$

be the differential of ψ_b. We let N_b be the normal sheaf of ψ_b, that is, the cokernel of the morphism $d\psi_b$ of sheaves on X_b. Equivalently, if we let

$$d\psi : TX \longrightarrow \psi^* T(B \times Y)$$

be the differential of ψ and $\mathcal{N} = \mathrm{Coker}(d\psi)$ the normal sheaf of ψ, then the normal sheaf N_b of ψ_b is the restriction of $\mathcal{N}$ to the fiber X_b, that is, $N_b = \mathcal{N} \otimes \mathcal{O}_{X_b}$. Note that if ψ_b is an immersion, then N_b will be locally free; more generally, if ψ_b is equidimensional onto its image then the sheaf N_b will have a torsion subsheaf supported exactly on the locus where $d\psi_b$ fails to be an injective bundle map.

We now describe the *Kodaira-Spencer map* of the family ψ of morphisms. This is a map $\Upsilon : T_b B \to H^0(X_b, N_b)$ that associates to any tangent vector $v \in T_b B$ a global section $\sigma = \Upsilon(v)$ of the normal sheaf, in such a way that the family is trivial — that is, the family $X \cong B \times X_b$ as B-schemes and the morphism $\psi = id_B \times \psi_b$ — if and only if $\Upsilon(v) = 0$ for every v. To define it, let $\pi : B \times Y \to B$ be the projection, so that we have an inclusion of bundles

$$\pi^* T_B \longrightarrow T_{B \times Y}.$$

We let $i : \psi^* \pi^* TB \to \psi^* T(B \times Y)$ be the corresponding inclusion of pullbacks to X, and let $\tilde{\Upsilon} : \psi^* \pi^* TB \to \mathcal{N}$ be the composition of i

with the surjection $\psi^*T_{B\times Y} \twoheadrightarrow \mathcal{N}$ as shown in the diagram

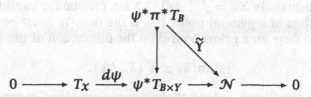

Restricting to X_b and taking global sections, we get a map

$$Y_b : T_b B \longmapsto H^0(X_b, \psi^*\pi^*T_B) \longrightarrow H^0(X_b, N_b),$$

which we'll call the *Kodaira-Spencer map* of the given family at b. Equivalently, we let Y be the pushforward of $\tilde{Y}$ to B, composed with the inclusion of T_B into $f_*\psi^*\pi^*T_B$: that is,

$$Y = f_*\tilde{Y} : T_B \longmapsto f_*\psi^*\pi^*T_B \longrightarrow f_*\mathcal{N}.$$

We will call Y the *global Kodaira-Spencer map* of the family; the maps Y_b are then the composition of the induced maps $T_b B \to (f_*\mathcal{N})_b$ on stalks with the natural maps $(f_*\mathcal{N})_b \to H^0(X_b, N_b)$.

There are two main facts to be stated in connection with this construction. The first, which can (and will) be left to you as an exercise, is simply that for any map $\psi_0 : X \to Y$ of smooth varieties, *the space of first-order deformations of ψ_0 is $H^0(X, N)$*, where N is the normal sheaf of ψ_0. The second, which is (as usual) substantially harder, is that *there exists a versal deformation space for ψ_0*, that is, there is a deformation $X \to B$ of $X \cong X_0$ and map $\psi : X \to B \times Y$ over B with $\psi|_{X_0} = \psi_0$, such that:

1. every deformation of the map ψ_0 is locally a pullback of (X, ψ), and,

2. the tangent space to B at 0 is the space of first-order deformations of ψ_0; that is, the Kodaira-Spencer map Y_0 of (X, ψ) at 0 is an isomorphism.

For proofs of these two facts, see [87] and [88] respectively.

EXERCISE (3.40) Let $\psi : X \to Y$ be any map of smooth varieties and let N be the normal sheaf of ψ. Show that the Kodaira-Spencer map gives an isomorphism of the space of first-order deformations of ψ with $H^0(X, N)$.

Dimension counts for plane curves

The most common application of these facts is a dimension count. It follows from the existence of the versal deformation space that if the family ψ of morphisms is nowhere isotrivial (that is, the restriction of ψ to the subfamily $X_{B_0} = f^{-1}(B_0) \subset X$ isn't trivial for any analytic arc $B_0 \subset B$), then at a general point $b \in B$ the map Y_b must be injective, so that we have an a priori bound on the dimension of the family:

$$\dim(B) \leq h^0(X_b, N_b).$$

Moreover, the Chern classes of the normal sheaf are in general readily calculated, so that in many cases it may be possible to estimate $h^0(X_b, N_b)$, giving us an upper bound on the dimensions of families of maps.

This is exactly what we need, for example, to estimate the dimension of the Severi variety $V_{d,g}$ of reduced plane curves of given degree and geometric genus. To set this up, let $\mathcal{C} \subset V_{d,g} \times \mathbb{P}^2$ be the universal curve, $X = \tilde{\mathcal{C}}$ the normalization of the total space $\mathcal{C}$, and $U \subset V_{d,g}$ the dense open subset of $V_{d,g}$ over which the map $X \to \mathcal{C} \to V_{d,g}$ is smooth; let $\psi : X \to U \times \mathbb{P}^2$. If $[C_0] \in U$ is a general point, so that X_0 is a smooth curve of genus g and $\psi_0 : X_0 \to C_0 \subset \mathbb{P}^2$ a birational embedding of X_0 as a plane curve of degree d, then the normal sheaf N_0 of ψ_0 is a rank 1 sheaf on the curve X_0, the degree of whose Chern class is

$$\deg(c_1(N_0)) = \deg(c_1(\psi_0^* T_{\mathbb{P}^2})) - \deg(c_1(T_{X_0}))$$

$$= 3d + 2g - 2$$

$$> 2g - 2.$$

We would thus expect that

$$\dim(U) \leq h^0(X_0, N_0)$$

$$= \deg(c_1(N_0)) - g + 1$$

$$= 3d + g - 1.$$

We cannot, however, conclude this yet. The difficulty arises from the possibility that ψ_0 isn't an immersion: if the differential $d\psi_0$ vanishes at points of X_0, the sheaf N_0 will have torsion there, and in this case the quotient $N_0/(N_0)_{\text{tors}}$ (and hence N_0 itself) may well be special. In such a case, the dimension $h^0(X_0, N_0)$ will indeed be larger than the naive estimate $3d + g - 1$ for the dimension of our family, and the method appears to fail.

Happily, there is a standard result, due to Arbarello and Cornalba [4], that deals with this situation. We have:

LEMMA (3.41) *Let* $X \rightarrow B$ *be a flat smooth proper family,* Y *a smooth variety and* $\psi : X \rightarrow B \times Y$ *a morphism of B-schemes; assume that* $\psi : X \rightarrow B \times Y$ *is birational onto its image. If* $b \in B$ *is a general point, then*

$$\mathrm{Im}(Y_b) \cap H^0(X_b, (N_b)_{\mathrm{tors}}) = 0.$$

Remarks. 1) If we don't assume the map ψ is birational onto its image, the conclusion of the lemma may well be false. In fact, it will fail when the map $\psi_b : X_b \rightarrow Y$ is multiple-to-one, with constant image but variable branch points.

2) While we won't introduce the definitions needed to make this precise, another way to express this lemma is to say that "the first-order deformation of the map ψ_b corresponding to a torsion section of N_b can never be equisingular". If $b \in B$ is general, the first-order deformations of ψ_b arising from the family $\psi : X \rightarrow B \times Y$ are necessarily equisingular; it follows that they can't be torsion.

PROOF.   Note first that, using the analytic topology, it's enough to prove the lemma in case B is one-dimensional: if we had $\mathrm{Im}(Y_b) \cap H^0(X_b, (N_b)_{\mathrm{tors}}) \neq 0$ at general $b \in B$, we could in an analytic neighborhood of b restrict to a curve whose tangent space was contained in $(Y_b)^{-1}(H^0(X_b, (N_b)_{\mathrm{tors}}))$ at each point.

We may thus assume that $\psi : X \rightarrow B \times Y$ is a one-parameter family of maps, the image of whose Kodaira-Spencer map Y_b at a general point is contained in $H^0(X_b, (\mathcal{N}_b)_{\mathrm{tors}})$. Let $Z = \psi(X) \subset B \times Y$ be the image of X and $p \in X$ a general point with image $\psi(p) = (b, q) \in B \times Y$. We're assuming that for any $v \in T_b B$, the image $Y_b(v)$ vanishes at p; that is, the tangent space $T_{(b,q)}Z$ is of the form

$$T_{(b,q)}Z = T_b B \times \Lambda_p$$

for some linear subspace $\Lambda_p \subset T_q Y$.

Now, let t be a local analytic coordinate on B near b, and let $(x, y_1, \ldots, y_n)$ be local coordinates on Y near q such that $\psi_b^* x$ is a local coordinate on X_b near p (so that the pair (t, x) give local coordinates on the surface X near p). We can write the map ψ locally as

$$y_i = f_i(t, x), \qquad i = 1, \ldots, n.$$

The tangent space $T_{(b,q)}Z$ is then the zero locus of the linear forms

$$dy_i - \frac{\partial f_i}{\partial t}dt - \frac{\partial f_i}{\partial x}dx$$

and the statement that $T_{(b,q)}Z = T_b B \times \Lambda_p$ for some linear subspace $\Lambda_p \subset T_q Y$ says that $\frac{\partial f_i}{\partial t}$ vanishes identically near p. We deduce that the image of ψ_b is constant, that is, that near (b, q) the image Z is equal to the product of a neighborhood of $b \in B$ with a neighborhood of $p \in X_b$.

This being true for general $p \in X$, it follows that $Z = B \times \psi_b(X_b)$ everywhere. Finally, since the map ψ is assumed birational, it follows that X is the normalization of Z; thus it's likewise a product, the map $\psi = id_B \times \psi_b$ and the Kodaira-Spencer map is identically zero. ∎

To restate the lemma, if we let

$$\overline{Y}_b : T_b B \longrightarrow H^0(X_b, N_b/(N_b)_{\text{tors}})$$

be the composition of Y_b with the natural map

$$H^0(N_b) \longrightarrow H^0(X_b, N_b/(N_b)_{\text{tors}}),$$

then the lemma implies that $\overline{Y}_b$ is an injection modulo the kernel of Y_b, that is, $\ker(\overline{Y}_b) = \ker(Y_b)$; in particular, if the family ψ is nowhere isotrivial then for general $b \in B$ the map $\overline{Y}_b$ is injective, and hence

$$\dim B \leq h^0(X_b, N_b/(N_b)_{\text{tors}}).$$

This is plenty to fix the argument given above for plane curves; in fact, it gives us a bit more. All we have to observe if that for any line bundle L of degree d on a smooth curve X of genus g, $h^0(X, L)$ is $d - g + 1$ if L is nonspecial, and at most g if L is special. Now, suppose that N is any rank 1 sheaf on X and $c_1(N) \geq 2g$. If the torsion part of N has length e, so that $\deg(N/N_{\text{tors}}) = c_1(N) - e$, we have

$$h^0(X, N/N_{\text{tors}}) \leq \max\{d - e - g + 1, g\} \leq c_1(N) - g + 1,$$

with equality holding if and only if N is torsion-free. Now, let $[C_b] \in V_{d,g}$ be a general point of any component of the Severi variety $V_{d,g}$ as before, $X_b \rightarrow C_b$ the normalization and N_b the normal sheaf of the map $\psi_b : X_b \rightarrow C_b \subset \mathbb{P}^2$. Applying Lemma (3.41) and the naive dimension estimate above, we have the:

COROLLARY (3.42) *The dimension of the Severi variety $V_{d,g}$ is $3d + g - 1$; and for $[C_b] \in V_{d,g}$ general, the map $X_b \rightarrow C_b \subset \mathbb{P}^2$ is an immersion and the Kodaira-Spencer map Y_b is onto.*

Now, to complete the proof of part 2 of Theorem (1.49), we simply need to establish the:

LEMMA (3.43) C_b *has no triple points or tacnodes.*

PROOF. The assertion of the lemma is that the map ψ_b is never three-to-one, and where $\psi_b(p) = \psi_b(q)$ for $p \neq q \in X_b$, the images of the differentials $d(\psi_b)_p$ and $d(\psi_b)_q$ aren't equal. This also follows from an analysis of the Kodaira-Spencer map. For the first, suppose that p, q and r are any three points of X_b mapping to the same point of C_b. If σ were a section of N_b vanishing at p and q, then under the corresponding first-order deformation of the map ψ_b, the point of intersection of the images of neighborhoods of p and q in X_b would be stationary; in order to preserve the triple point of C_b, then, σ would have to vanish at r as well.

Thus, to show that a general C_b has no triple points it's enough to show that if p, q and r are any three points of X_b then there exists a section of N_b vanishing at p and q but not at r. But this is immediate: N_b is a line bundle of degree $3d+2g-2 > 2g+1$, so of course the three points p, q and r impose independent conditions on $H^0(X_b, N_b)$.

Similarly, to show that a general C_b has no tacnodes it's enough to show that if $p, q \in X_b$ are points mapping to the same point in $\mathbb{P}^2$, there exists a section of the sheaf N_b vanishing at p but not at q, which follows from the same argument. ∎

Deformations of maps with tangency conditions

Like the deformation theory of varieties, the deformation theory of maps admits many variations. We will illustrate this by extending here the results obtained in the preceding subsection to deformations of a map $X \dashrightarrow \mathbb{P}^2$ that preserve tangency conditions with respect to a fixed line $L \subset \mathbb{P}^2$. This choice of topic is motivated in part by future applications: these extended dimension counts turn out to be crucial in the proof of part 3 of Theorem (1.49) (that is, the irreducibility of the Severi varieties), which we'll carry out in Section 6.E.

The key question here is: if $\psi : X \rightarrow \mathbb{P}^2$ is a map that has a point of tangency with L — that is, a point $q \in X$ such that the pullback $\psi^*(L)$ has multiplicity m at q — then, in the space of all deformations of the map ψ, can we identify the subspace of those preserving the tangency condition? In particular, can we describe the tangent space to this subspace as a subspace of $H^0(X, N)$?

To set this up, let $X \rightarrow B$ be as above a smooth family of curves over a reduced base B, $\psi : X \rightarrow B \times \mathbb{P}^2$ a morphism of B-schemes, and $Q \subset X$ a section of X over B such that the pullback divisor $\psi^*(L)$ contains the section Q with multiplicity exactly m. Let $b \in B$ be a general point and $q = X_b \cap Q$; suppose $\psi(q) = p \in L$. Let $v \in T_b B$, $\sigma = Y_b(v) \in H^0(X_b, N_b)$ the corresponding first-order deformation, and $\overline{\sigma} = \overline{Y}_b(v) \in H^0(X_b, N_b/(N_b)_{\text{tors}})$. Suppose finally that the differential $d\psi_b$ vanishes to order $l - 1$ at q, so that the image $\psi_b(\Delta)$ of a small neighborhood Δ of $q \in X_b$ will have multiplicity l at p. We then have the:

LEMMA (3.44) $\overline{\sigma}$ *vanishes to order at least* $m - l$ *at* q, *and cannot vanish to order exactly* k *for any* k *with* $m - l < k < m$. *Moreover, if* $\psi|_Q$ *is constant,* $\overline{\sigma}$ *vanishes to order at least* m *at* q.

PROOF. It will be sufficient to do this in case B is one-dimensional. Next, since B is reduced and $b \in B$ is general, B is smooth at b. Finally, since again $b \in B$ is general we may assume that the divisor $\psi_b^* L$ on X_b contains the point q with multiplicity exactly m as well.

Now, choose coordinates (x, y) in an analytic neighborhood of $p = \psi(q)$ so that the line L is given simply as the zero locus of y. Let then $\frac{\partial}{\partial x}$ and $\frac{\partial}{\partial y}$ be the generators of the rank 2 bundle TY at p; we'll abuse notation and write $\frac{\partial}{\partial x}$ and $\frac{\partial}{\partial y}$ also for the corresponding sections of $\psi_b^* T_Y$.

The first thing we'll show is that the image of $\frac{\partial}{\partial x}$ in $N_b / (N_b)_{\text{tors}}$ vanishes to order $m - l$ at q.

We treat the case $l < m$ first for simplicity, and leave the case $l = m$ for later. Let t be an mth root of $\psi_b^* y$ in a neighborhood of $q \in X_b$, then t will be a local coordinate on X_b near q and the map ψ_b will be given as

$$\psi_b : t \longmapsto (t^l + c_{l+1} t^{l+1} + \cdots, t^m)$$

so that the differential $d\psi_b$ is given by

$$d\psi_b : \frac{\partial}{\partial t} \longmapsto (l t^{l-1} + (l+1) c_{l+1} t^l + \cdots) \frac{\partial}{\partial x} + m t^{m-1} \frac{\partial}{\partial y}$$

$$= t^{l-1} \left((l + (l+1) c_{l+1} t + \cdots) \frac{\partial}{\partial x} + m t^{m-l} \frac{\partial}{\partial y} \right)$$

Set

$$\tau(t) := (l + (l+1) c_{l+1} t + \cdots) \frac{\partial}{\partial x} + m t^{m-l} \frac{\partial}{\partial y} .$$

The torsion subsheaf $(N_b)_{\text{tors}} \subset N_b$ is isomorphic to $\mathcal{O}_{X_b} / \mathfrak{m}_q^{l-1}$, and is generated by the section $\tau(t)$. Moreover, the quotient

$$N_b / (N_b)_{\text{tors}} = \mathcal{O}_{X_b} \{ \frac{\partial}{\partial x}, \frac{\partial}{\partial y} \} / \langle \tau \rangle$$

is generated by the image of the section $\frac{\partial}{\partial y}$. Note finally that modulo the subsheaf generated by τ,

$$\frac{\partial}{\partial x} \sim \frac{m t^{m-l}}{l + (l+1) c_{l+1} t + \cdots} \cdot \frac{\partial}{\partial y}$$

so that the image of the section $\frac{\partial}{\partial x}$ in $N_b / (N_b)_{\text{tors}}$ vanishes to order exactly $m - l$ at q.

Now let t and ε be local coordinates on X, with ε a local coordinate on B. A general deformation ψ of the map ψ_b over the base B may be given modulo ε^2 by

$$\psi(t, \varepsilon) = (\varepsilon;\ t^l + c_{l+1}t^{l+1} + \cdots + \varepsilon(\alpha_0 + \alpha_1 t + \cdots),\ t^m + \varepsilon(\beta_0 + \beta_1 t + \cdots)).$$

The condition that the divisor $\psi^*((y)) = mQ$ near q says that we can take t to be an mth root of the pullback $\psi^*(y)$ not just on X_b, but in a neighborhood of q in X. This means that a deformation satisfying the hypotheses of the lemma may be written modulo ε^2 as

$$\psi(t, \varepsilon) = (\varepsilon;\ t^l + c_{l+1}t^{l+1} + \cdots + \varepsilon(\alpha_0 + \alpha_1 t + \cdots), t^m).$$

From the definitions, the image $Y_b(\frac{\partial}{\partial \varepsilon}) \in H^0(X_b, N_b)$ of the tangent vector $\frac{\partial}{\partial \varepsilon} \in T_b B$ under the Kodaira-Spencer map will be given as the image in N_b of

$$\sigma := Y_b\left(\frac{\partial}{\partial \varepsilon}\right) = (\alpha_0 + \alpha_1 t + \cdots)\frac{\partial}{\partial x},$$

whose image $\overline{\sigma}$ in $N_b / (N_b)_{\text{tors}}$, as we've seen, vanishes to order at least $m - l$ at q. Moreover, since $b \in B$ is general, the differential $d\psi_\varepsilon$ will vanish to order $l - 1$ at $X_\varepsilon \cap Q$ for all ε near b; that is, $t^{l-1} | d\psi_\varepsilon$. This implies that

$$\alpha_1 = \alpha_2 = \ldots = \alpha_{l-1} = 0;$$

or in other words, the order of vanishing of $\overline{\sigma}$ at p can't equal $m - l + 1, \ldots, m - 1$. To complete the proof in case $m > l$, the further condition that $\psi|_Q$ is constant says that $\alpha_0 = 0$, which further implies that $\overline{\sigma}$ vanishes to order at least m at q.

The case $m = l$ is completely analogous. As before we write the map ψ_b as

$$\psi_b : t \longmapsto (t^n + c_{n+1}t^{n+1} + \cdots, t^m)$$

where now $n \geq m$. We leave it to you to check that the same argument yields that if $\psi|_Q$ is constant, the section $\overline{\sigma}$ vanishes to order at least m at q. ∎

We will apply Lemma (3.44) to obtain an estimate on the dimension of the varieties parameterizing plane curves of given degree and genus satisfying certain tangency conditions with respect to a line. First, some definitions. We fix again a line $L \subset \mathbb{P}^2$, and also a finite subset $S = \{p_1, \ldots, p_k\} \subset L$. For any positive integer β, we define the *generalized Severi variety* $V_{d,g}^\beta$ to be the closure, in the space $\mathbb{P}^N$ of all plane curves of degree d, of the locus of reduced, irreducible plane curves $C \subset \mathbb{P}^2$ of geometric genus g such that if $\psi_0 : X \to C$ is the normalization map, then

$$\#(\psi_0^{-1}(L \setminus S)) \leq \beta.$$

The variety $V_{d,g}^\beta$ will in general be very reducible: for example, in the simplest nontrivial case $\beta = d-1$, the general member of a component of $V_{d,g}^\beta$ may either pass through a point of S, or be simply tangent to L at a general point. In the case $\beta = d-2$, there are six possibilities. A general point $[C_0] \in V_{d,g}^\beta$ may correspond to a curve that has a node at a point of S, is tangent to L at a point of S, passes through two points of S, passes through one point of S and is tangent to L at a general point, is tangent to L at two general points, or has a flex along L. In general, though, as the dimension estimates we derive here will show, the dimension will depend only on β.

LEMMA (3.45) *The generalized Severi variety $V_{d,g}^\beta$ has dimension $2d + g - 1 + \beta$ everywhere. Moreover, if $[C]$ is a general point of any component of $V_{d,g}^\beta$ and $\psi_0 : X \to C \subset \mathbb{P}^2$ is the normalization map, then*

1) *ψ_0 is an immersion;*

2) *the only singularities of C away from S are nodes;*

3) *C is smooth along $L \setminus S$; and*

4) *$\#(\psi_0^{-1}(L \setminus S)) = \beta$.*

PROOF.   To begin with, it follows from a straightforward dimension count that $V_{d,g}^\beta$ has dimension at least $2d + g - 1 + \beta$ everywhere.

We thus have to show that $\dim V_{d,g}^\beta \leq 2d + g - 1 + \beta$ everywhere. Let $\mathcal{C} \subset V_{d,g}^\beta \times \mathbb{P}^2$ be the universal curve over $V_{d,g}^\beta$, $\psi : \tilde{\mathcal{C}} \to \mathcal{C}$ the normalization of the total space, $U \subset V_{d,g}^\beta$ the open subset over which the map $\tilde{\mathcal{C}} \to V_{d,g}^\beta$ is smooth, and X the inverse image of U in $\tilde{\mathcal{C}}$, so that $\pi : X \to U$ is a family of smooth curves of genus g.

Let $[C] \in U$ be a general point, $X \to C \subset \mathbb{P}^2$ the normalization and N the normal sheaf of the map $\psi_0 = \psi|_X : X \to C \subset \mathbb{P}^2$. Write

$$\psi_0^{-1}(L) = \sum_{i=1}^\beta m_i \cdot q_i + \sum_{i=1}^\beta n_i \cdot r_i$$

where $\psi_0(q_i) \in S$. By the definition of $V_{d,g}^\beta$ and the fact that $[C] \in U$ is general, we have, in an analytic neighborhood of $[X]$, collections of sections $\{Q_i\}$ and $\{R_i\} \subset X$ such that

$$\psi(Q_i) \subset S$$

(so that in particular $\psi|_{Q_i}$ is constant),

$$\psi^*(L) = \sum_{i=1}^\beta m_i \cdot Q_i + \sum_{i=1}^\beta n_i \cdot R_i.$$

and

$$q_i = Q_i \cap X \quad \text{and} \quad r_i = R_i \cap X.$$

We may assume that the points $\{q_i\}$ and $\{r_i\}$ are distinct: if not, then the component V of $V_{d,g}^\beta$ in which the curve $[C]$ is general would also be a component of a Severi variety $V_{d,g}^{\beta'}$ for some $\beta' < \beta$. Our argument will then show that the dimension of V is bounded above by $2d + g - 1 + \beta' < 2d + g - 1 + \beta$.

We need to introduce one more bit of notation. We denote by $l_i - 1$ the order of vanishing of the differential $d\psi_0$ at the point r_i. We then define divisors D and $D_0 \in \mathrm{Div}(X)$ by

$$D = \sum_{i=1}^\beta m_i \cdot q_i + \sum_{i=1}^\beta (n_i - 1) \cdot r_i$$

and

$$D_0 = \sum_{i=1}^\beta (l_i - 1) \cdot r_i.$$

Note that

$$\deg(D) = d - \beta$$

and that

$$\deg((\psi_0^* \mathcal{O}_{\mathbb{P}^2}(1))(-D)) \ge 0.$$

Note also that

$$\deg(c_1(N_{\mathrm{tors}})) \ge \deg(D_0),$$

with equality holding if and only if ψ_0 is an immersion away from $\{r_i\}$. Hence

$$\deg(c_1(N/N_{\mathrm{tors}})) \le \deg(c_1(N)) - \deg(D_0),$$

again with equality holding if and only if ψ_0 is an immersion away from $\{r_i\}$.

Finally, let D_1 be the effective part of $D - D_0$.

Now, applying Lemma (3.41) and Lemma (3.44), we see that

$$\dim U \le h^0(X, (N/N_{\mathrm{tors}})(-D_1)).$$

We have

$$\deg((N/N_{\mathrm{tors}})(-D_1)) \le \deg(c_1(N)) - \deg(D)$$

and since

$$c_1(N) = \psi_0^* \mathcal{O}_{\mathbb{P}^2}(3) \otimes \omega_X$$

we see that the line bundle

$$(c_1(N)(-D)) \otimes \omega_X^{-1} = ((\psi_0^* \mathcal{O}_{\mathbb{P}^2}(1))(-D)) \otimes \psi_0^* \mathcal{O}_{\mathbb{P}^2}(2)$$

has strictly positive degree. We may thus conclude that

$$\dim U \le h^0(X, (N/N_{\text{tors}})(-D_1))$$

$$\le \deg(c_1(N)(-D)) - g + 1$$

$$= (3d + 2g - 2 - \deg(D)) - g + 1$$

$$= 2d + g - 1 + \beta.$$

This completes the proof of the dimension statement in Lemma (3.45).

Notice that the argument above implies that the image of the Kodaira-Spencer map can be identified as follows:

$$\text{Im}(\overline{Y}_{[C]}) = H^0(X, (N/N_{\text{tors}})(-D_1)).$$

To prove the second half of Lemma (3.45), we start by establishing what is perhaps the subtlest point: that the map ψ_0 is indeed an immersion. In fact, much of this has already been accomplished in the proof of the first half: since the line bundle $(c_1(N)(-D) \otimes \omega_X^{-1}$ on X has degree at least 2, we may deduce that

$$(N/N_{\text{tors}})(-D_1) = c_1(N)(-D)$$

so that $D_1 = D - D_0$ and

$$N/N_{\text{tors}} = c_1(N)(-D_0)$$

and hence ψ_0 is an immersion away from $\{r_i\}$.

To see that ψ_0 is an immersion at the point r_i, we observe that the line bundle $(c_1(N)(-D)) \otimes \omega_X^{-1}$ has degree at least 4 on X_0, so that there exists a section $\overline{\sigma}$ of $c_1(N)(-D) = (N/N_{\text{tors}})(-D_1)$ vanishing to order exactly 1 at r_i, and this section must be in the image of the Kodaira-Spencer map

$$Y_{[X]} : T_{[X]}V \longrightarrow H^0(X, (N/N_{\text{tors}})).$$

But the multiplicity of r_i in the divisor $D_1 = D - D_0$ is

$$(n_i - 1) - (l_i - 1) = n_i - l_i,$$

and it follows that $\overline{\sigma}$, viewed as a section of N/N_{tors}, vanishes to order exactly $n_i - l_i + 1$ at r_i. By Lemma (3.44), then, we must have $l_i = 1$; that is, ψ_0 must be an immersion at r_i.

Next, to show that X has only nodes as singularities away from S, we have to show it has no triple points and that no two branches are tangent to each other. This is simply a variant of the argument given in the proof of Lemma (3.43), replacing N by $N(-D)$: for example, to show there are no triple points, it's enough to show that for any three points $p, q, r \in X$ there exists a section of $N(-D)$ vanishing at p and q but not at r, which follows immediately by degree considerations. Similarly, to establish part 3, we simply have to argue that for any two points p and q of $\psi_0^{-1}(L \setminus S)$, there is a first-order deformation that varies $\psi_0(p)$ but not $\psi_0(q)$.

Finally, the fact that $\#(\psi_0^{-1}(L \setminus S)) = \beta$ is an immediate consequence of the dimension count in the first part of the lemma. ∎

We mention, for future application, that the result of Lemma (3.45) doesn't depend on the hypothesis that the general member of $V_{d,g}^\beta$ is irreducible. In fact, we can derive the following more general statement as a corollary.

COROLLARY (3.46) *Let L in $\mathbb{P}^2$ be a line and let S be any finite subset. Let $V \subset \mathbb{P}^N$ be any locally closed subset of the space $\mathbb{P}^N$ of plane curves of degree d, and let $[C] \in V$ be a general point. Suppose that the curve C is the image of an abstract nodal curve X of geometric genus g under a map $\eta : X \to C$ that isn't constant on any connected component of X and such that the inverse image $\eta^{-1}(L \setminus S)$ contains at most β points. Then*

$$\dim V \leq 2d + g + \beta - 1,$$

and if equality holds, then C is a nodal curve smooth at its points of intersection with $L \setminus S$.

PROOF. This follows by simply applying Lemma (3.45) to each component of C in turn. ∎

C Stable reduction

Results

It's a basic fact, quoted without proof above, that the moduli space $\overline{\mathcal{M}}_g$ of stable curves of genus g is compact and separated. According to the valuative criterion for properness, the compactness property implies that any regular map from the complement of a point on a smooth curve to $\overline{\mathcal{M}}_g$ admits an extension to a regular map on the whole curve. Likewise, the separability of $\overline{\mathcal{M}}_g$ can be viewed as asserting that this extension is unique.

As remarked in Chapter 1, a map of a smooth curve B, punctured or not, to a coarse moduli space $\overline{\mathcal{M}}_g$ corresponds (possibly after a base change to rigidify) to a family of stable curves over a branched cover of B. It should therefore follow that:

PROPOSITION (3.47) (STABLE REDUCTION) *Let B be a smooth curve,
0 a point of B and B* = B \ {0}. Let X → B* be a flat family of stable curves of genus g ≥ 2. Then there exists a branched cover B' → B
totally ramified over 0 and a family X' → B' of stable curves extending the fiber product X ×_{B*} B'. Moreover, any two such extensions are
dominated by a third. In particular, their special fibers — those over
the preimage of 0 in B' — are isomorphic.*

The process of finding the family $X' \to B'$ is called *stable reduction*.
It arises quite frequently in practice, since even geometrically smooth
families of curves (e.g., linear systems of curves on a surface, families of branched covers) are apt to specialize to nonstable curves —
curves with a cusp or worse singularity, or curves with multiple components. In this circumstance, we're assured in the abstract that we
can, in any one-parameter subfamily with smooth base, replace the
unstable fibers with stable ones by making a base change and birational modification. For many purposes, we need to know not just
that this can be done, but what stable curves actually appear as limits
of one-parameter subfamilies tending to an unstable curve.
Sometimes the geometry of the specialization gives us a good idea
of what special fiber to expect. For example, in a general pencil of
plane quartics degenerating to a double conic, we might guess that
the special fiber will be a (smooth) hyperelliptic curve (this is worked
out in the following subsection). In other cases (many examples are
given below), what the stable reduction will be is far from clear.
One warning we should offer here is that we cannot necessarily take
the total space of the family $X' \to B'$ to be smooth. It will, however,
have only a very limited range of possible singularities: by the description above of the versal deformation space of a node, any singularity
of X' will be given locally by the equation $xy - t^k$ for some k (in the
usual terminology, will be an A_{k-1} *singularity*). These can be resolved
by blowing up to obtain an exceptional divisor consisting of a chain of
$k - 1$ rational curves each appearing with multiplicity 1 in the fiber of
the blown-up surface over the origin $0 \in B$. Recalling from Section 2.C
that a connected curve C is *semistable* if it has only nodes as singularities and if each rational component of the normalization $\tilde{C}$ of C
contains at least 2 (as opposed to 3) points lying over nodes of C, we
then have the:

PROPOSITION (3.48) (SEMISTABLE REDUCTION) *Let B be a smooth
curve, 0 a point of B and B* = B \ {0}. Let X → B* be a flat family of semistable curves of genus g ≥ 2. Then there exists a branched
cover B' → B totally ramified over 0 and a family X' → B' of semistable
curves extending the fiber product X ×_{B*} B' and having smooth total
space X'. Any two such extensions are dominated by third and so have*

special fibers whose stable models — obtained by contracting smooth rational components meeting the rest of the curve in fewer than three points — are isomorphic.

This will be useful in a number of situations: for example, when we have a line bundle L on X that we want to extend to some compactification X'.

Finally, a third variant will be useful when we wish to resolve the indeterminacy of a rational map or to replace a divisor in X, finite over B, with a collection of disjoint sections. This can be done by making analogous modifications to X' (either blowups or base changes and blowups) without reintroducing multiple components of the special fiber. Since we may have to blow up smooth points of the special fiber X_0, producing smooth rational components of X_0 meeting the rest of X_0 only once, we can't assert that the resulting family is semistable, but we can ensure that it's a family of nodal curves.

PROPOSITION (3.49) (NODAL REDUCTION) *Let B be a smooth curve, 0 a point of B and $B^* = B \setminus \{0\}$. Let $X \rightarrow B^*$ be a flat family of nodal curves of genus g, $\psi : X \rightarrow Z$ any morphism to a projective scheme Z, and $D \subset X$ any divisor finite over B^*. Then, there exists a branched cover $B' \rightarrow B$ and a family $X' \rightarrow B'$ of nodal curves extending the fiber product $X \times_{B^*} B'$ with the following properties:*

1) *The total space X' is smooth.*

2) *The morphism $\pi_X \circ \psi : X \times_{B^*} B' \rightarrow Z$ extends to a regular morphism on all of X'.*

3) *The closure of the inverse image $\pi_X^{-1}(D)$ in X' is a disjoint union of sections of $X' \rightarrow B'$.*

Any two such extensions are dominated by a third and so have special fibers whose stable models are isomorphic.

Remark. The last property makes it easy to prove analogous results for families of stable pointed curves, which we leave you to formulate precisely.

There are now two things to do. We should prove the propositions and we should give some examples of how the process of finding the extension $X' \rightarrow B$ of a given family is carried out in practice. You can by now easily guess which one we'll actually do. In fact, proofs of all three results follow, in outline, rather closely the steps illustrated by the examples below. After we've gone through these, we'll indicate what additional ingredients enter into the proof. (A more detailed proof can be found in [11].) We should, however, at least remark that the way we've motivated these results in the first paragraph above is somewhat misleading. Historically, these theorems preceded the discovery of $\overline{\mathcal{M}}_g$ and their proof did not depend on knowing that such a compactification exists. Quite the reverse: stable reduction is a key ingredient in the proof of the existence of $\overline{\mathcal{M}}_g$.

Examples

Glueing a constant section to a moving one

We start with one of the simplest examples. Suppose that C is a fixed smooth curve of genus $g - 1$, $p \in C$ is a point, and for each point $q \neq p \in C$, C_q is the stable curve of genus g obtained by identifying the points p and q on C. Fixing p and letting q vary, it's clear that these fit together to form a family $\mathcal{X}$ of stable curves over the punctured curve $C - \{p\}$. We ask: what happens as q approaches p as indicated in Figure (3.50)? At the level of the fibers, we want to know how we can

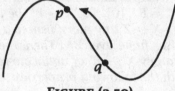

FIGURE (3.50)

complete the family of stable curves shown in Figure (3.51) (in which q approaches p as we move from left to right) to one defined over all of C? In other words, what stable curve should replace the question mark at the right? There is an obvious way to complete this to a family

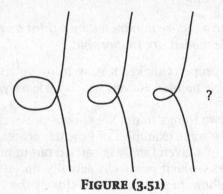

FIGURE (3.51)

of curves: in the product $C \times C$, we can simply identify the two cross-sections Δ and Γ_p, where Δ is the diagonal and $\Gamma_p = \{p\} \times C$ is the horizontal cross-section as in Figure (3.52). This, however, will yield

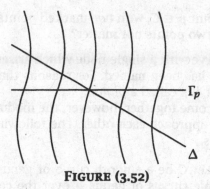

FIGURE (3.52)

a family whose fiber over $p \in C$ is a cuspidal curve. This is perhaps the first case in which it's unclear what the stable central fiber will be.

To find out, we do the next obvious thing: we blow up the original family $C \times C$ at the offending point (p, p) before making the identifications. This yields the family in Figure (3.53), and now we can

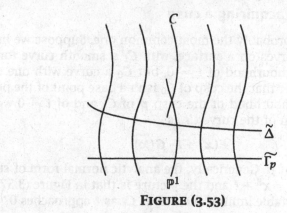

FIGURE (3.53)

make the identification of the proper transforms $\tilde{\Delta}$ and $\tilde{\Gamma}_p$ to arrive at a family whose fiber over p is the stable curve shown in Figure (3.54).

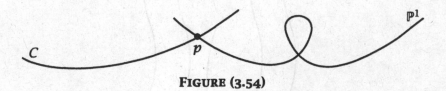

FIGURE (3.54)

EXERCISE (3.55) Show that the stable limit of the two-pointed curve (C, p, q) as q approaches p is the union of C and a copy of the line $\mathbb{P}^1$

attached at the point $p \in C$, with two marked points on $\mathbb{P}^1$. Why does the choice of the two points not matter?

The rational curve with a single node which arises here appears so frequently that it has been named. For reasons that should be clear from Figure (3.54), it's called a *pigtail*.

If more points come together, however, the moduli of the limit will depend how they approach each other. The following exercise shows an example of this.

EXERCISE (3.56) Let C be a smooth curve of genus $g - 2$. Consider the family of stable curves of genus g over the complement of the diagonal in the fourfold product C^4 whose fiber over a point (p, q, r, s) is the curve obtained by identifying p with q and r with s. What are the stable limits of curves in this family? Does this family extend to a family of stable curves over all of C^4? Over what blowup of C^4 does it extend?

Smooth curves acquiring a cusp

This example is probably the most common one. Suppose we have a pencil $\{C_t\}$ of curves on a surface, with C_t a smooth curve for t in a punctured neighborhood of $t = 0$, but C_0 a curve with one cusp. Suppose moreover that the cusp of C_0 isn't a base point of the pencil, so that in a neighborhood of the cusp p of C_0 and of $t = 0$ we can write the equation of the curve C_t as

$$F(x) + t \cdot G(x)$$

with G nonzero at p. Generically, the analytic normal form of such a pencil will be $y^2 = x^3 + t$ and the picture is that in Figure (3.57). We ask: what is the stable limit of the curves C_t as t approaches 0?

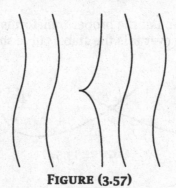

FIGURE (3.57)

This problem is substantially more subtle than the preceding one; for example, here we'll definitely have to make a base change to carry out a stable reduction. Before we do this, however, we should clean up the problem a little by getting rid of the cusp in the special fiber. Indeed, by blowing up the total space $X_0 \subset \mathbb{P}^1 \times S$ of the original pencil we can always arrive at a family whose fiber over the origin is supported on a nodal curve — the problem is that this fiber will have multiple components — and we do this first.[4]

This takes three blowups. Note first of all that, because of the hypothesis that the cusp p of C_0 isn't a base point of the pencil, the surface X_0 is smooth (this is in fact the only aspect of this process that involves the fact that the original family is a pencil). Starting with the family in Figure (3.57) and blowing up once, we arrive at the family in Figure (3.58) whose special fiber consists of the normalization $\tilde{C}$ of

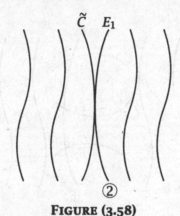

FIGURE (3.58)

C and an exceptional divisor $E_1 \cong \mathbb{P}^1$ appearing with multiplicity 2 in the fiber. Here, and in the sequel, *we indicate component multiplicities greater than 1 by circled integers.*

Next, we blow up a second time to arrive at the family in Figure (3.59) whose special fiber consists of $\tilde{C}$, the proper transform of E_1 (which we'll continue, by abuse of notation, to call E_1), and a new exceptional divisor E_2 appearing with multiplicity 3 in the fiber, all smooth and intersecting at a common point p. Finally, we blow up the point p, introducing another exceptional divisor E_3 appearing with multiplicity 6 in the fiber as in Figure (3.60). This last blowup separates the

---

[4]If you're familiar with proofs of the stable reduction theorem you may find this procedure somewhat surprising since these usually begin by making all necessary base changes and only afterwards perform blowups as needed to smooth. However, the procedure we adopt in this example is, in fact, fairly typical. We will see shortly that making base changes only after having obtained a (nonreduced) nodal fiber greatly simplifies the bookkeeping of component multiplicities and intersection numbers in the special fiber: this is a task which is superfluous in an existence proof but essential when we want to identify a particular stable limit.

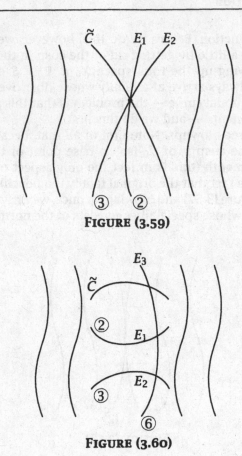

$$\tilde{C} \quad E_1 \quad E_2$$

③ ②

FIGURE (3.59)

$$E_3$$

$$\tilde{C}$$

② E_1

③ E_2

⑥

FIGURE (3.60)

proper transforms of the components in the fiber of the previous family (which we continue to call $\tilde{C}$, E_1 and E_2); they now all intersect E_3 at distinct points.

We have thus arrived at a family whose reduced special fiber has only nodes as singularities. The problem, of course, is that the fiber is highly nonreduced, having components of multiplicities 2, 3 and 6. This we deal with by making a *base change*: we take the fiber product $X' = X \times_B B'$ of our existing family $X \to B$ with a branched cover $B' \to B$ of the curve B ramified over 0 — almost always, the map given locally for some m by $t \mapsto t^m$.

Here is where additional practical considerations arise. The base change will generally introduce new singularities in the surface X, and the bookkeeping is simplified if we package each base change with the normalization of the resulting surface. In practice, it's also usually best to perform only base changes with prime exponent m, if necessary factoring the base change into several stages. Of course, since the operation of base change is associative, we could just make a single base change of composite order, but doing so makes it much harder to keep track of what is going on.

In the present case, since we have components of multiplicity divisible by 2 and 3, we'll want to make base changes of these orders. To start with, we make a base change of order 2; that is, we take the fiber product of X with a double cover B' of B given locally by the map $t \mapsto t^2$. This is equivalent to taking the double cover of X branched along the divisor $t = 0$, so that the local equation of the resulting surface will be $u^2 = t$ everywhere. If $D \subset X$ is a component of multiplicity m in the special fiber, then in a neighborhood of a point p of D, t is the mth power of a local coordinate z on X, so that the local equation of X' will be

$$u^2 = z^m.$$

Of course, if $m > 1$, this will be singular along the inverse image of D. However, the normalization process will smooth this locus, replacing u by a local coordinate $v = u/z^{\lfloor m/2 \rfloor}$, so that the local equation of the normalization will be either $v^2 = z$ or $v^2 = 1$ depending on whether m is odd or even. This suggests the following definition.

DEFINITION (3.61) *For any divisor*

$$D = \sum a_i \cdot D_i$$

on a surface and $n \in \mathbb{Z}$, define the divisor $D_{\equiv n}$ (called the divisor D reduced mod n) to be the divisor

$$D_{\equiv n} = \sum \bar{a}_i \cdot D_i$$

where $0 \le \bar{a}_i \le n - 1$ and $\bar{a}_i \equiv a_i \bmod n$.

In these terms, we can summarize the discussion in the preceding paragraph by saying that *the effect of the base change of prime order p followed by normalization is to take the branched cover $\tilde{X}$ of X branched along the divisor (t) reduced mod p.* The simplicity of this description is the main reason for factoring any base change into a succession of base changes of prime order.

EXERCISE (3.62) Compute the result of performing a base change of order 6 on X and normalizing the resulting surface and show that the description above can't be extended to base changes of composite order.

In the present circumstance, the divisor (t) reduced mod 2 is simply the sum of the components E_2 and $\tilde{C}$ of the special fiber that are shown thickened in Figure (3.63). Since this branch divisor is smooth, the resulting surface will be smooth as well. The inverse images of

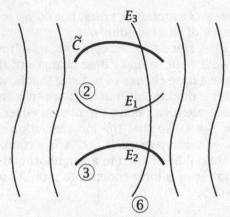

FIGURE (3.63)

E_2 and $\tilde{C}$ will of course be curves mapping isomorphically to them. Since E_3 meets the branch locus in two points, its inverse image will be a double cover of $E_3 \cong \mathbb{P}^1$ branched at two points, which is to say, a single rational curve that we shall continue to call E_3. On the other hand, E_1 is disjoint from the branch locus, so that its inverse image will be an unramified cover of $E_1 \cong \mathbb{P}^1$; that is, two disjoint rational curves that we'll call E_1' and E_1''.

The multiplicities of the various components in the special fiber are not hard to calculate either. Briefly, the pullback to $\tilde{X}$ of the divisor (t) on X is simply the sum of the components of the inverse image of the special fiber in X, with multiplicities unchanged from that of the corresponding component of (t) on X for components not contained in the branch divisor, and with multiplicity doubled for components in the branch divisor. Thus,

$$\eta^* = 2\tilde{C} + 6E_2 + 2E_1' + 2E_1'' + 6E_3.$$

But the special fiber (u) of the new family $\tilde{X} \to B'$ is exactly one-half of this divisor: thus

$$(u) = \tilde{C} + 3E_2 + E_1' + E_1'' + 3E_3,$$

and the picture of our new surface is shown in Figure (3.64) with all components smooth rational curves (except for $\tilde{C}$, of course), and all multiplicities 1 unless marked.

Again, we can apply the same principles to a base change of any prime order p except that the multiplicities of components of the special fiber X_t in the branch divisor are then multiplied by p in the inverse image of X_t and the new special fiber is $(\frac{1}{p})^{\text{th}}$ of this inverse image. As an example, we take the logical next step of making a base change of order 3 and normalizing. We must form the cyclic triple

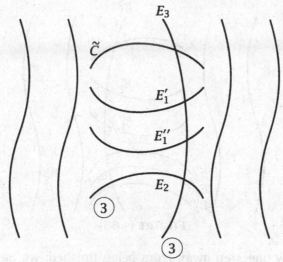

E_3

$\tilde{C}$

E_1'

E_1''

E_2

③

③

FIGURE (3.64)

cover of our surface branched over the special fiber reduced mod 3, which is to say the curve $\tilde{C} \cup E_1' \cup E_2''$ shown thickened in Figure (3.65). The inverse images of $\tilde{C}$, E_1' and E_1'' are copies of themselves. Since E_2

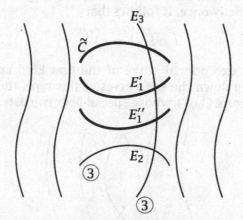

E_3

$\tilde{C}$

E_1'

E_1''

E_2

③

③

FIGURE (3.65)

is disjoint from the branch divisor, its inverse image is a disjoint union of three rational curves, which we'll call E_2', E_2'' and E_2'''. Finally, the inverse image of E_3 will be a triple cover of E_3 branched over the three points where E_3 meets $\tilde{C}$, E_1' and E_1'' — that is, by Riemann-Hurwitz, an elliptic curve, which we'll continue to call simply E_3. Figure (3.66) shows the picture we finally arrive at. Since in $\eta^*(t)$ all components now have multiplicity 3, the new special fiber will be reduced.

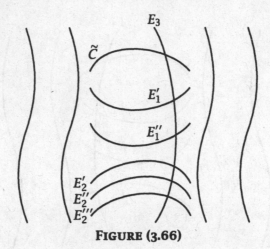

FIGURE (3.66)

We're now one step away from being finished: we have a family whose special fiber is a reduced curve with only nodes as singularities. It's not semistable, however, because of the presence of the five E_1 and E_2 curves, which are rational curves meeting the rest of the fiber only once. This is, in fact, exactly what allows us to get rid of them. The intersection number of each of any component E of the special fiber with the whole special fiber is 0. Therefore, if E meets the rest of the special fiber exactly once, it follows that

$$E \cdot E = -1.$$

Hence, E is an exceptional curve of the first kind and can be contracted. Blowing down the five curves of this type, then, we arrive at the family in Figure (3.67) whose special fiber consists of the union of

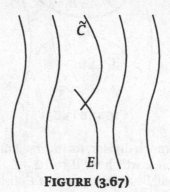

FIGURE (3.67)

the normalization $\tilde{C}$ of the original curve C together with the elliptic curve E (called an *elliptic tail*), joined at the point of $\tilde{C}$ lying over the cusp of C. This, finally, is the stable reduction of the original family.

EXERCISE (3.68) At each stage of this process, calculate the self-intersection of each component of the special fiber.

EXERCISE (3.69) Consider the family $F + t \cdot G$ above, with F a quartic with an ordinary cusp and G a general quartic. Carry out the process of stable reduction globally, making base changes of orders 2 and 3 by taking covers $\mathbb{P}^1 \to \mathbb{P}^1$ of degrees 2 and 3 branched over the point $t = 0$ and one other general point of $\mathbb{P}^1$. What are the numerical invariants of the resulting surface?

There is one amusing (and significant) point to be made here. We haven't really specified in the description above the moduli point in $\overline{\mathcal{M}}_g$ of the special fiber $\overline{X}_0$ because we haven't said *which* elliptic curve E arises. Looking back at the reduction process we see that E appeared in the process of making the base change of order 3 as a triple cover of $\mathbb{P}^1$ totally ramified over three points. Since any three points on $\mathbb{P}^1$ are projectively equivalent, this description completely specifies E. In fact, it shows that E is Galois with Galois group $\mathbb{Z}/3\mathbb{Z}$ and so has an automorphism group of order 3. Thus, the associated lattice is spanned by 1 and $e^{2\pi i/3}$ and the elliptic curve E *always* has j-invariant 0! To see what is funny about this, consider a two-parameter family of curves $C_{(a,b)}$ of genus $g \geq 2$ given locally by

$$y^2 = x^3 + a \cdot x + b$$

(which is the versal deformation space of a cusp singularity); assume that for $(a, b) \neq (0, 0)$, the curve $C_{(a,b)}$ is stable. Figure (3.70) shows the type of singularity in the fiber over each point of the base of this family. What the analysis above shows is that if we approach the origin via a *general* one-parameter family in this plane not tangent to the a-axis at the origin, we'll get the elliptic curve with j-invariant 0 in the limit. What happens if we approach along the discriminant Δ? This is essentially the situation of the first example treated above and we saw there that the limit is always the *pigtail* for which $j = \infty$.

Where are the other elliptic curves hiding? That generic directions lead to tails E with j-invariant 0 shows what the answer must be. Associating to a point of the (a, b)-plane $\mathbb{C}^2_{a,b}$ other than the origin the isomorphism class $[C_{(a,b)}]$ of the curve $C_{(a,b)}$ defines a rational map $\mathbb{C}^2_{a,b} \dashrightarrow \overline{\mathcal{M}}_g$. Now, not only is this map not regular at the origin; but what we've seen is that even after we blow up the origin in the (a, b)-plane, the map doesn't extend to a regular map. Indeed, what we've seen is that if $\widetilde{\mathbb{C}}^2_{a,b}$ is the blowup of $\mathbb{C}^2_{a,b}$ at the origin, $E \subset \widetilde{\mathbb{C}}^2_{a,b}$ the exceptional divisor, then the map φ extends to a regular map on the complement in $\widetilde{\mathbb{C}}^2_{a,b}$ of the point $p \in E$ corresponding to the line $(b = 0)$, constant on $E \setminus \{p\}$: for any arc in $\mathbb{C}^2_{a,b}$ with tangent line at $(0, 0)$ not equal to $(b = 0)$, the total space of the corresponding

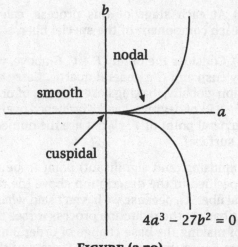

FIGURE (3.70)

one-parameter family of curves will be smooth, and the same analysis shows that the stable limit will have an elliptic tail with j-invariant 0.

In fact, a sequence of three blowups exactly analogous to those needed to desingularize the total space $\mathcal{X}$ in the preceding example is required to regularize this map. Their effect is shown schematically in Figure (3.71). At the final stage shown on the bottom left, we get

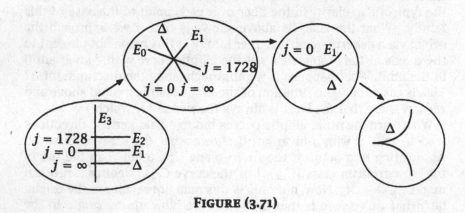

FIGURE (3.71)

a map to $\overline{\mathcal{M}}_g$ that sends the points of E_3 to the joins of $\widetilde{C}$ to elliptic tails having every j-invariant. This map blows down the curves E_1, E_2 and Δ respectively to the three joins with tails having j-invariants 0, 1728 and ∞.

Smooth curves acquiring a triple point

For another (and in fact easier) example, consider a family of curves, generically smooth, tending to a curve with a triple point. As in the previous case, we'll assume that the family has no infinitesimal base locus so that we have a smooth total space $X \to B$. (As the last example illustrated, this last assumption is quite a restrictive one.) Locally such a family might have equation of the form $x^3 + y^3 + g(x,y) + t$ where g vanishes to order at least 4 at the origin and look like that in Figure (3.72). As before, the first order of business is to reduce to

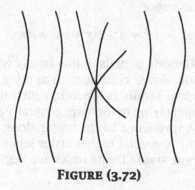

FIGURE (3.72)

the case where the reduced special fiber has only nodes. This can be done in one step by blowing up the triple point of the special fiber: we arrive at the family in Figure (3.73) in which the special fiber consists of the normalization $\tilde{C}$ of the original curve C with a triple point, plus an exceptional divisor E_1 meeting $\tilde{C}$ in the three points lying over the triple point of C and appearing with multiplicity 3 in the fiber.

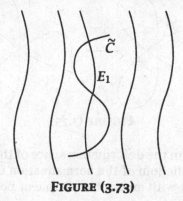

FIGURE (3.73)

Now, to get rid of the multiplicity we have to make a base change of order 3, followed by a normalization. As before, this amounts to taking a cyclic triple cover of the total space of our family branched over the curve $\tilde{C}$. The inverse image of E_1 is then a cyclic triple cover of E_1 totally branched over the three points of intersection of E_1 with $\tilde{C}$. Once again, this is the elliptic curve E of j-invariant 0. Since the points of intersection of E with $\tilde{C}$ are the fixed points of the automorphisms of E, they form a subgroup of order 3 of E. We summarize this example: the stable limit of our family is the normalization $\tilde{C}$ of the original curve C with an elliptic curve E of j-invariant 0 attached by identifying the points of $\tilde{C}$ lying over the triple point of C with the points of a subgroup of order 3 in E.

Again, we can now ask what variations on this reduction will appear over the deformation space

$$x^3 + y^3 + a + bx + cy + dxy$$

of the triple point. Generic pencils in this family will lead to the limit above. The derivation above continues to apply as long as the total space X of the original family is smooth, which will be true here as long as we avoid pencils lying in (or more generally arcs tangent to) the hyperplane $a = 0$. Approaches to the origin along special directions, or more generally, with special higher-order jets will lead to a whole menagerie of different stable limits which we begin to explore in the exercises.

EXERCISE (3.74) 1) What is the stable limit of a family of curves with three nodes degenerating to a curve with an ordinary triple point as shown below? What conditions must the jet of an arc in the deformation space of the triple point satisfy to yield a family of curves of this type?

FIGURE (3.75)

2) Construct an arc in the deformation space of the triple point whose stable reduction is the join of the normalization $\tilde{C}$ at the points lying over the triple point with an arbitrary triple of points on an arbitrary elliptic curve.

3) (Harder) Describe the regularization of the rational map to moduli from the base of the deformation space of the triple point.

Plane quartics specializing to a double conic

Here is a final example with a somewhat different flavor: we consider a family of plane quartic curves specializing to a double conic. To set this up, say $Q(x, y)$ is the equation of a smooth conic in the plane $\mathbb{P}^3$, and $F(x, y)$ a general quartic, and consider the pencil given by the equation $Q^2 + t \cdot F$ which is sketched in Figure (3.76).

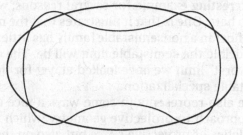

FIGURE (3.76)

Assuming the conic $Q = 0$ and the quartic $F = 0$ intersect transversely at eight points p_i as in the figure, we may take local coordinates x and y around p_i so that $Q = x$ and $F = y$. The local equation of the family is then $x^2 + yt = 0$. In particular we see that the total space $X \subset \mathbb{P}^2 \times \Delta$ of our family necessarily has an A_1 singularity at the points $(p_i, 0)$. This we can deal with by blowing up once to get the family in Figure (3.77) whose special fiber is the conic curve C given by $Q = 0$ (appearing with multiplicity 2 in the fiber), plus eight copies E_i of $\mathbb{P}^1$ attached at the points $p_i \in C$. Note that the E_i will have self-intersection -2 and will appear with multiplicity 1 in the fiber.

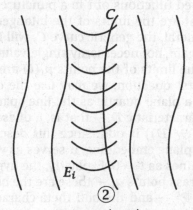

FIGURE (3.77)

The next step is to make a base change of order 2, i.e., take a double cover of the total space branched over the union of the curves E_i.

The inverse image of C is then the double cover of C branched at the eight points p_i, which is a hyperelliptic curve C_0 of genus 3. At the same time, the inverse image of each E_i is then just a rational curve mapping isomorphically to E_i. These will appear with multiplicity 1 in the fiber and will have self-intersection -1 so they can be blown down to arrive at a family whose special fiber is simply the smooth hyperelliptic curve C_0.

This is an interesting example for several reasons; we should mention at least two here. One is that it illustrates that the nastiness of the nonsemistable fiber in a nonsemistable family has little to do with how singular or reducible the semistable limit will be. The double conic is certainly the "worst" limit we have looked at, yet for the first time we get a smooth stable specialization.

This example also represents in some ways a nice example of the 20th century approach to projective geometry, which is to focus not only on subvarieties of projective space but also on the abstract varieties associated to them. Thus, in the original family of plane curves, the limiting curve is a double conic, which has relatively little structure (e.g., all double conics are isomorphic, and all points on a double conic look the same). If we think of the family as a family of abstract curves of genus 3, however, the natural limiting object is a smooth, hyperelliptic curve of genus 3 — an object that does have a great deal of structure.

To illustrate how this notion can be used in practice, we consider the following classical problem. In the family of curves $C_t \subset \mathbb{P}^2$ given by $Q^2 + t \cdot F = 0$, the curve C_t will, for each small $t \neq 0$, be smooth and so will have exactly 28 bitangent lines. We will denote these, somewhat abusively, by $L_1(t), \ldots, L_{28}(t)$ although these aren't necessarily, a priori, single-valued functions of t in a punctured neighborhood of $t = 0$. We ask: what are the limits of the bitangents $L_i(t)$ as $t \to 0$? Similarly, if F is general, the generic curve C_t will have 24 flex points $p_1(t), \ldots, p_{24}(t)$ (again, not necessarily single-valued functions of t).[5] We may ask what the limits of the points $p_i(t)$ are as $t \to 0$.

To answer the first question, we may use the characterization of a bitangent line to a plane quartic as the line spanned by the points of an odd theta-characteristic D — that is, a divisor $D \geq 0$ such that $2D \sim K$ and $h^0(C, \mathcal{O}_C(D))$ is odd. Since this description doesn't explicitly involve the plane embedding, it serves as well to characterize the limits of these lines as $t \to 0$. Explicitly, the hyperelliptic curve C_0 will have 8 Weierstrass points q_i — these are the branch points of the double cover $C_0 \to \mathbb{P}^1$ — and the odd theta-characteristics of C_0 are

---

[5] In this example, it turns out that the lines $L_i(t)$ are single-valued while the points $p_i(t)$ are defined only over $\mathcal{O}_\Delta[t^{1/3}]$.

just the pairwise sums $q_i + q_j$ of distinct Weierstrass points. Since the points $q_i \in C_0$ map to the points p_i of intersection of the plane conic C with the quartic curve $F = 0$, we conclude that *the limits of the* 28 *bitangent lines to the curves C_t are the lines joining pairs of points of intersection of C with the curve $F = 0$.*

Again, the point is that the original problem — finding the limits of bitangents to a family of plane curves — doesn't involve in any way the stable limit of the family. If we look only at the plane curves, however, it's difficult to see where the bitangents should go (though it's a worthwhile exercise to try and work out the problem rigorously without invoking stable reduction).

EXERCISE (3.78) In the example above, show that as $t \to 0$ the 24 lines to which the general fiber X_t flexes approach the eight tangent lines to the conic at the base points of the pencil and that each tangent line occurs with multiplicity 3 in the limit.

EXERCISE (3.79) 1) Let $\{C_t\}$ be a general pencil of plane sextics specializing to a double cubic curve $C_0 = 2E$. What will be the stable limit of the family?

2) A general smooth sextic plane curve will have 324 bitangent lines. What are the limiting positions of the bitangent lines to $\{C_t\}$ as t approaches 0?

EXERCISE (3.80) 1) Show that a general pencil of plane sextics specializing to a triple conic will have as stable limit a smooth cyclic trigonal curve of genus 10.

2) Make a dimension count that shows that other trigonal curves of genus 10 must be limits of (special pencils of) plane sextics as well.

3) Show that if Q, R and S are a sufficiently generic quadric, quartic and sextic respectively, then the pencil $Q^3 + tQR + t^2 S$ has stable limit a noncyclic trigonal curve. What does "sufficiently generic" mean in this example?

4) Exactly which trigonal curves arise as limits of pencils of sextics?

Here are a few more exercises that treat stable reductions that come up frequently in applications.

EXERCISE (3.81) Find the stable limit of a general pencil $\{C_t\}$ of plane quartics specializing to:

1) the union of a smooth cubic plane curve C and a transverse line L

2) the union of a smooth cubic plane curve C and a line L simply tangent to C

3) the union of a smooth cubic plane curve C and a flex line L to C

4) the union of a cuspidal cubic C and a general line L through the cusp of C

5) the union of a cuspidal cubic C and the "tangent line" (i.e., the reduced projective tangent cone) to C at the cusp.

EXERCISE (3.82) 1) If $\{C_t\}$ is a general pencil of curves, smooth for $t \neq 0$, specializing to a curve C with an ordinary fourfold point, show that the stable limit of the family will be the union of the normalization $\tilde{C}$ with a curve B of genus 3 meeting $\tilde{C}$ at the four points $p_1, \ldots, p_4 \in \tilde{C}$ lying over the fourfold point of C.

2) Show that if $\{C_t\}$ is any such family, the stable limit will be the union of $\tilde{C}$ with a stable 4-pointed curve $(B; q_1, \ldots, q_4)$ of genus 3 obtained by identifying p_i with q_i.

3) Show that not all such stable 4-pointed curves $(B; q_1, \ldots, q_4)$ arise, by showing that the versal deformation space of a fourfold point has dimension 9 and the space of pointed curves $(B; q_1, \ldots, q_4)$ as above has dimension 10.

4) *Harder*: Naively, the locus of B's that arise should have the same dimension as the projectivization of the tangent space at the origin to the deformation space: in the example above, we expect an eight-dimensional family. Prove or disprove the

CONJECTURE (3.83) *The stable 4-pointed curve* $(B; q_1, \ldots, q_4)$ *of genus 3 arises above if and only if* $q_1 + \cdots + q_4$ *is a canonical divisor on B.*

EXERCISE (3.84) Partially generalize the previous exercise to the case of families of curves C_t that for $t \neq 0$ are smooth but for which the special fiber C has an ordinary n-fold point p by showing:

1) Every stable limit will be the join of the normalization $\tilde{C}$ of C at the n points lying over P with some stable n-pointed curve $(B; q_1, \ldots, q_n)$ of genus $h = \binom{n-1}{2}$.

2) The B that consists of a union of n lines each meeting the others and $\tilde{C}$ at one point always appears as a limit.

3) The B that arises as the stable limit of the generic such family is the n-sheeted cover of $\mathbb{P}^1$ totally ramified at the n points corresponding to the tangent directions of the branches to C at the n-fold point.

PROBLEM (3.85) Is it always the case that the locus in $\overline{\mathcal{M}}_{h,n}$ of curves B that arise as limits in this way has dimension 1 less than that of the versal deformation space of the n-fold point?

Substantial progress has been made recently on this and related questions by Hassett [85].

Outline of the proof of existence of stable reductions

As we remarked before looking at examples, the proof of the existence of stable reductions follows, in outline, the stages we have repeatedly carried out above.

We start with the case where we have a family $\pi : X \to B$ over a smooth one-dimensional base B smooth over $B \setminus \{0\}$ and with otherwise arbitrary special fiber X_0 over the point $0 \in B$. We will proceed in stages but continue to use notation like $\pi : X \to B$ to denote the family that results after each step of the process.

The first step is to apply resolution of singularities to the pair (X, X_0): by blowing up, we arrive a family such that X is smooth, and X_0 has set-theoretic normal crossings, that is, the reduced scheme $(X_0)_{\text{red}}$ is nodal. At this point, the map π will be given by an equation of the form $t = x^a y^b$ in terms of a local coordinate t on B and local coordinates x and y on X.

Stage two is to perform a base change. If m is the least common multiple of the multiplicities of the components of the special fiber X_0, we make the base change $t \mapsto t^m$ and normalize the resulting total space. A local calculation then shows that the special fiber X_0 has reduced normal crossings and the map π has local equation of the form either $t^n = x$ or, at nodes of the special fiber, $t^n = xy$ where t is again a local coordinate on the base B. In the latter case, the total space X will be smooth at the node if and only if $n = 1$. If $n > 1$, there is an A_{n-1} singularity at the node. Note that, in both cases, the special fiber is now reduced and nodal.

Third, we minimally resolve the A_{n-1} singularities that arise. This has the effect of replacing each singularity by a chain of $(n-1)$ rational curves. We have now arrived at a family with smooth total space and reduced, nodal special fiber.

What we do next depends on which of the three variants we're after. To obtain the semistable reduction, we simply blow down any exceptional curves of the first kind in the total space X: these are just the smooth rational components of the special fiber $X - 0$ meeting the rest of X_0 only once. To obtain the stable reduction, we further blow-down all semistable chains: that is, chains of smooth rational curves of self-intersection (-2). To obtain the nodal reduction, it may be necessary to go in the other direction, blowing up some points of X to extend the map ψ. This causes no trouble unless we need to blow up a node of X_0 in which case the exceptional divisor will be a nonreduced component of X_0 and we need to repeat step two.

Finally, the case where the general fiber is singular requires a few extra operations. By a base change, we may assume that the monodromy acts trivially on the branches of the nodes of the general fiber. Then, we apply the steps above to the normalization $\tilde{X}$ of the total space.

Next, at the end, we make further blowups and base changes as necessary to ensure that closures of the inverse images of the nodes of the general fiber are disjoint sections of $\tilde{\mathcal{X}}$. Finally, we reidentify these appropriately.

As for the uniqueness assertions in all the variants, these all follow by applying the facts about B-isomorphisms discussed at the end of Section A.

Flat completion of embedded families

Before we leave stable reduction behind, we wish to mention a problem raised by Carel Faber that roughly involves understanding the inverse of the stable reduction process: Given a family of stable curves $\mathcal{X} \to \Delta^*$ over the punctured disc together with a line bundle $\mathcal{L}$ on $\mathcal{X}$ whose sections give embeddings

$$X_t \to Y_t \subset \mathbb{P}^n$$

of the curves in the family as curves in $\mathbb{P}^n$, describe all the curves Y_0 in $\mathbb{P}^n$ that can arise as flat limits of the Y_t's. In other words, describe all the embedded families that (up to base change) have $\mathcal{X} \to \Delta^*$ as their stable reduction. We call this problem the *flat completion* problem.

The first interesting example to look at is a constant family of smooth plane quartics. That is, each X_t for $t \neq 0$ is isomorphic to a fixed smooth plane quartic curve and the line bundle $\mathcal{L}$ is the pullback of $\mathcal{O}_{\mathbb{P}^2}(1)$ (equivalently, the relative dualizing sheaf $\omega_{\mathcal{X}/\Delta^*}$). This means that the images D_t of all the X_t's are obtained from any one — say, D_1 — by projectivities with coefficients that depend on t.

EXERCISE (3.86) 1) Show that if we obtain D_t from D_1 by the diagonal projectivity $\mathrm{diag}(1, 1, t)$, then the limiting curve D_0 is either a fourfold line or the union of a triple line with another line depending on whether the fixed point $(0, 0, 1)$ lies off or on the curve D_1.

2) Show that the projectivity $\mathrm{diag}(1, t, t)$ yields as limit the sum of four concurrent lines (not necessarily distinct).

3) How can you obtain the union of the cuspidal cubic $y^2 z - x^3 = 0$ and its tangent line $y = 0$ at the cusp?

Aluffi and Faber [2] show that the set of possible limits that can arise in isotrivial families like the examples above can depend on the *intrinsic* geometry of the general fiber D_1. For example, the curve $y^3 z - x^4 = 0$ appears if and only if the plane model of D_1 has a hyperflex. Almost nothing general is known, even about the isotrivial case, but the problem is yet another that clearly merits study.

D   Interlude: calculations on the moduli stack

> "OF course, here I'm working with the moduli stack rather than with the moduli space. For those of you who aren't familiar with stacks, don't worry: basically, all it means is that I'm allowed to pretend that the moduli space is smooth and that there's a universal family over it."

Who hasn't heard these words, or their equivalent, spoken in a talk? And who hasn't fantasized about grabbing the speaker by the lapels and shaking him until he says what — exactly — he means by them? But perhaps you're now thinking that all that is in the past, and that at long last you're going to learn what a stack is and what they do.

Fat chance. Unless you've picked up this book for the first time and have opened it at random to this page, you must know better. But, we're not going to evade the issue entirely. Briefly, here is the situation:

One of the basic techniques we'll be applying in our study of moduli is to find relations among the various cycle classes on the moduli space $\overline{\mathcal{M}}_g$. (Some of these classes were introduced in Section 2.D; others will appear below). Most of the ways we do this — the Grothendieck-Riemann-Roch formula, which we'll discuss in the following subsection, is a perfect example — don't a priori give such relations. Rather, they give relations among corresponding divisor classes on the base of any family $X \to B$ of stable curves. In order to get results on the cohomology or Picard group of $\overline{\mathcal{M}}_g$, then, we have to translate these statements.

There is a perfect vehicle for doing this: the language of stacks. As we indicated in Section 2.A, the category of stacks is a relatively slight enlargement of the category of schemes — slight enough, at any rate, that we can meaningfully extend to the category of stacks the definitions of such things as the Picard group of a scheme, the cohomology ring of a scheme, and intersection numbers between line bundles and curves. But the category of stacks *is* just large enough that, in it, the functor of families of stable curves is representable: in other words, there is a stack $\overline{\mathcal{M}}_g^{\text{fun}}$ such that for any scheme B we have a natural identification between the set of families of stable curves over B and the set of morphisms of B to $\overline{\mathcal{M}}_g^{\text{fun}}$. Moreover, since there is (again, in the category of stacks) a universal stable curve over $\overline{\mathcal{M}}_g^{\text{fun}}$, there is a natural map $\pi : \overline{\mathcal{M}}_g^{\text{fun}} \to \overline{\mathcal{M}}_g$. Now, the results we obtain by applying Grothendieck-Riemann-Roch and other formulas to families of stable curves yield directly relations among classes on the stack $\overline{\mathcal{M}}_g^{\text{fun}}$; and applying the map π we derive in turn results about the cohomology and Picard groups of $\overline{\mathcal{M}}_g$.

Well, that's how it goes in theory, anyway. In practice, there is one respect in which the language of stacks isn't wholly perfect: it's difficult to understand even the definition of a stack (we're speaking strictly for ourselves here). Actually, once you've absorbed the basic definitions, the rest is not so bad; but there's no question that the initial learning curve is steep, not to say vertical.

What are we going to do in this book? Basically, we're going to be guided by two principles. First, we aren't going to define a stack; and, given that we're not going to define them, we won't use any definition or result that relies on the definition of a stack. Second, we *will* provide a reasonably self-contained logical framework for the divisor-class calculations that form an essential part of our study of $\overline{\mathcal{M}}_g$. We've therefore chosen to make purely local and ad-hoc definitions of the objects we'll be using. But, we'll also try, parenthetically, to indicate how the definitions we make relate to those in the theory of stacks, so that you can relate our calculations to those that appear elsewhere in the literature. If you want to explore further, a good place to begin is Vistoli's article [149].

Divisor classes on the moduli stack

We start with the definition of a rational divisor class. On the moduli space, this is straightforward: since $\overline{\mathcal{M}}_g$ has only finite quotient singularities, every codimension 1 subvariety of $\overline{\mathcal{M}}_g$ is $\mathbb{Q}$-Cartier, so that we have an equality

$$A^1(\overline{\mathcal{M}}_g) \otimes \mathbb{Q} = \text{Pic}(\overline{\mathcal{M}}_g) \otimes \mathbb{Q}.$$

We will call an element of this group a *rational divisor class on the moduli space*. At the same time, it'll be helpful to introduce the notion of a divisor class on the moduli stack.

DEFINITION (3.87) *By a* rational divisor class on the moduli stack *we'll mean an association* γ *to each family* $\rho : \mathcal{X} \to B$ *of stable curves of a rational divisor class* $\gamma(\rho) \in \text{Pic}(B) \otimes \mathbb{Q}$ *on the base of the family, such that for any fiber square*

$$
\begin{array}{ccc}
\mathcal{X}' \cong B' \times_B \mathcal{X} & \longrightarrow & \mathcal{X} \\
{\scriptstyle \rho'} \downarrow & & \downarrow {\scriptstyle \rho} \\
B' & \longrightarrow & B
\end{array}
$$

the class $\gamma(\rho')$ associated to the morphism $\rho' : X' \to B'$ is the pullback of the class $\gamma(\rho)$ associated to $\rho : X \to B$. The group of rational divisor classes on the moduli stack will be denoted $\mathrm{Pic}_{\mathrm{fun}}(\overline{\mathcal{M}}_g) \otimes \mathbb{Q}$.

We may define analogously *rational cohomology classes on the moduli stack.* We should emphasize again that terms like "rational divisor class on the moduli stack" and "rational cohomology class on the moduli stack" are to be taken as self-contained and atomic: remember that we do not and will not define a stack. Moreover, the definition above does *not even suggest* the correct definition of a line bundle on a stack: a line bundle on the moduli stack is officially something that associates to every family $X \to B$ of stable curves a line bundle on the base of the family, and for every fiber square *specifies an isomorphism* between the line bundle associated to the morphism $X' \to B'$ and the pullback of the line bundle associated to $X \to B$. It will turn out, however, that in the present circumstance the two definitions yield the same group of bundles.

Of the various divisors and classes we've discussed so far, some seem naturally to be rational divisor classes on the moduli stack: the class λ, for example. Others, like δ, by contrast, are easy to describe as rational Cartier divisors (and hence rational line bundles) on the moduli space $\overline{\mathcal{M}}_g$; but it may not at first be apparent how to define a corresponding rational divisor class on the moduli stack: what do we do, for example, with a family $X \to B$ of stable curves, all of which are singular?

Our first task, then, is to show that, in fact, a rational divisor class on the moduli stack is the same thing as a rational divisor class on the moduli space, that is, an element of $\mathrm{Pic}(\overline{\mathcal{M}}_g) \otimes \mathbb{Q}$. Once we've established this isomorphism, we'll be free to define and deal with our divisor classes in whatever way seems best suited to the class at hand. We will also give an explicit description of the rational divisor class on the moduli stack associated to a divisor in $\overline{\mathcal{M}}_g$, and in particular answer the question at the end of the last paragraph. In any event, the first step is the:

PROPOSITION (3.88)

$$\mathrm{Pic}_{\mathrm{fun}}(\overline{\mathcal{M}}_g) \otimes \mathbb{Q} \cong \mathrm{Pic}(\overline{\mathcal{M}}_g) \otimes \mathbb{Q}.$$

PROOF. We will give inverse maps between the two. In one direction this is straightforward: given a rational divisor class $\nu \in \mathrm{Pic}(\overline{\mathcal{M}}_g)$, some multiple $m\nu$ of ν may be represented by a line bundle L on $\overline{\mathcal{M}}_g$. Now, to any family $\rho : X \to B$ of stable curves with induced map $\varphi : B \to \overline{\mathcal{M}}_g$ we associate $\frac{1}{m}$ times the pullback line bundle: in other words, we set

$$\gamma(\rho) = \frac{1}{m} \varphi^*(L).$$

This defines a rational divisor class γ on the moduli stack, and thus a map

$$\pi^* : \mathrm{Pic}(\overline{\mathcal{M}}_g) \otimes \mathbb{Q} \to \mathrm{Pic}_{\mathrm{fun}}(\overline{\mathcal{M}}_g) \otimes \mathbb{Q}.$$

To define the inverse isomorphism, it will be useful to have a basic lemma:

LEMMA (3.89) *1) There exists a family $\rho : X \to \Omega$ of stable curves such that the induced map $\varphi : \Omega \to \overline{\mathcal{M}}_g$ is surjective and finite.*
2) For any point $[C] \in \overline{\mathcal{M}}_g$, there exists a family $X \to \Omega$ of stable curves over a smooth base Ω such that the induced map $\Omega \to \overline{\mathcal{M}}_g$ is surjective, generically finite, and finite over $[C]$.

We should point out that this is a very weak form of a theorem of Looijenga and Pikaart [108], who have shown (by explicit construction) that we can take Ω to be *simultaneously* smooth and finite over $\overline{\mathcal{M}}_g$. It is also true for more general moduli problems; see Kollár [99]. This simple statement is enough for our purposes, however, and (more importantly) within our ability to prove, so we'll leave it at that. The proof will be deferred to the next subsection. The lemma, as stated, also follows easily from the existence of universal curves with suitably defined level structure, as in Popp ([131], [132]).

Given the lemma, we can define a map from the group of rational divisor classes on the moduli stack to $\mathrm{Pic}(\overline{\mathcal{M}}_g)$ as follows. Given $\gamma \in \mathrm{Pic}_{\mathrm{fun}}(\overline{\mathcal{M}}_g) \otimes \mathbb{Q}$, we define a divisor class on $\overline{\mathcal{M}}_g$ by choosing any tautological family $\rho : X \to \Omega$ as in the lemma, letting $D = \gamma(\rho) \in \mathrm{Pic}(\Omega) \otimes \mathbb{Q}$ be the divisor class associated to this family by γ, taking the norm (or pushforward) of D under the map $\varphi : \Omega \to \overline{\mathcal{M}}_g$, and finally dividing the result by the degree of the map φ: in other words, we define a map

$$\pi_* : \mathrm{Pic}_{\mathrm{fun}}(\overline{\mathcal{M}}_g) \otimes \mathbb{Q} \to \mathrm{Pic}(\overline{\mathcal{M}}_g) \otimes \mathbb{Q}$$

by

$$\pi_*(\gamma) = \frac{1}{\deg(\varphi)} \mathrm{Norm}(\gamma(\rho)).$$

This is independent of the choice of tautological family $\rho : X \to \Omega$ (any two tautological families are in turn covered by their fiber product over $\overline{\mathcal{M}}_g$), and gives a two-sided inverse to π^* above. ∎

Note that since we haven't defined the notion of stack, morphism of stacks, or the pullback of a rational divisor class under a morphism of stacks, we really shouldn't use the symbols "π^*" and "π_*" to denote the maps above; something neutral, like "F" and "G" would have been more honest. But the fact is, there are such things as stacks, morphisms of stacks, and the pullback of a rational divisor class under a morphism of stacks; there is a natural morphism π from the moduli stack to the moduli space, and the map π^* above is simply the pullback under this morphism.

The following exercise is helpful in thinking about divisor classes on the moduli stack.

EXERCISE (3.90) 1) Show that any rational divisor class on the moduli stack — say $\gamma \in \mathrm{Pic}_{\mathrm{fun}}(\overline{\mathcal{M}}_g)$ — is determined by its values $\gamma(\rho)$ on families $\rho : X \to B$ with smooth, one-dimensional base B.

2) Let $\Sigma \subset \overline{\mathcal{M}}_g$ be any proper subvariety of $\overline{\mathcal{M}}_g$. Extending the result of the first part, show that any rational divisor class $\gamma \in \mathrm{Pic}_{\mathrm{fun}}(\overline{\mathcal{M}}_g)$ on the moduli stack is determined by its values $\gamma(\rho)$ on families $\rho : X \to B$ such that B is smooth and one-dimensional, and such that the image $\varphi(B) \subset \overline{\mathcal{M}}_g$ of B under the induced may $\varphi : B \to \overline{\mathcal{M}}_g$ isn't contained in Σ.

3) Finally, in view of the fact that $\mathrm{Pic}(\overline{\mathcal{M}}_g)$ is discrete, show that any rational divisor class $\gamma \in \mathrm{Pic}_{\mathrm{fun}}(\overline{\mathcal{M}}_g)$ on the moduli stack is determined by the *degrees*

$$\deg(\gamma(\rho)) \in \mathbb{Q}$$

of its values on families $\rho : X \to B$ with smooth, one-dimensional base B and $\varphi(B) \not\subseteq \Sigma$.

Thus, we may think of a rational divisor class $\gamma \in \mathrm{Pic}_{\mathrm{fun}}(\overline{\mathcal{M}}_g)$ on the moduli stack as a gadget that, for any one-parameter family of stable curves, measures the nontriviality of the family, for example by counting the number of fibers of a certain type (e.g., the divisor class δ counts the number of singular fibers). This in turn suggests another way of associating to a divisor on $\overline{\mathcal{M}}_g$ a divisor on the moduli stack:

Let $\Sigma \subset \overline{\mathcal{M}}_g$ be any closed subvariety of codimension 1. Then to any one-parameter family $\rho : X \to B$ of stable curves of genus g we may associate a number, or more generally a divisor class $\sigma(\rho) \in \mathrm{Pic}(B) \otimes \mathbb{Q}$ (naively, the "number of members of the family lying in Σ"), as follows:

Case 1: *Only a finite number of fibers X_b of the family correspond to points of Σ.*

In this case, we assign to each fiber X_b lying in Σ a multiplicity $\mathrm{mult}_b(\sigma)$ and add these up, setting $\sigma(\rho) = \sum_b \mathrm{mult}_b(\sigma) \cdot b$. To define the multiplicity, let $\widetilde{\Sigma}$ be the inverse image of Σ in the versal deformation space $\mathrm{Def}(X_b)$ of X_b; since the versal deformation space is smooth this is a Cartier divisor. Now, for any $b \in B$ we have a natural map from a neighborhood of $b \in B$ to the versal deformation space of X_b, and we define $\mathrm{mult}_b(\sigma)$ to be the multiplicity of the pullback under this map of the divisor $\widetilde{\Sigma}$.

Case 2: *All the fibers X_b belong to Σ.*

Here, we describe a line bundle L on B as follows: in a neighborhood of each point $b \in B$, we take L to be the pullback of the normal space to $\widetilde{\Sigma}$ in $\mathrm{Def}(X_b)$. Combining the openness of versality [Exercise (3.39)] with the uniqueness of the map to $\mathrm{Def}(X_b)$, we see that this is indeed a well-defined line bundle. We then define the divisor class $\sigma(\rho)$ of curves belonging to Σ to be the line bundle L.

With all this said, we have the:

PROPOSITION (3.91) *For any codimension 1 subvariety $\Sigma \subset \overline{\mathcal{M}}_g$, there exists a rational divisor class σ on the moduli stack whose values on one-parameter families $\rho : \mathcal{X} \to B$ are given as above.*

Note that by Exercise (3.90), γ is determined by its agreement with the degrees specified in Case 1 above, so we can avoid the seemingly trickier calculations invoked in Case 2. Actually, as we'll see in the explicit description in Lemma (3.94) of the divisor class on the moduli stack associated to the boundary $\Delta \subset \overline{\mathcal{M}}_g$, the calculation in Case 2 is often quite straightforward to carry out.

We now have two ways of passing from an irreducible, codimension 1 subvariety $\Sigma \subset \overline{\mathcal{M}}_g$ to a rational divisor class on the moduli stack: we can apply the "pullback" map π^* of Proposition (3.88) to the divisor class $[\Sigma] \in \mathrm{Pic}(\overline{\mathcal{M}}_g) \otimes \mathbb{Q}$ to arrive at a rational divisor class on the moduli stack; or we can define a rational divisor class σ on the moduli stack as in Proposition (3.91). The relationship between the two is straightforward: we have the:

PROPOSITION (3.92) *Let $\Sigma \subset \overline{\mathcal{M}}_g$ be an irreducible, codimension 1 subvariety, and $\sigma \in \mathrm{Pic_{fun}}(\overline{\mathcal{M}}_g)$ the divisor class on the moduli stack associated to Σ as in Proposition (3.91). Let $[C] \in \Sigma$ be a general point. Then*

$$\sigma = \frac{1}{\#\mathrm{Aut}(C)} \pi^*[\Sigma].$$

PROOF. The proof of this proposition is immediate and is simultaneously a proof of the last proposition. ∎

Applying Exercise (2.28), we see in particular that the divisor class on the moduli stack associated by Proposition (3.92) to an effective divisor $D \subset \overline{\mathcal{M}}_g$ coincides with the the class $\pi^*([D]) \in \mathrm{Pic_{fun}}(\overline{\mathcal{M}}_g) \otimes \mathbb{Q}$ associated to $[D] \in \mathrm{Pic}(\overline{\mathcal{M}}_g)$ by the isomorphism in Proposition (3.88) except in the cases of genus 2, of the divisor $H_3 \subset \overline{\mathcal{M}}_3$ of hyperelliptic curves of genus 3, and of the divisor Δ_1 in general.

Now, all of the above may seem somewhat like hairsplitting: why introduce rational divisor classes on the moduli stack at all if they are so closely related to rational divisor classes on the moduli space?

The answer is a practical one. As we will see very amply illustrated in the calculations that follow, the rational divisor classes on the moduli stack are the ones that we can calculate readily: typically, if we want to find the degree of a divisor class on a one-parameter family of curves, it's easier to calculate directly the degree of the corresponding divisor class on the moduli stack. Likewise, if we want to find relations among divisor classes in $\text{Pic}(\overline{\mathcal{M}}_g)$, we will typically find relations among the degrees of corresponding divisor classes on the moduli stack, and then translate the result back into terms of divisor classes on the space $\overline{\mathcal{M}}_g$. Explicitly, we'll invoke the:

BASIC PROPOSITION (3.93) *Let* Γ *in* $\text{Pic}(\overline{\mathcal{M}}_g) \otimes \mathbb{Q}$ *and* $\gamma = \pi^*(\Gamma)$ *in* $\text{Pic}_{\text{fun}}(\overline{\mathcal{M}}_g) \otimes \mathbb{Q}$ *be divisor classes that correspond under the isomorphism of Proposition (3.88). Let* $\Sigma_1, \ldots, \Sigma_k \subset \overline{\mathcal{M}}_g$ *be irreducible codimension 1 subvarieties, and* $\sigma_1, \ldots, \sigma_k \in \text{Pic}_{\text{fun}}(\overline{\mathcal{M}}_g) \otimes \mathbb{Q}$ *the rational divisor classes on the moduli stack associated to the subvarieties* Σ_i *as in Proposition (3.91). Let* $[C_i]$ *be a general point of* Σ_i *and let* a_i *be the order of the automorphism group of* C_i. *The following statements are equivalent:*

1) *For any family* $\rho : X \to B$ *of stable curves of genus* g, *we have the relation among divisor classes on* B

$$\sum c_i \cdot \sigma_i(\rho) = \gamma(\rho) \in \text{Pic}(B).$$

2) *We have the relation among divisor classes on* $\overline{\mathcal{M}}_g$

$$\sum \frac{c_i}{a_i} \cdot [\Sigma_i] = \Gamma \in \text{Pic}(\overline{\mathcal{M}}_g) \otimes \mathbb{Q}.$$

Moreover the second statement follows if we know the first for families $X \to B$ *whose general member does not actually belong to any* Σ_i.

The notion of the degree of a rational divisor class γ on the moduli stack on a one-parameter family $\rho : X \to B$ of stable curves extends naturally to families that are only generically stable, that is, whose general member is stable: by the degree of γ on such a family we'll mean the degree of γ on any family $\rho' : X' \to B'$ obtained from $\rho : X \to B$ by semistable reduction, divided by the degree of the base change morphism $B' \to B$ involved. Note that the first statement of Proposition (3.93) follows for generically stable families as well from the second statement.

At this point we should do a fundamental example: the description of the rational divisor class δ on the moduli stack associated by Proposition (3.91) to the codimension 1 subvariety $\Delta = \overline{\mathcal{M}}_g \setminus \mathcal{M}_g \subset \overline{\mathcal{M}}_g$.

To set this up, suppose that $\rho : X \to B$ is a family of stable curves over a smooth, one-dimensional base with local parameter t. We consider two cases:

Suppose first that the general fiber X_t of the family is smooth, and that the fiber X_0 over 0 in B has exactly one node, at $p \in X_0$. By our description of the versal deformation space of a node, we can choose local coordinates x and y on X in a neighborhood of the point p so that $xy = t^k$ for some k. Then, by our description of the versal deformation space of a nodal curve we see that, in the versal deformation space of X_0, the image of B will be a curve with contact of order k with the (smooth) hypersurface of singular deformations. Therefore the image of $\varphi(B) \in \overline{\mathcal{M}}_g$ will be tangent to Δ to order k — or, more accurately (we don't know that φ is one-to-one), the pullback to B of the defining equation of $\Delta \subset \overline{\mathcal{M}}_g$ will vanish to order exactly k at $t = 0$. Note in particular the equivalence: the map $\varphi : B \to \overline{\mathcal{M}}_g$ will be transverse to Δ at $t = 0$ if and only if X_0 has a single node p and X is smooth at p.

Generalizing this, suppose now that the fiber X_b over b in B has exactly n nodes, at the points $p_1, p_2, \ldots, p_n$. In the versal deformation space of X_0, the hypersurface of singular deformations will have n smooth sheets corresponding naturally to the nodes p_i of X_0. (If the curve X_0 has no automorphisms, the same will be true in $\overline{\mathcal{M}}_g$: the divisor Δ will have n smooth sheets in a neighborhood of the point $[X_0]$). Moreover, if the local equation of X at p_i is $xy - t^{k_i}$, then the pullback to B of the defining equation of the branch of the discriminant hypersurface corresponding to p_i will vanish to order k_i. All in all, we see that the pullback to B of the defining equation of the discriminant hypersurface in the versal deformation space of X_0 vanishes to order $k_1 + k_2 + \cdots + k_n$ at 0; or, in other words, the multiplicity

$$\mathrm{mult}_b(\delta) = k_1 + k_2 + \cdots + k_n.$$

Now, we turn to the second case. That is, we consider a family of stable curves $\rho : X \to B$, still over a smooth, one-dimensional base, but now where the general fiber is singular. To simplify things, suppose for the moment that each fiber X_b of ρ has a single node p. The crucial fact here is Proposition (3.31), which says that the normal space to the discriminant hypersurface in the versal deformation space of each fiber X_b is naturally identified with the tensor product of the tangent spaces to the branches of C at the node. To apply this, we introduce the normalization $\nu : \tilde{X} \to X$ of the total space of X. Then, we let $\tilde{\rho} = \rho \circ \nu : \tilde{X} \to B$ be the composition (so that in particular $\tilde{\rho}$ will be a smooth map), and $\tilde{\Gamma} \subset \tilde{X}$ be the locus of points lying over the nodes of the fibers X_b of the original map. The map $\tilde{\Gamma} \to B$ is unramified of degree 2; after a base change we may assume that $\tilde{\Gamma}$ consists simply of two disjoint sections $\tilde{\Gamma}_1$ and $\tilde{\Gamma}_2$ of $\tilde{X} \to B$. In these terms, what Proposition (3.31) is telling us is that the normal space of the discriminant hypersurface, in the versal deformation space of each fiber X_b, is the tensor product of the tangent spaces to the fiber $\tilde{X}_b$ at the two points of $\tilde{\Gamma}$ lying over b. Equivalently, *the line bundle associated to the family* $\mathcal{X} \to B$ *by the divisor class* δ *on the moduli stack is the tensor product of the normal bundles to the components of* $\tilde{\Gamma}$ *in* $\tilde{\mathcal{X}}$:

$$\delta(\rho) = N_{\tilde{\Gamma}_1/\tilde{\mathcal{X}}} \otimes N_{\tilde{\Gamma}_2/\tilde{\mathcal{X}}}.$$

Needless to say, the same analysis can be carried out for a family $\mathcal{X} \to B$ of stable curves whose general fiber has any number of nodes.

Combining these two cases, we arrive at the following description of δ. Suppose $\rho : \mathcal{X} \to B$ is a family of stable curves of genus g over a smooth, one-dimensional base B with parameter t, whose general fiber has n nodes; let $\tilde{\mathcal{X}} \to \mathcal{X}$ be the normalization of the total space of $\mathcal{X}$ and $\tilde{\rho} : \tilde{\mathcal{X}} \to B$ the composition. For each singular point $p \in \mathrm{sing}(\tilde{\rho})$ of the map $\tilde{\rho}$, let $k = k(p)$ be the unique integer such that there exist local coordinates x, y, t on $\tilde{\mathcal{X}}$ near p satisfying $xy = t^k$. Let $\Gamma \subset \mathcal{X}$ be the positive-dimensional components of the singular locus of ρ, and $\tilde{\Gamma} \subset \tilde{\mathcal{X}}$ the inverse image of Γ in $\tilde{\mathcal{X}}$. After making a base change, if necessary, we may suppose that $\tilde{\Gamma}$ consists simply of $2n$ disjoint sections $\tilde{\Gamma}_i$.

LEMMA (3.94) *Let* $\rho : \mathcal{X} \to B$ *be as above, and let* δ *be the divisor class on the moduli stack associated to the boundary* $\Delta \subset \overline{\mathcal{M}}_g$. *Then*

$$\delta(\rho) = \bigotimes_{i=1}^{2n} N_{\tilde{\Gamma}_i/\tilde{\mathcal{X}}} \otimes \mathcal{O}_B \Big(\sum_{p \in \mathrm{sing}(\tilde{\rho})} k(p) \cdot \tilde{\rho}(p) \Big).$$

In particular, the degree of δ *is given by*

$$\deg(\delta(\rho)) = (\tilde{\Gamma})^2 + \sum_{p \in \mathrm{sing}(\tilde{\rho})} k(p).$$

We can give similar descriptions of the divisor classes δ_i on the moduli stack associated to the divisors $\Delta_i \subset \overline{\mathcal{M}}_g$. Note that as a consequence of Proposition (3.92), the pullback $\pi^*([\Delta])$ of the divisor class $[\Delta] \in \mathrm{Pic}(\overline{\mathcal{M}}_g) \otimes \mathbb{Q}$, as defined in Proposition (3.88), is *not* the divisor δ on the moduli stack. Rather, since the general point $[C] \in \Delta_1 \subset \overline{\mathcal{M}}_g$ corresponds to a curve with automorphism group of order 2 (there is an involution on the elliptic tail fixing the point of attachment), we have:

COROLLARY (3.95)

$$\pi^*([\Delta]) = \delta_0 + 2\delta_1 + \delta_2 + \cdots + \delta_{\lfloor \frac{g}{2} \rfloor}.$$

A warning about notation. In view of the potential ambiguity, we have to be careful to distinguish between the class $[\Delta]$ of the boundary

$\Delta \subset \overline{\mathcal{M}}_g$ and the divisor δ on the moduli stack, and more generally between the class of any divisor $\Sigma \subset \overline{\mathcal{M}}_g$, a component of whose support lies in the locus of curves with automorphisms, and the associated divisor class σ on the moduli stack. In the case of the divisor classes λ and κ, which are defined only as classes and aren't naturally associated to any particular divisor, there is no such confusion. Therefore, we've yielded to temptation, and used the symbols λ and κ to denote both the divisor classes on $\overline{\mathcal{M}}_g$ and their pullbacks $\pi^*\lambda$ and $\pi^*\kappa \in \mathrm{Pic}_{\mathrm{fun}}(\overline{\mathcal{M}}_g) \otimes \mathbb{Q}$.

Existence of tautological families

To complete this section, we want to give a proof of Lemma (3.89).

PROOF. We observe first that for any point $[C] \in \overline{\mathcal{M}}_g$, there is a Zariski open neighborhood U of the point $[C]$ in $\overline{\mathcal{M}}_g$, a finite map $\Omega_U \rightarrow U$ and a family $Y_U \rightarrow \Omega_U$ of stable curves inducing the map $\Omega_U \rightarrow U \subset \overline{\mathcal{M}}_g$. To see this, we look at the locally closed subset K of the Hilbert scheme parameterizing m-canonically embedded stable curves and take Ω a linear section of K transverse to the locus of curves isomorphic to $[C]$ and $Y \rightarrow \Omega$ the restriction to Ω of the universal family over the Hilbert scheme. The induced map $\Omega \rightarrow \overline{\mathcal{M}}_g$ will then be finite over some neighborhood U of $[C]$ in $\overline{\mathcal{M}}_g$; we simply take Ω_U and Y_U the inverse images of U in Ω and Y respectively.

Next, suppose we're given two families $Y_U \rightarrow \Omega_U$ and $Y_V \rightarrow \Omega_V$ of stable curves whose associated maps $\Omega_U \rightarrow \overline{\mathcal{M}}_g$ and $\Omega_V \rightarrow \overline{\mathcal{M}}_g$ are finite and surjective onto open sets U and V in $\overline{\mathcal{M}}_g$; we want to construct a family $Y \rightarrow \Omega$ whose associated map $\overline{\mathcal{M}}_g$ is finite onto $U \cup V$. The construction is reasonably straightforward: very briefly, we'll extend Ω_U and Ω_V to finite covers of $\overline{\mathcal{M}}_g$, take their fiber product, pull back the families $Y_U \rightarrow \Omega_U$ and $Y_V \rightarrow \Omega_V$ to the inverse images of Ω_U and Ω_V in the product, and then make a further base change in order to make them agree over the inverse image of $U \cap V$ so we can paste them together to form a single family over the union $U \cup V$. The details, however, will sound somewhat complicated as we trace through them.

To start with, let $\overline{\Omega}_U$ and $\overline{\Omega}_V$ be the normalizations of $\overline{\mathcal{M}}_g$ in the function fields of Ω_U and Ω_V respectively (these are normal varieties containing Ω_U and Ω_V as open sets, to which the maps of Ω_U and Ω_V to $\overline{\mathcal{M}}_g$ extend). Let Ω' be the fiber product of $\overline{\Omega}_U$ and $\overline{\Omega}_V$ over $\overline{\mathcal{M}}_g$, and let

$$\pi_U : \Omega' \longrightarrow \overline{\Omega}_U$$

$$\pi_V : \Omega' \longrightarrow \overline{\Omega}_V$$

and

$$\pi : \Omega' \longrightarrow \overline{\mathcal{M}}_g$$

be the projections. Note that π, being a composition of finite maps, is again finite. Finally (for now!), let $\Omega'_U = (\pi_U)^{-1}(\Omega_U) = \pi^{-1}(U)$ and $\Omega'_V = (\pi_V)^{-1}(\Omega_V) = \pi^{-1}(V)$ be the inverse images of U and V in Ω'.

Via the projection $\pi_U : \Omega'_U \to \Omega_U$ we can pull back the family of stable curves $Y_U \to \Omega_U$ to obtain a family $Y'_U = Y_U \times_{\Omega_U} \Omega'$; we define a family $Y'_V \to \Omega'_V$ likewise. We now want to patch together the families Y'_U and Y'_V to form a single family over the inverse image $\Omega'_U \cup \Omega'_V = \pi^{-1}(U \cup V) \subset \Omega'$. This, however, requires a further base change: even though for each point p of $\pi^{-1}(U \cap V) \subset \Omega'$ the fibers of Y'_U and Y'_V over p are isomorphic, there may be no set of choices of such isomorphisms for each p which glue to give an isomorphism between the inverse images of $U \cap V$ in $Y_{U'}$ and $Y_{V'}$.

To overcome this problem, we introduce the variety Z of pairs (p, ψ) where p lies in $\Omega'_U \cap \Omega'_V$ and $\psi : (Y'_U)_p \to (Y'_V)_p$ is an isomorphism. Let Z_0 be any irreducible component of Z dominating $\Omega'_U \cap \Omega'_V$, and take Ω'' the normalization of Ω' in the function field of Z_0. Note that the projection $\Omega'' \to \Omega'$, and hence the projection $\Omega'' \to \overline{\mathcal{M}}_g$, are once more finite maps. Now, on the inverse images Ω''_U and Ω''_V of Ω'_U and Ω'_V in Ω'', we have pullback families $Y''_U = Y'_U \times_{\Omega'_U} \Omega''_U \to \Omega''_U$ and $Y''_V = Y'_V \times_{\Omega'_V} \Omega''_V \to \Omega''_V$; and their restrictions to the overlap $\Omega''_U \cap \Omega''_V$ are isomorphic. We may thus patch them together to form a single family $Y = Y''_U \cup Y''_V \to \Omega = \Omega''_U \cup \Omega''_V$ over the union Ω of Ω''_U and Ω''_V, whose associated structure map $\Omega \to U \cup V$ is finite. This, at last, is the family we're after over $U \cup V$.

Finally, since we can cover $\overline{\mathcal{M}}_g$ with a finite number of open sets U admitting such families $Y_U \to \Omega_U$, this glueing step completes the proof of the Lemma. ∎

It's worth noting that each step in this construction is already needed in the case of the moduli space $\overline{\mathcal{M}}_1$ of curves of genus 1: i.e., the affine line with coordinate j. To begin with, in order to have a family of smooth curves over a neighborhood U of the point 0, we have to make a base change $\Omega_U \to U$ of order 3, ramified at 0; and similarly to have a smooth family near 1728, we have to make a base change $\Omega_V \to V$ of order 2, ramified at 1728. To cover both, we then have to take the fiber product of these covers, giving us a six-sheeted cover of $\overline{\mathcal{M}}_1$. But this is still not enough: in order to patch these families together to form a single family covering all of $\overline{\mathcal{M}}_1$, we have to make a further base change of order 2, arriving (for example) at the relatively familiar family of curves of genus 1, with j-function of degree 12, given as a pencil of plane cubics.

We note one corollary of Lemma (3.89), which we may think as of stable reduction for families of curves over higher-dimensional bases, and whose proof we leave as an exercise. To state this corollary, we need one more bit of terminology. Suppose we're given a family $\rho : X \rightarrow B$ of curves, and we want to apply to this family the base change associated to a generically finite map $B' \rightarrow B$ with B' irreducible. If, in fact, the fiber dimension of ρ does jump, then it may happen that the fiber product $X \times_B B'$ is no longer irreducible. In this case, we'll simply disregard the components of the fiber product that fail to dominate B: we define the *essential pullback* $X' \rightarrow B'$ of our family to be the unique irreducible component X' of the fiber product $X \times_B B'$ dominating B', equipped with the restriction of the projection map to B'. We then have the:

COROLLARY (3.96) (STABLE REDUCTION OVER GENERAL BASES) *For any morphism $f : X \rightarrow B$ of integral varieties whose general fiber is a smooth curve of genus $g \geq 2$, there exists a generically finite map $B' \rightarrow B$, a family of stable curves $X' \rightarrow B'$ and a birational isomorphism of X' with the essential pullback to B' of the family $X \rightarrow B$.*

We should note (though it's not really within the purview of this book) that an analogue of the basic stable reduction theorem Proposition (3.47) holds for families $X \rightarrow B$ of higher-dimensional varieties over a one-dimensional base B: after base change and birational modifications, we can arrive at a family all of whose fibers are scheme-theoretic normal crossings. The situation for families of higher-dimensional varieties over higher-dimensional bases, however, is much less clear; in particular, the analogue of Corollary (3.96) doesn't seem to hold. For the best statement we know in this direction, see Abramovich and Karu [1].

E Grothendieck-Riemann-Roch and Porteous

Grothendieck-Riemann-Roch

The classical Riemann-Roch formula expresses the holomorphic Euler characteristic of a vector bundle E on a complex manifold X in terms of topological invariants of the bundle and of the manifold. A more naive interpretation is as a solution to the initial problem of giving a formula for the dimension $h^0(X, E)$ of the space of global sections of E in terms of topological invariants. The difficulty is that $h^0(X, E)$ isn't a topological invariant — it need not even be constant in holomorphically varying families of bundles E. On the other hand, if we throw in "error terms" $\pm h^i(X, E)$ coming from the higher cohomology groups of E, we arrive at the holomorphic Euler characteristic $\chi(E)$, which is a topological invariant, and which is expressed by the Riemann-Roch formula.

The Grothendieck form of the Riemann-Roch formula is, in these terms, just its extension to the relative case. We have, instead of a single variety or scheme X, a family of schemes $\{X_b\}$ (that is, a morphism $\pi : X \to B$ of schemes) with smooth connected base B, and a family of vector bundles on that family of schemes (that is, a vector bundle E on X). We can then try to form the spaces $H^0(X_b, E)$ into a bundle on B, and attempt to describe that bundle. Of course, this cannot, in general, be done. But, as a first approximation to such a bundle, we can take the direct image sheaf $\pi_* E$ on B — after all, if E is flat over B, then at any point $b \in B$ in the open subset $U \subset B$ where the dimension $h^0(X_b, E)$ assumes its generic value, $\pi_* E$ will in fact be locally free with fiber $(\pi_* E) \otimes k_b = H^0(X_b, E)$.

This said, we should next specify what information we're looking for. To begin with, we would obviously like to know the rank of the sheaf $\pi_* E$ — this is just the information given to us by the classical Riemann-Roch formula applied to the restriction of E to the general fiber X_b of π. Beyond this, however, we would like to understand the twisting of the sheaf $\pi_* E$, as measured by its Chern classes. (Indeed, we can think of the rank of a sheaf as just the 0th graded piece of its Chern character, an object which we would like to understand completely.) The problem, then, is to find a formula for the Chern class or Chern character $c(\pi_* E)$.

The difficulty now is that, in analogy with the classical case, even the topology of the sheaf $\pi_* E$ isn't a topological invariant of the bundle E and the map π. Instead, we have to throw in "error terms" as before — in this case, the higher direct image sheaves $R^i \pi_* E$ (which have fibers $H^i(X_b, E)$ at general points $b \in B$). The way to do this that minimizes the amount of bookkeeping required is to take the alternating sum of the Chern characters $\operatorname{ch}(R^i \pi_* E)$ in the K-group $K_0(B)$.

We now briefly review how this is set up and carried out. To start with, we fix, as above, a proper morphism $\pi : X \to B$ of projective varieties and a vector bundle E on X. (Typically, we're going to view this data as describing a family of vector bundles E_b on the fibers X_b of X over B, but we'll see below that there are interesting results to be had even in the case where the map π is the inclusion of a point X in a variety B.) For the rest of this section, we'll assume that B *is smooth*. In fact, in the applications, we'll want to take B to be the (singular) moduli space $\mathcal{M}_g$ or $\overline{\mathcal{M}}_g$ but we'll postpone dealing with this extra complication.

Our formula will be an equality in the Grothendieck group $K_0(B)$ of coherent sheaves on B (which, since B is smooth, is naturally isomorphic to the Grothendieck group $K^0(B)$ of vector bundles on B). Recall that, by definition, $K^0(B)$ is the quotient of the free abelian group generated by the vector bundles F on B by the subgroup generated by all elements $F - E - G$ for which there is an exact sequence

(3.97) $$0 \longrightarrow E \longrightarrow F \longrightarrow G \longrightarrow 0.$$

If $c(F)$ denotes the total Chern class

$$c(F) = 1 + c_1(F) + c_2(F) + \cdots,$$

the Whitney product formula says that, when such an exact sequence exists,

(3.98) $$c(F) = c(E)\,c(G).$$

To define the Chern *character*, we introduce a formal factorization:

$$c(E) = \prod_i (1 + \alpha_i(E)).$$

We avoid the need to define the $\alpha_i(E)$'s themselves by working only with symmetric functions of them: these are all expressible as polynomials in the elementary symmetric functions of the $\alpha_i(E)$'s — that is, as polynomials in the Chern classes $c_i(E)$. The most important example of such a function is the Chern character, which we define as the formal sum

$$\mathrm{ch}(E) = \sum_i e^{\alpha_i(E)}.$$

If E has rank r and we expand each exponential as a formal series in the α_i's and then group terms of like degree, we obtain

$$\mathrm{ch}(E) = \sum_i \left(1 + \alpha_i(E) + \frac{\alpha_i(E)^2}{2!} + \frac{\alpha_i(E)^3}{3!} + \cdots \right)$$

$$= (1 + 1 + \cdots + 1) + (\alpha_1 + \alpha_2 + \cdots + \alpha_r)$$

(3.99) $$+ \frac{(\alpha_1^2 + \alpha_2^2 + \cdots + \alpha_r^2)}{2} + \frac{(\alpha_1^3 + \alpha_2^3 + \cdots + \alpha_r^3)}{6}$$

$$+ \cdots$$

$$= \mathrm{rank}(E) + c_1(E) + \frac{\left(c_1^2(E) - 2c_2(E) \right)}{2} + \cdots.$$

EXERCISE (3.100) The formula for the quadratic term above comes from the formula

$$\sigma_1^2 = n_2 + 2\sigma_2$$

in which σ_i denotes the ith elementary symmetric function and n_i the ith Newton symmetric function (i.e., the sum of the ith powers). Use the analogous formula for σ_3 to derive the cubic term in the expansion of the Chern character above.

The exponentiation in the definition of the Chern character gives it two convenient properties. First, given an exact sequence like (3.97), the identity (3.98) yields

$$\text{ch}(F) = \text{ch}(E) + \text{ch}(G).$$

Second, for any bundles E and F

$$\text{ch}(E \otimes F) = \text{ch}(E)\,\text{ch}(F).$$

Together, these are equivalent to the statement that the Chern character defines a ring homomorphism

$$\text{ch} : K^0(X) \to H^*(X, \mathbb{Q}).$$

This identity can be used to give a characterization of the Chern character avoiding the need to introduce the formal roots α_i.

EXERCISE (3.101) Use the Whitney splitting principle to show that the map that associates to a line bundle L in $K(X)$ the class $e^{c_1(L)}$ in $H^*(X, \mathbb{Q})$ extends to a ring homomorphism. Then show that this homomorphism $K^0(X) \to H^*(X, \mathbb{Q})$ equals the Chern character.

Next, we introduce the *Todd class* $\text{td}(E)$ of E. This is defined in terms of the α_i's by

$$\text{td}(E) = \prod_i \frac{\alpha_i}{1 - e^{-\alpha_i}} \in H^*(X, \mathbb{Q}).$$

As this, like the Chern character, is symmetric in the α_i's, it must be expressible in terms of the Chern classes of E. The expansion

$$\frac{\alpha}{1 - e^{-\alpha}} = 1 + \frac{\alpha}{2} + \frac{\alpha^2}{12} + \cdots$$

yields, after expanding and rewriting the product defining $\text{td}(E)$, the expansion

(3.102) $$\text{td}(E) = 1 + \frac{c_1(E)}{2} + \frac{c_1^2(E) + c_2(E)}{12} + \cdots.$$

Following custom, we write $\text{td}(X)$ for the Todd class of the tangent bundle of X.

EXERCISE (3.103) 1) Show that the Todd class is multiplicative in the sense that, given an exact sequence $0 \to E \to F \to G \to 0$, we have $\mathrm{td}(F) = \mathrm{td}(E) \cdot \mathrm{td}(G)$.

2) Find the degree 3 term in the expression of $\mathrm{td}(E)$ in terms of the Chern classes of E.

Finally, recall that the *shriek* of E by π is given by

$$\pi_!(E) = \sum_i (-1)^i R^i(\pi_*(E)).$$

While the definitions are lengthy, the formula that relates them is beautifully succinct.

THEOREM (3.104) (GROTHENDIECK-RIEMANN-ROCH) *If* $\pi : X \to B$ *is a proper morphism with smooth base B, then*

$$\mathrm{ch}(\pi_!(E)) \cdot \mathrm{td}(B) = \pi_*(\mathrm{ch}(E) \cdot \mathrm{td}(X)).$$

EXERCISE (3.105) 1) Show that, when B is a point, $\pi_!(E) = \chi(E)$ and $\mathrm{td}(B) = 1$ and use this to reduce the formula above to the Hirzebruch-Riemann-Roch formula

$$\chi(E) = (\mathrm{ch}(E) \cdot \mathrm{td}(X))[X],$$

where $[X]$ denotes integration over the fundamental class of X.

2) Show that, if X is a curve, then

$$\chi(E) = c_1(E) + (\mathrm{rank}(E) \cdot (1 - g))$$

and, in particular, if we take E to be a line bundle L of degree d, then

$$\chi(L) = d - g + 1;$$

i.e., we recover the most basic Riemann-Roch formula.

Of course, when B has larger dimension, interpreting the Grothendieck-Riemann-Roch formula isn't so easy. We will see how to do so in practice in the next sections by applying it in three circumstances to the universal family of curves over the moduli space $\mathcal{M}_g$.

Chern classes of the Hodge bundle

Our first example of the use of the Grothendieck-Riemann-Roch formula will be a calculation already referred to in the general discussion of cohomology classes on moduli spaces in Chapter 2: the expression of the Chern classes λ_i of the Hodge bundle in terms of the tautological classes κ_i.

To recall the circumstances, the *Hodge bundle* Λ on $\mathcal{M}_g$ is the bundle whose fiber at a point $[C] \in \mathcal{M}_g$ is the space $H^0(C, K_C)$ of holomorphic differentials on C. More precisely, let $\pi : \mathcal{C}_g \to \mathcal{M}_g$ be the universal curve, $\omega = \omega_{C/\mathcal{M}}$ be the relative dualizing sheaf, and $\gamma = c_1(\omega)$; we define

$$\Lambda = \pi_*(\omega).$$

Note that this bundle exists only away from the locus of curves with automorphisms. This is for the most part not a serious problem: since the locus of curves with automorphisms has codimension $g - 2$, computations involving the Chern classes $c_i(\Lambda)$ with $i < g - 2$ will still be valid. Alternately, the bundle Λ exists on a finite cover of $\mathcal{M}_g$, and we can define the Chern class of Λ (with rational coefficients) to be the pushforward of the class of this bundle, divided by the degree of the cover. We will address these issues more systematically in a moment.

In any event, the Grothendieck-Riemann-Roch formula gives us a simple expression for these classes. To begin with, note that the conormal bundle of the map π — that is, the difference $T_C - T_{\mathcal{M}}$ in $K(\mathcal{C}_g)$ — is simply minus the relative tangent bundle, which is the dual of the relative dualizing sheaf. Hence,

$$\frac{\mathrm{td}(\mathcal{C}_g)}{\pi^* \mathrm{td}(\mathcal{M}_g)} = \mathrm{td}(\omega^\vee) = 1 - \frac{\gamma}{2} + \frac{\gamma^2}{12} - \cdots.$$

Grothendieck-Riemann-Roch thus says that

$$\mathrm{ch}(\pi_! \omega) = \pi_* \left(\left(1 - \frac{\gamma}{2} + \frac{\gamma^2}{12} + \cdots \right) \cdot \left(1 + \gamma + \frac{\gamma^2}{2} + \frac{\gamma^3}{6} + \cdots \right) \right)$$

$$= \pi_* \left(1 + \frac{\gamma}{2} + \frac{\gamma^2}{12} + \cdots \right).$$

To evaluate the left-hand side of this equation, note that the higher direct image $R^1 \pi_* \omega$ is the structure sheaf $\mathcal{O}_{\mathcal{M}_g}$. Thus $\mathrm{ch}(R^1 \pi_* \omega) = 1$ and the higher direct images vanish, giving

$$\mathrm{ch}(\Lambda) - 1 = \pi_* \left(\frac{\gamma}{2} + \frac{\gamma^2}{12} + \cdots \right).$$

To evaluate the right-hand side, note first that γ has degree $2g - 2$ on a fiber of $\mathcal{C}_g$ over $\mathcal{M}_g$; thus

$$\mathrm{rank}(\Lambda) - 1 = g - 1,$$

which of course is no surprise. Next, we have

$$(3.106) \qquad c_1(\Lambda) = \mathrm{ch}_1(\Lambda) = \pi_* \left(\frac{\gamma}{2} \right) = \frac{\kappa}{12},$$

where $\kappa = \kappa_1$ is the first tautological class. Similarly, to find $\mathrm{ch}_2(\Lambda)$ we write

$$(3.107) \qquad c_2(\Lambda) = \frac{\mathrm{ch}_1(\Lambda)^2}{2} - \mathrm{ch}_2(\Lambda) = \frac{\kappa^2}{288}$$

since $\mathrm{ch}_2(\Lambda) = 0$. In general, it's clear that the Grothendieck-Riemann-Roch in this case expresses each of the Chern classes of the Hodge bundle as a polynomial (with rational coefficients) in the tautological classes κ_i, and that the polynomial may be worked out explicitly in any given case. Note in particular that, while the λ_i are polynomials in the κ_i, the above examples already show that the converse is not true.

Next, we consider how this computation — at least in the case of the codimension 1 classes in $\overline{\mathcal{M}}_g$ — may be extended over all of the stable compactification $\overline{\mathcal{M}}_g$. Here we'll see the discussion of Section D used in practice. First of all, to define our terms, we will denote by ω the relative dualizing sheaf of $\overline{\mathcal{C}}_g$ over $\overline{\mathcal{M}}_g$, and call the direct image $\pi_* \omega$ on $\overline{\mathcal{M}}_g$ the Hodge bundle Λ. Note that the problem we were able to gloss over above has now become more serious: the universal curve now fails to be universal over a codimension 1 locus (all the points $[C] \in \Delta_1 \subset \overline{\mathcal{M}}_g$ correspond to curves with automorphisms). But now we have an alternative: by Proposition (3.93), in order to derive or prove any relation among divisor classes on the moduli space *we simply have to verify the corresponding relation among the associated divisor classes on the base B of any family $X \to B$ of stable curves with smooth, one-dimensional base and smooth general fiber.*

To do this, let $\rho : X \to B$ be any such one-parameter family of stable curves. We will use t to denote a local coordinate on the base B of the family. We make one modification: we let $\mu : \mathcal{Y} \to X$ be a minimal resolution of the singularities of the total space X, and let $\nu = \rho \circ \mu : \mathcal{Y} \to B$ be the composition. This has the effect, for each node p of a fiber of $X \to B$ with local coordinates x, y, t satisfying $xy = t^k$, of replacing the point p by a chain of $k - 1$ rational curves. In this way we arrive at a family $\nu : \mathcal{Y} \to B$ of semistable curves, with smooth total space and having k nodes lying over each node of a fiber of X with local equation $xy - t^k$. To relate the invariants of the new family $\nu : \mathcal{Y} \to B$ to those of the original, we have the:

EXERCISE (3.108) 1) Show that the relative dualizing sheaf of the new family is trivial on the exceptional divisor of the map μ, and hence that it's simply the pullback of the relative dualizing sheaf of $\rho : X \to B$, i.e.,

$$\omega_{\mathcal{Y}/B} = \mu^* \omega_{X/B}.$$

2) Deduce that their direct images are equal, that is,

$$\nu_* \omega_{\mathcal{Y}/B} = \rho_* \omega_{X/B}$$

and likewise that

$$\nu_* (c_1(\omega_{\mathcal{Y}/B})^2) = \rho_* (c_1(\omega_{X/B})^2).$$

3) Let $Z \subset \mathcal{Y}$ be the locus of nodes of fibers of $\mathcal{Y} \to B$. Show that

$$\nu_*([Z]) = \delta(\rho).$$

With all this said, pretty much the same calculation is made in this setting, simply replacing $\overline{C}_g$ and $\overline{\mathcal{M}}_g$ by $\mathcal{Y}$ and B respectively. Only one thing has changed: the relative tangent bundle $T_{\mathcal{Y}} - \rho_* T_B$ (viewed as an element of the K-ring $K(\mathcal{Y})$) is no longer the dual of the relative dualizing sheaf. To see what it is, let C be a stable curve with a node p, let t be a coordinate on B near the point $[C]$, and let (x, y) be coordinates on $\mathcal{Y}$ near p in terms of which the map to B is given by $t = xy$. Working with the cotangent bundles rather than tangent bundles, the pullback map gives an injection

$$\pi^* T_B^{\vee} \to T_{\mathcal{Y}}^{\vee}$$

which we may view more concretely as the map

$$\mathcal{O}_{\mathcal{Y}}\langle dt \rangle \to \mathcal{O}_{\mathcal{Y}}\langle dx, dy \rangle$$

sending dt to $x\,dy + y\,dx$. The cokernel is the relative cotangent bundle

$$\Omega_{\mathcal{Y}/B} = \frac{\mathcal{O}_{\mathcal{Y}}\langle dx, dy \rangle}{\langle x\,dy + y\,dx \rangle}.$$

Note that this is locally free of rank 1 everywhere except along the locus Z of nodes of fibers of $\mathcal{Y}$ over B.

Now, the relative dualizing sheaf $\omega = \omega_{\mathcal{Y}/B}$ may be characterized as the unique invertible sheaf whose restriction to the locus $\mathcal{Y} \setminus Z$ of smooth points of fibers of π is isomorphic to the relative cotangent bundle. It follows that we can write

$$\omega = \mathcal{O}_C \langle \alpha \rangle$$

where

$$\alpha = \frac{dx}{x} - \frac{dy}{y}.$$

Note that $x\alpha = 2\,dx$, while $y\alpha = -2\,dy$, so that

$$\Omega := \Omega_{\mathcal{Y}/B} = \mathcal{I}_Z \otimes \omega.$$

We can use this to calculate the Todd class of the relative cotangent bundle Ω in terms of ω. To begin with, to calculate the Chern character of the ideal sheaf of Z, we apply Grothendieck-Riemann-Roch to the inclusion $i : Z \to Y$. We have

$$\text{ch}(i_*\mathcal{O}_Z) = i_* \left(\text{ch}(\mathcal{O}_Z) \cdot \text{td}(T_Z - i^*T_Y) \right) = i_*(\eta)$$

where η denotes the class of the locus Z.

From the exact sequence

$$0 \to \mathcal{I}_Z \to \mathcal{O}_Y \to \mathcal{O}_Z \to 0$$

we have

$$\text{ch}(\mathcal{I}_Z) = 1 - \text{ch}(\mathcal{O}_Z) = 1 - \eta$$

and so finally:

$$\text{ch}(\Omega) = \text{ch}(\omega) \cdot \text{ch}(\mathcal{I}_Z)$$

(3.109)
$$= (1 + \gamma + \gamma^2 + \cdots) \cdot (1 - \eta + \cdots)$$

$$= 1 + \gamma + \left(\frac{\gamma^2}{2} - \eta \right) + \cdots.$$

Thus, $c_1(\Omega) = c_1(\omega) = \gamma$ — no news here, since we are modifying ω only on a codimension two locus. Further,

$$c_2(\Omega) = \frac{1}{2}\text{ch}_1(\Omega)^2 - \text{ch}_2(\Omega) = \eta$$

so that

$$\text{td}(Y/B) = 1 - \frac{\gamma}{2} + \frac{\gamma^2 + \eta}{12} + \cdots.$$

Plugging this into the Grothendieck-Riemann-Roch, we arrive at

$$c_1(\nu_*(\omega_{Y/B})) = \nu_* \left(\frac{c_1(\omega_{Y/B})^2 + \eta}{12} \right)$$

and finally invoking Exercise (3.108) we have the relation among the divisor classes on B associated to the family $\rho : X \to B$

(3.110)
$$\lambda = \frac{\kappa + \delta}{12}.$$

Applying the translation Proposition (3.93) and Corollary (3.95), we arrive at the corresponding formula

$$12\lambda - \kappa = [\Delta_0] + \frac{1}{2}[\Delta_1] + [\Delta_2] + \cdots + [\Delta_{\lfloor \frac{g}{2} \rfloor}] \in \text{Pic}(\overline{\mathcal{M}}_g) \otimes \mathbb{Q}.$$

EXERCISE (3.111) Carry out the calculation for the relation (3.110) above without introducing the resolution Y of the total space X of the original family: that is, describe the relationship between the relative dualizing sheaf and the relative cotangent bundle on a family $X \to B$ of stable curves near a point with local equation $xy - t^k$, and use this to calculate the Todd class of X/B directly.

Chern class of the tangent bundle

For our final example we'll calculate the canonical class of $\overline{\mathcal{M}}_g$ in terms of the standard generators of its Picard group. Before we can even begin to try to compute this class, we have to make sense of its definition. On the smooth open sublocus $\overline{\mathcal{M}}_g^0$, we mean, as usual, the bundle generated by the holomorphic differential forms of top degree $(3g - 3)$. However, this definition doesn't make sense at singular points of $\overline{\mathcal{M}}_g$. We solve this problem by defining "the canonical bundle on $\overline{\mathcal{M}}_g$" to be the unique rational line bundle on $\overline{\mathcal{M}}_g$ extending the canonical bundle on its smooth locus.

Having thus defined the canonical bundle, the computation of its class turns out to be very similar to the previous two computations. The connection is provided by the characterization of the tangent space to the versal deformation space of a nodal curve C. As we saw in (3.30), this is just the global Ext group $\text{Ext}^1(\Omega_C, \mathcal{O}_C)$. Applying duality, then, the cotangent space to the moduli stack at a point C will be the space

$$T^\vee = H^0(C, \Omega_C \otimes \omega_C)$$

of global sections of the tensor product of the dualizing sheaf and the sheaf of differentials.

Accordingly, we'll introduce what we will call the *canonical class of the moduli stack*: this will be the divisor class K on the moduli stack that associates to any family $\rho : X \to B$ of stable curves the class

$$K(\rho) = \rho_*(\Omega_{X/B} \otimes \omega_{X/B}).$$

Again, the phrase "canonical class of the moduli stack" should be treated as atomic: we haven't defined a stack, let alone the canonical class of one. The bundle on the right is simply the bundle that associates to each point $b \in B$ the top exterior power of the cotangent space to the versal deformation space of the fiber X_b.

We will express the class K in terms of the usual generators λ and δ_i of $\text{Pic}_{\text{fun}}(\overline{\mathcal{M}}_g) \otimes \mathbb{Q}$, and then use this to derive an expression for the canonical bundle of the moduli space $\overline{\mathcal{M}}_g$ in terms of the generators λ and $[\Delta_i]$ of $\text{Pic}(\overline{\mathcal{M}}_g) \otimes \mathbb{Q}$.

The actual calculation of the class K is completely straightforward, given what we have already done. We let $\rho : X \to B$ be any family of stable curves with smooth, one-dimensional base and smooth general fiber and apply Grothendieck-Riemann-Roch to find the first Chern class of the direct image $\rho_*(\Omega_{X/B} \otimes \omega_{X/B})$: since the higher direct images are all zero, we have

$$\mathrm{ch}(\rho_*(\Omega_{X/B}\otimes\omega_{X/B}))$$

$$= \pi_*\left(\left(1 + 2\gamma + (2\gamma^2 - \eta)\right)\left(1 - \frac{\gamma}{2} + \frac{\gamma^2 + \eta}{12}\right)\right)$$

$$= \pi_*\left(1 + \frac{3}{2}\gamma + \left(\frac{13}{12}\gamma^2 - \frac{11}{12}\eta\right)\right)$$

$$= (3g - 3) + \left(\frac{13}{12}\kappa_1 - \frac{11}{12}\delta\right)$$

$$= (3g - 3) + (13\lambda - 2\delta).$$

Hence, in particular, the canonical class K of the moduli stack is given by

(3.112) $$\hspace{4cm} K = 13\lambda - 2\delta.$$

How do we use this to get a formula for the canonical class of the moduli space? We're actually pretty close: after all, at any point $[C] \in \overline{\mathcal{M}}_g$ corresponding to a curve without automorphisms, the fiber of the canonical bundle $K_{\overline{\mathcal{M}}_g}$ at $[C]$ is again the top exterior power of the cotangent space to the versal deformation space of the fiber X_b, so the pullback of $K_{\overline{\mathcal{M}}_g}$ to the moduli stack should be just K — or it would be if it were not for the presence of the divisor Δ_1. For a general point $[C] \in \Delta_1$, the versal deformation space of C is a two-sheeted cover of its image in $\overline{\mathcal{M}}_g$, ramified along Δ_1. It follows that

$$\pi^* K_{\overline{\mathcal{M}}_g} = K + \delta_1 \in \mathrm{Pic}_{\mathrm{fun}}(\overline{\mathcal{M}}_g)\otimes\mathbb{Q},$$

that is, we must subtract δ_1 from the canonical class of the moduli stack to obtain the pullback to the moduli stack of the canonical class $K_{\overline{\mathcal{M}}_g}$ of $\overline{\mathcal{M}}_g$. (In the language of stacks, we would say that "the moduli stack is ramified over the moduli space along the locus Δ_1" — a pretty nifty trick, considering that the map π from the moduli stack to the moduli space is finite of degree 1). We have then

$$K - \delta_1 = 13\lambda - 2\delta - \delta_1$$

and so we find, by again applying our dictionary, that on $\overline{\mathcal{M}}_g$ itself

(3.113)
$$K_{\overline{\mathcal{M}}_g} = 13\lambda - 2[\Delta_0] - \frac{3}{2}[\Delta_1] - 2[\Delta_2] - \cdots$$

$$= 13\lambda - 2[\Delta] + \frac{1}{2}[\Delta_1].$$

Porteous' formula

One further tool that is often of use in analyzing the geometry of moduli spaces is Porteous' formula, which expresses the class of the locus where the rank of a map between vector bundles is less than or equal to a given bound. Applications of this formula are already abundant in the theory of a single curve. To cite one example, the Riemann-Roch formula for divisors on a curve C says that a divisor D of degree d moves in a linear series of dimension at least r if and only if the rank of the evaluation map

$$H^0(K_C) \longrightarrow H^0(K_C/K_C(-D))$$

is $d - r$ or less. As D varies, the target and domain spaces of this map give vector bundles over the symmetric product C_d of C, and applying Porteous to the corresponding bundle map we arrive at a formula for the class of the locus in C_d of divisors D such that $r(D) \geq r$. In particular, observing that this class is nonzero (when its codimension is $d - r$ or less) gives the first proof of the existence of special linear series on an arbitrary curve whenever the Brill-Noether number $\rho \geq 0$.

For such reasons, the subject of Porteous' formula and its application to curves is already discussed at reasonable length in [7, Chapters 2 and 8]. We will simply state the formula here, assuming a familiarity with its derivation and the applications to a fixed curve, and concentrating on giving further applications of the formula to the study of the geometry of families of curves.

To state this, we need to recall two additional notations. The first is the Chern polynomial. For a vector bundle E, this is just the formal polynomial

$$c_t(E) = \sum_i c_i(E) t^i.$$

By the Whitney product formula, this extends to a group homomorphism from $K(X)$ (which we write additively) to the multiplicative group of units of the formal power series ring $H^*(X)[[t]]$.

Next, for any formal series $c_t = \sum_i c_i t^i$, any integer a and any positive integer b, we define $M_{a,b}(c_t)$ to be the $b \times b$ matrix whose $(i,j)^{\text{th}}$ entry is c_{a+j-i}. Finally, we set $\Delta_{a,b}(c_t) = \det(M_{a,b}(c_t))$. In these terms, Porteous' formula is:

THEOREM (3.114) (PORTEOUS' FORMULA) *Let $\varphi : E \to F$ be a homomorphism between vector bundles of respective ranks m and n on a smooth variety X. Let*

$$X_k(\varphi) = \{x \in X \mid \text{rank}(\varphi_x) \leq k\}.$$

and let $[X_k(\varphi)]$ be the fundamental class of $X_k(\varphi)$. If $X_k(\varphi)$ is either empty, or of the expected codimension $(m - k)(n - k)$, then

$$[X_k(\varphi)] = \Delta_{n-k,m-k}\left((c_t(F - E))\right)$$

$$= \Delta_{m-k,n-k}\left((c_t(E^\vee - F^\vee))\right)$$

$$= (-1)^{(e-k)(f-k)}\Delta_{m-k,n-k}\left((c_t(E - F))\right).$$

The hyperelliptic locus in $\mathcal{M}_3$

The first problem to which we'll apply Porteous' formula is the following. In the moduli space $\mathcal{M}_3$ of smooth curves of genus 3, let $H = H_3$ denote the locus of hyperelliptic curves. Since H is a closed subvariety of codimension 1, it has a class in $\operatorname{Pic}(\mathcal{M}_3)\otimes\mathbb{Q}$. We ask now what that class is.

This will be another example of how these sorts of computations are most naturally carried out on the moduli stack, rather than the moduli space. The steps by now should be familiar: we'll introduce the divisor class h on the moduli stack associated to the subvariety $H \subset \mathcal{M}_3$ as in Proposition (3.91); we calculate the degree $\deg(h(\pi))$ of h on a one-parameter family $\pi : X \to B$ of curves of genus 3 — that is, the number of hyperelliptic curves in such a family — in terms of the degree $\lambda(\pi)$, and finally use this to deduce a relation between the classes $[H]$ and $\lambda \in \operatorname{Pic}(\overline{\mathcal{M}}_g)$. The part that is new is the middle part, the calculation of the number of hyperelliptic curves in a one-parameter family, which will involve an application of Porteous.

To carry this out, we first need a good characterization of hyperelliptic curves. There are, of course, many: the canonical map isn't an embedding; we have an involution with $2g + 2$ fixed points; there is a degree 2 map to $\mathbb{P}^1$; and so on. The one that is most useful here, however, is the characterization via Weierstrass points: a smooth curve C is hyperelliptic if and only if it contains a point $p \in C$ such that $2p$ fails to impose two independent conditions on the canonical series, i.e., with

$$h^0\left(K_C(-2p)\right) = 2.$$

To globalize this, suppose that $\pi : X \to B$ is any smooth family of curves of genus 3, not all hyperelliptic. We can define two vector bundles on the total space X of the family, as follows. First, we let E be the bundle whose fiber at a point (b, p) (where $b \in B$ and p is a point of X_b) is the space of sections $H^0(X_b, K_{X_b})$. This is just the pullback of the Hodge bundle:

$$E = \pi^*(\pi_*\omega_{X/B});$$

or, if we let $X_2 = X \times_B X$ be the fiber product of X with itself over B and π_i the projection maps to X,

$$E = (\pi_1)_* (\pi_2^* \omega_{X/B}).$$

Next, we take F to be the bundle whose fiber over (b, p) is the space $H^0(K_C/K_C(-2p))$ of differentials in a neighborhood of p in X_b, modulo those vanishing to order 2 at p: that is, if X_2 is as above and $\Delta \subset X_2$ is the diagonal

$$F = (\pi_1)_* \left(\pi_2^* \omega_{X/B} \otimes \mathcal{O}_{X_2}/\mathcal{I}_\Delta^2 \right).$$

We then have a natural evaluation map $\varphi : E \to F$, sending each global holomorphic differential on X_b to its truncated Taylor series at p. Abstractly, this is just the pushforward under π_1 of the restriction map

$$\pi_2^* \omega_{X/B} \longrightarrow \pi_2^* \omega_{X/B} \otimes \mathcal{O}_{X_2}/\mathcal{I}_\Delta^2$$

on X_2. The key point is then that the locus Ω of points $(b, p) \in X$ such that p is a hyperelliptic Weierstrass point of X_b is exactly the locus where the map φ fails to be surjective. Porteous' formula (in a simple case at that) will give us a formula for the class of Ω, and, since the generic hyperelliptic X_b has exactly eight such p's, the class $h(\pi)$ will be simply $(\frac{1}{8})^{\text{th}}$ of the pushforward of this class.

To make the calculation, we need to know the Chern classes of E and F. The class of E we've already calculated: it's the pullback to X of the class

$$\lambda = \frac{\kappa}{12} = \frac{\pi_*(\gamma^2)}{12},$$

where $\gamma = c_1(\omega_{X/B})$ is the first Chern class of the dualizing sheaf. On the other hand, for F we have a two-term filtration

(3.115) $$0 \longrightarrow F_2 \longrightarrow F \longrightarrow F_1 \longrightarrow 0$$

where the fiber of F_1 at (b, p) is $H^0 (K_{X_b}/K_{X_b}(-p))$ and the fiber of F_2 is $H^0 (K_{X_b}(-p)/K_{X_b}(-2p))$. Now, the line bundle F_1 is just the relative dualizing sheaf itself. Similarly, the bundle F_2 is just the square of the relative dualizing sheaf. It follows that

$$c(F) = (1 + \gamma)(1 + 2\gamma) = 1 + 3\gamma + 2\gamma^2.$$

We now apply Porteous' formula (3.114) to conclude that the locus where φ fails to be surjective has class

$$[\Omega] = c_2(E^\vee - F^\vee).$$

(Since here we have $n = m = 2$ and $k = 1$, the matrix M has a single entry, which we've chosen to express using the second of the three forms in (3.114).) Modulo terms of codimension 2 in B, we have

$$c(E^\vee) = 1 - \lambda,$$

and from the above,

$$c(-F^\vee) = (1 - 3\omega + 2\omega^2)^{-1} = 1 + 3\omega + 7\omega^2,$$

so that

$$c(E^\vee - F^\vee) = 1 + (3\omega - \lambda) + (7\omega^2 - 3\omega\lambda).$$

Thus,

$$[\Omega] = 7\omega^2 - 3\omega\lambda$$

and hence

$$\pi_*([\Omega]) = 7\kappa - 12\lambda = 72\lambda.$$

We conclude that

$$h = 9\lambda \in \text{Pic}_{\text{fun}}(\mathcal{M}_3) \otimes \mathbb{Q}$$

and hence by our Basic Proposition (3.93) that the class $[H]$ of the locus H is given by

$$[H] = 18\lambda \in \text{Pic}(\mathcal{M}_3) \otimes \mathbb{Q}.$$

Here are some exercises about hyperelliptic and related loci.

EXERCISE (3.116) The calculations above make the implicit assumption that the scheme Ω — defined as the determinantal scheme associated to the map φ — is reduced. Verify this by writing φ explicitly for a family of curves whose associated arc in $\mathcal{M}_3$ is transverse to B — for example, the stable reduction of the general pencil of quartics specializing to a double conic, as described above.

EXERCISE (3.117) Find the class in $\text{Pic}(\mathcal{M}_3)$ of the locus of curves C with a point p such that $4p \sim K_C$, or equivalently the union of the locus of plane quartics with a hyperflex and the hyperelliptic locus.

EXERCISE (3.118) More generally, for each semigroup S of nonnegative integers having index g (that is, such that $\#(\mathbb{Z}^{\geq 0} \setminus S) = g$), let W_S denote the locus of curves C with genus g possessing a Weierstrass point with semigroup S. For $g = 3$ and 4, determine when this locus is reduced and of the expected codimension, and when it is, calculate its class in $A(\mathcal{M}_3)$. (A discussion of these loci in moduli spaces of curves of general genus g can be found in Section 5.D.)

EXERCISE (3.119) Let $\mathcal{W} \subset \mathcal{M}_g$ be the locus of curves C with a *subcanonical point* — that is, a point p such that $(2g - 2)p \sim K_C$. What is the expected dimension of $\mathcal{W}$? Assuming that $\mathcal{W}$ has this dimension, what is its class?

EXERCISE (3.120) Prove that the locus of curves in $\mathcal{M}_4$ that have only one pencil of degree 3 is a divisor — the general curve of genus 4 has two such pencils — and find the class of this divisor.

Relations amongst standard cohomology classes

Note that a variant of this calculation gives us relations in the co-homology ring of $\mathcal{M}_g$. To see an example in genus 3, consider the restriction map from the pullback E of the Hodge bundle to the bun-dle $F_1 \cong \omega$ in (3.115) above. Since the canonical series of a smooth curve of positive genus never has a base point, this map is surjective, from which it follows that $c_3(E^\vee \otimes \omega) = 0$ in $H^*(\mathcal{C}_3)$. Calculating this out yields

$$0 = c_3(E^\vee \otimes \omega)$$

$$= \gamma^3 + \gamma^2 c_1(E^\vee) + \gamma c_2(E^\vee) + c_3(E^\vee)$$

$$= \gamma^3 - \gamma^2 \cdot \frac{\kappa}{12} + \gamma \cdot \frac{\kappa^2}{288} + \cdots .$$

(We have written κ for the pullback to $\mathcal{C}_3$ of the class $\kappa = \kappa_1$ and, in the final line, omitted the last term, which is the pullback of a class in codimension 3 in $\mathcal{M}_3$). Pushing this forward to $H^4(\mathcal{M}_3)$, we have

$$0 = \kappa_2 - \frac{\kappa^2}{12} + 4\frac{\kappa^2}{288},$$

or, after simplifying,

$$72\kappa_2 = 5\kappa^2 .$$

In general, similar constructions can be applied to give relations among the generators κ_i of $H^*(\mathcal{M}_g)$ for all g. For example, as above we can use the fact that the canonical series is base point free to deduce that $c_g(E^\vee \otimes \omega) = 0$ in $\mathcal{M}_g$, where E is as before the pullback to $\mathcal{C}_g$ of the Hodge bundle. As above, pushing this forward gives us a relation in $H^{2g-2}(\mathcal{M}_g)$. At the same time, it's equally true that no differential on a smooth curve of genus g vanishes to order $2g - 1$ at any point. This says that the bundle map

$$E^\vee \longrightarrow J_{2g-1}$$

is injective, where J_k is the bundle on $\mathcal{C}_g$ whose fiber at a point (C, p) is the space $H^0(C, K_C/K_C(-kp))$. This, by Porteous' formula (3.114), yields a relation in degree g among the classes γ and κ_i in $H^*(\mathcal{C}_g)$, and hence a relation among the classes κ_i in degree $g - 1$ in $H^*(\mathcal{M}_g)$. Although, in view of Looijenga's vanishing theorem [Theorem (2.52)], we might expect all terms in this relation to be 0, Faber's work shows that such apparently useless relations can in fact have important im-plications.

EXERCISE (3.121) Find this relation in $H^2(\mathcal{M}_3)$.

We can generalize this further by defining E^m to be the pullback to $\mathcal{C}_g$ of the bundle of m-canonical differentials and $\mathcal{J}_k^m$ to be the bundle of kth-order jets of such differentials. (The previous example is just the case $m = 1$.) If C is any smooth curve, p is a point of C and $m \geq 2$, then

> the multiple $((m-1)(2g-2) - 1) \cdot p$ imposes independent conditions on $|mK_C|$,

and,

> no m-fold differential vanishes to order $m(2g-2)+1$ at p.

For each m, this yields two more relations among the classes κ_i and y in degrees $g + 1$ in $H^*(\mathcal{C}_g)$, and, correspondingly, relations in degree g in $H^*(\mathcal{M}_g)$. Unfortunately, writing down almost any of these infinitely many relations explicitly (at least by hand) is essentially impossible. If we note, however, that the coefficient of each κ_i in the mth relation is polynomial in m, we do get one simpler relation.

EXERCISE (3.122) Find the leading term in m of the sequence of relations defined above to derive an explicit polynomial of degree $g + 1$ in κ_i and y vanishing in $H^{2g+2}(\mathcal{C}_g)$.

Divisor classes on Hilbert schemes

Throughout this section, we've been applying the Grothendieck-Riemann-Roch and Porteous formulas to derive relations among classes on families of stable curves in the abstract, and thereby relations among classes on the moduli space. We can also apply them to find relations among classes on Hilbert schemes parameterizing curves in projective space.

To set this up, let $\mathcal{H}_0$ be a component of the restricted Hilbert scheme parameterizing curves $C \subset \mathbb{P}^r$ of degree d and genus g, and let $\mathcal{H}$ be the open subset of $\mathcal{H}_0$ parameterizing smooth curves. There are ways of extending these calculations to larger subsets of $\mathcal{H}_0$ — see [33], for example, for analogous computations in case $r = 2$ — but this comes at the expense of greater technical complications, and since we are here trying to indicate simply what calculations are possible, we'll skip this.

We first introduce the basic divisor classes on $\mathcal{H}$. As on page 64, we let $X \subset \mathcal{H} \times \mathbb{P}^r$ be the universal curve; we let $\omega = \omega_{X/\mathcal{H}}$ be the relative dualizing sheaf of X over $\mathcal{H}$, and we let $\mathcal{O}_X(1)$ be the pullback of $\mathcal{O}_{\mathbb{P}^r}(1)$ via the projection $X \to \mathbb{P}^r$. We then denote by η and ξ the Chern classes of the line bundles ω and $\mathcal{O}_X(1)$, and set

$$A = \pi_*(\xi^2), \quad B = \pi_*(\xi \cdot \eta) \quad \text{and} \quad C = \pi_*(\eta^2)$$

where $\pi : X \to \mathcal{H}$ is the projection. Note that by what we've already established, the class C is simply the pullback of the class $\kappa = 12\lambda$ under the induced map $\mathcal{H} \to \mathcal{M}_g$.

In addition to these abstractly defined classes, we have a number of other divisors and line bundles that may be defined in terms of the geometry of the projective curves parametrized by $\mathcal{H}$, and we may ask whether we can describe their classes as linear combinations of the classes A, B and C. We will give these as a series of exercises.

EXERCISE (3.123) For any codimension 2 linear space $\Lambda \cong \mathbb{P}^{r-2} \subset \mathbb{P}^r$, let

$$I = I_\Lambda = \{[C] \in \mathcal{H} : C \cap \Lambda \neq \varnothing\}$$

be the locus of curves meeting Λ. Show that the class of I is simply A.

EXERCISE (3.124) For any hyperplane $H \subset \mathbb{P}^r$, let $T = T_H$ be the locus of curves tangent to H. Show that the class of T is simply B.

EXERCISE (3.125) Let $S \subset \mathcal{H}$ be the locus of curves C that possess a *hyperstall*, that is, a hyperplane having contact of order $r + 2$ or more with C at a point $p \in C$. (In classical terminology, a point $p \in C$ whose osculating hyperplane has contact of order $r + 1$ or more with C at p — what we now call a ramification point — was called a *stall*.) Show that S has pure codimension 1 in $\mathcal{H}$, and find its class.

EXERCISE (3.126) Similarly, let $F \subset \mathcal{H}$ be the locus of curves C that possess a point p whose osculating $(r - 2)$-plane has contact of order r or more with C at p. Show that F has pure codimension 1 in $\mathcal{H}$, and find its class. (You can check that S and F are the only two divisors in $\mathcal{H}$ defined by ramification conditions.)

EXERCISE (3.127) To check the preceding two exercises, let $R \subset X$ be the locus of ramification points. Find its class, and find the branch divisor of the projection $R \rightarrowtail X \to \mathcal{H}$.

EXERCISE (3.128) Now let $r = 3$, and let $B \subset \mathcal{H}$ be the locus of curves possessing a bitangent line. Again, show that B is a divisor, and find its class.

EXERCISE (3.129) Suppose that $nd > 2g - 2$, and that $E_n = \pi_* \mathcal{O}_X(n)$ is the vector bundle on $\mathcal{H}$ whose fiber at each point $[C]$ is the vector space $H^0(C, \mathcal{O}_C(n))$. Find the first Chern class of E_n.

EXERCISE (3.130) Suppose now that $d = g+r > 2g-2$ and let $D \subset \mathcal{H}$ be the locus of degenerate curves, that is, curves C lying in some hyperplane in $\mathbb{P}^r$. Using the result of the preceding exercise, show that D is a divisor, and find its class.

EXERCISE (3.131) Let now $r = 3$, $d = 6$, and $g = 3$. In this case, the locus of curves $C \subset \mathbb{P}^3$ lying on a quadric forms a divisor in $\mathcal{H}$; find its class. (*Hint*: again apply Exercise (3.129).) What does this have to do with the calculation of the class of the hyperelliptic locus in $\mathcal{M}_3$?

EXERCISE (3.132) Suppose now that $d = 2g - 2$ and $r = g - 2$. In this case, the locus of curves that aren't linearly normal (i.e., that are projections of canonical curves from $\mathbb{P}^{g-1}$; or, equivalently, such that $\mathcal{O}_C(1) \cong \omega_C$) form a divisor. Once more, find its class.

Finally, here is a general challenge:

PROBLEM (3.133) Consider two larger open subsets of the Hilbert scheme containing $\mathcal{H}$: the locus of stable curves embedded in $\mathbb{P}^r$, and larger still, the locus of nodal curves. Can you carry out any or all of the above computations on these loci?

Note that to do this you'll have to introduce as well the classes of the divisors corresponding to singular curves, which may be complicated: for example, the loci of curves with two components C_1 and C_2 meeting at a point will have to indexed by both the genera g_1 and $g_2 = g - g_1$ and the degrees d_1 and $d_2 = d - d_1$ of the components. And, even if we start with $d \gg g$, the degrees and genera of some of the components of curves appearing in the boundary may not satisfy any such inequality.

F Test curves: the hyperelliptic locus in $\overline{\mathcal{M}}_3$ begun

We will now complete the calculation of the class of the locus H of hyperelliptic curves in the moduli space of curves of genus 3. By "complete", we mean the following: we let $\overline{H} = \overline{H}_3$ be the closure in $\overline{\mathcal{M}}_3$ of the closed subvariety $H \subset \mathcal{M}_3$ and find the class of $\overline{H}$ in terms of the three generators λ, $[\Delta_0]$ and $[\Delta_1]$ of $\text{Pic}(\overline{\mathcal{M}}_3)\otimes\mathbb{Q}$. Equivalently, if we let $\overline{h}$ denote the divisor class on the moduli stack associated to the subvariety $\overline{H} \subset \overline{\mathcal{M}}_3$ as in Proposition (3.91), our goal is to express $\overline{h}$ as a linear combination of the three generators λ, δ_0 and δ_1 of $\text{Pic}_{\text{fun}}(\overline{\mathcal{M}}_3)\otimes\mathbb{Q}$.

We will refer to $\overline{H}$ as the locus of hyperelliptic stable curves of genus 3. In general, whenever we refer to a subvariety of the moduli space of stable curves characterized by a property normally ascribed to smooth curves (e.g., hyperelliptic curves, trigonal curves, plane quintics of genus 6, etc.) we'll mean the closure in $\overline{\mathcal{M}}_g$ of the locus of smooth curves with this property. Thus we *define* a hyperelliptic stable curve of genus g to be a stable curve that is the limit of smooth hyperelliptic curves.

Warning. $\overline{H}$ is *not* the locus of curves that possess a linear series of degree 2 and dimension 1 because *any* reducible curve has such a series. Just take a line bundle of large degree on one component to get the sections and use a bundle of negative degree on the other to make the total degree come out right.

How do we calculate the class of $\overline{H}$? Trying to to extend the application of Porteous' formula is the most obvious approach. The problem here is that one of the bundles in question — the bundle $F = J_2$ whose fiber at a point (C, p) is the space $H^0(C, K_C/K_C(-2p))$ — cannot be extended to a vector bundle over the nodes of fibers of the family of curves. At a node P, there is an entire one-parameter family of degree 2 divisors supported at P and none of these is singled out. More formally, in a family $\mathcal{X} \to B$, F can be defined as

$$\pi_{1*}\left(\pi_2^*(\omega_{\mathcal{X}/B}) \otimes \left(\mathcal{O}_{\mathcal{X} \times_B \mathcal{X}}/\mathcal{I}_\Delta^2\right)\right).$$

This definition certainly extends, but, near nodes, the diagonal Δ isn't a local complete intersection, and so has no Cartier divisor structure. This means that the direct image by π_1 is only a coherent sheaf and not a vector bundle.

What we would need, therefore, to carry out this strategy is a version of the Porteous formula for coherent sheaves. Such a formula, even with strong restrictions on the sheaves involved, would be extremely useful in many contexts; but no one has, as present, been successful in producing one. Nonetheless, we pose the:

QUESTION (3.134) Is there a Porteous type formula for maps of torsion-free coherent sheaves? That is, given a map $\varphi : E \to F$ of such sheaves on X, can we give the locus

$X_r :=$ closure of $\{p \mid E$ and F are locally free at p and $\mathrm{rank}_p(\varphi) \leq r\}$

a scheme structure, and express its class in terms of the Chern classes of E and F and of local contributions at points where E and F aren't locally free?

Fortunately, there is an alternative, if less direct, approach to the problem. We know there exists a relation expressing the divisor class

$\overline{h}$ on the moduli stack as a linear combination of the generators

$$\overline{h} = a \cdot \lambda + b \cdot \delta_0 + c \cdot \delta_1.$$

Indeed, we know already that the coefficient $a = 9$ and the problem is simply to determine b and c. We will do this by taking explicit one-parameter families of stable curves of genus 3 and evaluating both sides of this equation on them to deduce two independent linear relations among the coefficients a, b and c. This is colloquially referred to as the *method of test curves*.

We start with probably the simplest family of (generically smooth) stable curves of genus 3: a general pencil of plane quartics; that is, the family of curves $\{C_t\}$ given by polynomials $F(X,t) = t_0 G(X) + t_1 H(X)$ for general G and H. We have to describe the degree, on the base $B = \mathbb{P}^1$ of this family, of the various divisor classes $\overline{h}$, λ, δ_0 and δ_1.

Two are easy. Since the pencil is general, it will consist entirely of smooth curves and irreducible curves with one node. Thus, all the curves C_t are stable, no stable reduction is necessary, and the family contains no curves corresponding to points in Δ_1. The family will also contain no hyperelliptic curves although this point is subtler: since we don't know what plane models singular elements of $\overline{H}$ might have, we need to check that none of the irreducible nodal curves in the pencil lies in $\overline{H}$. The birational map from the $\mathbb{P}^{14}$ of plane quartics to $\mathcal{M}_3$ takes the discriminant locus D *onto* Δ. Therefore, the locus in D of curves mapping to $\overline{H}$ is proper and a generic pencil avoids it. The upshot is that

$$\deg(\overline{h}(\rho)) = \deg(\delta_1(\rho)) = 0.$$

Next we want to count the points (p,t) at which C_t is singular. Each of the forms $\partial F/\partial x$, $\partial F/\partial y$ and $\partial F/\partial z$ must vanish at (p,t). Since each is bihomogeneous of type $(3,1)$ in (X,t), what we want is the intersection of three divisors of type $(3,1)$ on $\mathbb{P}^2 \times \mathbb{P}^1$. Letting α be a divisor of type $(1,0)$ and β be one of type $(0,1)$, this is

$$(3\alpha + \beta)^2 = 3(3\alpha)^2\beta = 27\alpha^2\beta = 27.$$

Thus, the discriminant locus has degree 27 and our pencil meets it transversely that many times. It's tempting to conclude directly that the image of the pencil in $\overline{\mathcal{M}}_3$ likewise meets Δ transversely 27 times. This is, in fact, true here but requires a further argument: specifically, we have to invoke Lemma (3.94). Since our pencil is generic, each of its 27 singular elements contains a single node. Moreover, none of the nodes is a base point of the pencil. Hence the total space of the associated family of curves is smooth and we may apply the lemma to conclude that

$$\deg(\delta_0(\rho)) = 27.$$

To find the degree of λ, we observe that if $\pi : \mathcal{X} \to B$ is a family of stable curves, $\omega = \omega_{\mathcal{X}/B}$, and $E = \pi_* \omega$, then $\deg_B(\lambda) = c_1(E)$. For a generic pencil, we can identify E by writing down a spanning set of sections (or equivalently of differentials on all curves in the pencil). On the curve in the pencil corresponding to any finite value of t, with affine equation $f(x, y) = 0$, three differentials are

$$\eta_1 = \frac{dx}{\partial f/\partial y}, \quad \eta_2 = \frac{x\, dx}{\partial f/\partial y} \quad \text{and} \quad \eta_3 = \frac{y\, dx}{\partial f/\partial y}.$$

Homogenizing by setting $x = \frac{X}{Z}$ and $y = \frac{Y}{Z}$, this gives three rational sections of E which are defined and independent except at $t = \infty$. This is even true at the nodes as ω_C is the restriction to C of $\mathcal{O}_{\mathbb{P}^2}(1)$ for all C in the pencil. As $t \to \infty$, each of the η's has a simple zero because the equation $F(t, X, Y, Z)$ involves t linearly. Thus, the differentials $t \cdot \eta_1$, $t \cdot \eta_2$, and $t \cdot \eta_3$, are nonvanishing and independent at infinity. This shows that

$$E = \mathcal{O}_{\mathbb{P}^2}(1) \oplus \mathcal{O}_{\mathbb{P}^2}(1) \oplus \mathcal{O}_{\mathbb{P}^2}(1)$$

and hence that

$$\deg(\lambda(\rho)) = 3.$$

Another approach to finding the degree of λ is to view $\mathcal{X}$ as a divisor of type $(1, 4)$ on $\mathbb{P}^1 \times \mathbb{P}^2$. By adjunction,

$$K_{\mathcal{X}} = K_{\mathbb{P}^1 \times \mathbb{P}^2} \otimes \mathcal{O}(\mathcal{X}) = \mathcal{O}_{\mathcal{X}}(-1, 1).$$

Since $\pi^*(K_B) = \mathcal{O}_{\mathcal{X}}(-2, 0)$, we find that

$$\omega_{\mathcal{X}/B} = K_{\mathcal{X}} \otimes \pi^*(K_B)^{-1} = \mathcal{O}_{\mathcal{X}}(1, 1).$$

Hence, if we denote by α a divisor of class $(1, 0)$ and by β one of class $(0, 1)$, we find that $\deg(\kappa(\rho))$ is the intersection number on $\mathbb{P}^1 \times \mathbb{P}^2$

$$(\alpha + 4\beta)(\alpha + \beta)(\alpha + \beta) = 9.$$

In view of the relation $\kappa + \delta = 12\lambda$ and $\deg(\delta(\rho)) = 27$, we again arrive at $\deg(\lambda(\rho)) = 3$.

Applying our assumed relation of classes

$$\overline{h} = a\lambda + b\delta_0 + c\delta_1$$

to the one-parameter family $\rho : \mathcal{X} \to B$ and taking degrees thus yields the relation of coefficients

$$0 = 3a + 27b + 0c,$$

which, since we know that $a = 9$, tells us that $b = -1$.

EXERCISE (3.135) Verify the calculations of a and b above by restricting to the families $\rho : \mathcal{X} \longrightarrow B$ obtained by

1) taking a pencil of hyperplane sections of a smooth quartic surface in $\mathbb{P}^3$; and

2) taking a pencil of plane quartics containing a double conic but otherwise generic. (This will require a semistable reduction, and is a substantially more delicate calculation. The answer, along with the analogous degrees for other pencils, is given in Exercise (3.166).)

What sort of pencil might we use to extract the final coefficient c? The unfortunate fact is that no pencil of quartics whose singular elements are stable as abstract curves will do: the only such curves containing an elliptic component will consist of an elliptic curve plus a rational component meeting it 3 times. Since none of these three nodes is a disconnecting one, such curves all lie in Δ_0 rather than in Δ_1! What is needed is to carry out the stable reduction of a pencil containing a cuspidal quartic member. This is possible but onerous since, during the entire sequence of blowups, base changes, normalizations and blow downs, we need to do the bookkeeping of the canonical class of $\mathcal{X}$, of the relative dualizing sheaf $\omega_{\mathcal{X}/B}$, and of all the related intersection numbers. Moreover, to treat values of g larger than 3 we'll clearly need other methods. As g increases, writing down generically smooth pencils becomes increasingly difficult and the accessible examples such as families of hyperelliptic or trigonal curves only cover small subvarieties in $\overline{\mathcal{M}}_g$.

The most common approach is to use families consisting entirely of singular curves. We will look at three examples of such families. Since we have uses for these families beyond the consideration of the hyperelliptic locus in $\mathcal{M}_3$, we consider them in all genera greater than 2.

EXAMPLE (3.136) Fix a curve D of genus $(g-1)$ and an elliptic curve E and attach a fixed point p of E to a varying point of D. In other words, the total space $\mathcal{X}$ of our family would be the disjoint union of $D \times D$ and $D \times E$ modulo the identification of the diagonal Δ of $D \times D$ with $D \times \{p\}$ in E as shown in Figure (3.138).

EXAMPLE (3.137) Fix a curve D of genus $(g-1)$ and identify a fixed point p of D with a varying point q of D. This gives a family lying in Δ_0. However, as we established in Section C, the fiber over p itself (that is, where q approaches p) is a copy of D joined by a disconnecting node to a rational curve with a node or "pigtail" and this family therefore meets Δ_1 once. (To see this, begin with $D \times D$ as in the diagram on the left of Figure (3.139), then blowup the point (p, p) obtaining the diagram on the right and finally identify the now disjoint sections $D \times \{p\}$ and Δ to get the bottom diagram.)

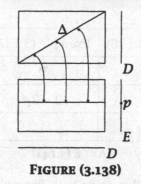

FIGURE (3.138)

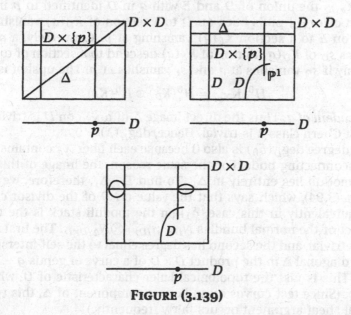

FIGURE (3.139)

EXAMPLE (3.140) Fix a curve D of genus $(g - 1)$ and identify a fixed point p of D with a base point q of a generic pencil of plane cubic curves E to obtain a family of stable curves of genus g over $\mathbb{P}^1$. As the elliptic curves degenerate, we again pick up a special fibers with a "pigtail".

The difficulty with all these families is that we have no way of telling which elements are hyperelliptic (i.e., lie in $\overline{H}_3$). For this, we'll need to introduce the notion of *admissible cover*, which will be the subject of the next section. Before we turn to this subject, we've recorded in Table (3.141) the intersection numbers of each with the standard classes. We haven't listed the degrees of the δ_i's for $i \geq 2$ because these are all clearly 0.

	Example (3.136)	Example (3.137)	Example (3.140)
$\deg(\lambda)$	0	0	1
$\deg(\delta_0)$	0	$2 - 2g$	12
$\deg(\delta_1)$	$4 - 2g$	1	-1

TABLE (3.141)

Let's begin to verify this table with Example (3.136) in which the fiber $X_{\bar{q}}$ is the union of D and E with q in D identified to p in E. A section of $\omega_{X/D}$ pulls back on D to a section of $K_D(q)$ vanishing at q and on E to a section $K_E(q)$ vanishing at p. Conversely, a pair of sections s_D of $K_D(q)$ and s_E of $K_E(q)$ descend to a section of $\omega_{X/D}$ if and only if s_D vanishes at q and s_E vanishes at p. The upshot is that

$$H^0(K_{X_q}) = H^0(K_D) \oplus H^0(K_E)$$

independent of q. Thus the direct image Λ of $\omega_{X/D}$ on D is trivial and its first Chern class λ is trivial. Hence, $\deg_D(\lambda) = 0$.

The degree $\deg_D(\delta_0)$ is also 0 because each fiber X_q contains a single disconnecting node. For the same reason, the image of this family in moduli lies entirely in Δ_1. To find $D \cdot \Delta_1$, therefore, we apply Lemma (3.94), which says that the value on D of the divisor class δ (or, equivalently in this case, δ_1) on the moduli stack is the tensor product of the normal bundles $N_{D \times \{p\}/D \times E} \otimes N_{\Delta/D \times D}$. The first factor here is trivial, and the second has degree equal to the self-intersection of the diagonal Δ in the product $D \times D$ of a curve of genus $g - 1$ with itself. This is just the topological Euler characteristic of D, which is $4 - 2g$. (Since test curves often lie in a component of Δ, this type of normal sheaf argument occurs fairly frequently.)

Example (3.137) illustrates this in a somewhat dual manner. Only the fiber X_p contains a disconnecting node and since the surface X is smooth at this point it follows that $\deg_D(\delta_1) = 1$. However the image of this family in moduli lies entirely in Δ_0, so we again need to compute the restriction to D of the normal bundle to Δ_0 in $\overline{\mathcal{M}}_g$ to evaluate $\deg_D(\delta_0)$. Here this bundle is the tensor product of the normal bundles to the proper transforms of Δ and of $D \times \{p\}$ on the blowup of $D \times D$ at (p, p). On $D \times D$, $\Delta^2 = 4 - 2g$ and $(D \times \{p\})^2 = 0$. Since each curve passes through (p, p), each self-intersection drops by one when we blow up, yielding $\deg_D(\delta_0) = 2 - 2g$.

To calculate $\deg_D(\lambda)$, we use the exact sequence on X_q

$$0 \longrightarrow H^0(K_D) \longrightarrow H^0(\omega_{X_q}) \xrightarrow{\text{res}_p} \mathbb{C} \longrightarrow 0.$$

The corresponding sequence of direct images is

$$0 \longrightarrow H^0(K_D)\otimes\mathcal{O} \longrightarrow \pi_*(\omega_{X/D}) \longrightarrow \mathcal{O} \longrightarrow 0$$

from which it's immediate that the first Chern class of $\pi_*(\omega_{X/D})$ is trivial and, hence, that $\deg_D(\lambda) = 0$.

Finally, we come to Example (3.140). Since the elliptic pencil here is general, there will be twelve singular elements, each a rational nodal curve or "pigtail". Moreover, the base point of the pencil will be smooth on all the elements of the pencil so the total space of the family is smooth. By Lemma (3.94), each intersection of the family with Δ_0 will be transverse yielding $\deg_D(\delta_0) = 12$.

EXERCISE (3.142) 1) Use a normal bundle argument to show that, in Example (3.140), $\deg_D(\delta_1)$ equals the self-intersection, on the rational elliptic surface associated to the pencil, of the section σ_q corresponding to the base point q and conclude that $\deg_{\mathbb{P}^1}(\delta_1) = -1$.
2) Verify that $\deg_{\mathbb{P}^1}(\lambda) = 1$.

EXERCISE (3.143) The case $g = 2$ is special: we have $\text{Pic}(\mathcal{M}_2)\otimes\mathbb{Q} = 0$ (prove this!), and so the class $\lambda \in \text{Pic}_{\text{fun}}(\overline{\mathcal{M}}_2)\otimes\mathbb{Q}$ must be expressible as a linear combination of the boundary classes δ_0 and δ_1. Use Table (3.141) to show that this relation is

$$\lambda = \frac{1}{10}\delta_0 + \frac{1}{5}\delta_1.$$

G Admissible covers

In order to complete the calculations begun in the preceding section (and for many other reasons as well), we have to face the question we ducked earlier: which stable curves are hyperelliptic? Or, more generally, which stable curves are limits of smooth d-gonal curves?

We will determine the answer by constructing a nice compactification of the Hurwitz scheme. Recall from Section 1.G that the classical Hurwitz scheme $\mathcal{H}_{d,g}$ is the scheme parameterizing pairs (C, π), where C is a smooth curve of genus g and $\pi : C \to \mathbb{P}^1$ a branched cover of $\mathbb{P}^1$ of degree d, branched over $b = 2d + 2g - 2$ distinct points $p_1, \ldots, p_b$. The scheme $\mathcal{H}_{d,g}$ maps naturally to the moduli space $\mathcal{M}_g$ of smooth curves of genus g, and its image contains an open subset of the locus in $\mathcal{M}_g$ of d-gonal curves. What we need is a compactification $\overline{\mathcal{H}}_{d,g}$ of $\mathcal{H}_{d,g}$ with the following somewhat informally expressed list of properties:

DESIRED PROPERTIES (3.144)   1) *The space $\overline{\mathcal{H}}_{d,g}$ should be relatively accessible. For example, its singularities should be reasonable; it should be possible to describe the components of the boundary $\overline{\mathcal{H}}_{d,g} - \mathcal{H}_{d,g}$; and so on.*

2) *It should be modular, i.e., its points should actually correspond to some sort of geometric object, preferably a branched cover.*

3) *It should admit a map to the moduli space $\overline{\mathcal{M}}_g$ of stable curves extending the map of $\mathcal{H}_{d,g}$ to $\mathcal{M}_g$, i.e., we should have a diagram.*

(3.145)

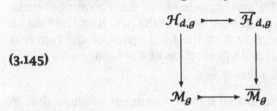

At first glance, this seems not unreasonable. Clearly, we have to understand what happens when branch points of a branched cover $\eta : X \to \mathbb{P}^1$ come together. For example, suppose we have a family of branched covers $\{C_t \to \mathbb{P}^1\}$ — that is, a family $\mathcal{C}^* \to \Delta^*$ of smooth curves over the punctured disc, with a map $\pi : \mathcal{C}^* \to \mathbb{P}^1$ such that $\pi|_{C_t}$ is simply branched at b points $p_1(t), \ldots, p_b(t)$ — we can assume after base change that these are single valued — and suppose that two of the branch points p_i and p_j both approach the point p as $t \to 0$.

What happens in this situation depends on how the simple transpositions that express the monodromy of the fibers around p_i and p_j are related. There are three possibilities indicated in Figure (3.146). The two transpositions could be equal, they could be noncommuting transpositions (that is, overlapping but distinct), or they could be disjoint transpositions. In the first case, we get in the limit a curve C_0 with a simple node. (Note that this curve need not be stable: it may be reducible with one component rational. This is, in fact, what happens if the stable limit of the curves C_t is smooth, but the pencil giving the map $C_t \to \mathbb{P}^1$ acquires a base point at $p_i(0) = p_j(0)$.) In the other two cases, we see no visible degeneration as the curve C_t approaches C_0 although the covering $C_t \to \mathbb{P}^1$ does degenerate. If the monodromies overlap, this cover has one triple ramification point instead of two simple ones; if they are disjoint, we see two simple ramification points aligned above one another in a single fiber of the covering. In all three cases, we can fill in the family $\mathcal{C}^*$ to a family $\mathcal{C} \to \Delta$ having only nodal fibers, with map $\pi : \mathcal{C} \to \mathbb{P}^1$ extending the given map π on $\mathcal{C}^*$ and expressing the fiber C_0 again as a d-sheeted branched cover of $\mathbb{P}^1$.

The problems start when more branch points coincide. What will the limiting branched cover look like when three or more branch points p_i

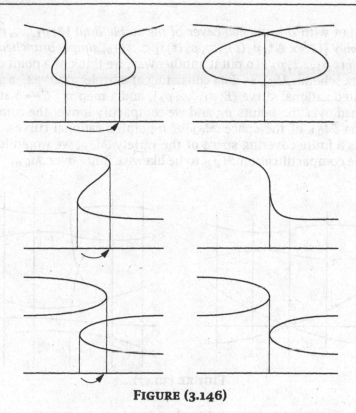

FIGURE (3.146)

approach a common limit p? Making sketches like those above of a few
examples will convince you that even in cases where you can complete
the family by throwing in a singular curve C_0 expressed as a branched
cover of $\mathbb{P}^1$, the limiting singularity will depend on the relative rates
of approach of the points $p_i(t)$ to p. This makes it unclear what the
space of such covers will look like. Moreover, since the singularities
of the special fiber X_0 can become extremely complicated, there is no
direct method for determining from the covering data what the *stable*
limit of the family of curves X_t will be. This in turn prevents us from
describing concretely the closure in $\overline{\mathcal{M}}_g$ of the image of $\mathcal{H}_{d,g}$.

What is the solution to these problems? The idea, which goes back
to Knudsen [97], is simple: *we never allow the branch points to coincide.* This may at first sound outlandish, but in fact we've already
seen other examples of how this can be carried out: the moduli space
$\overline{\mathcal{M}}_{g,n}$ of stable n-pointed curves $(C; p_1, \ldots, p_b)$ of genus g is compact, even though the points p_i are never allowed to come together.
In fact, we adopt exactly the same strategy here. The target space of
a simple branched cover will be a $\mathbb{P}^1$ with b marked points. Given a
family of branched covers $\mathcal{C}^* \to \Delta^*$ with maps $C_t \to \mathbb{P}^1$ as above, we
can try to fill in the family not with a possibly nastily branched cover

of $\mathbb{P}^1$, but with *a branched cover of the stable limit* $(B; p_1, \ldots, p_b)$ *of the family* $\{(\mathbb{P}^1 \times \Delta^*; p_1(t), \ldots, p_b(t))\} \subset \overline{\mathcal{M}}_{0,b}$, *simply branched over the points* $p_1, \ldots, p_b$. To put it another way, we think of a point of the Hurwitz scheme $\mathcal{H}_{d,g}$ as data consisting of a triple: a curve C, a stable b-pointed rational curve $(B; p_1, \ldots, p_b)$, and a map $\pi : C \to B$ simply branched over the points p_i; and we compactify it over the compactification $\overline{\mathcal{M}}_{0,b}$ of the space $\mathcal{M}_{0,b}$ of b-pointed rational curves. Since $\mathcal{H}_{d,g}$ is a finite covering space of the variety $\mathcal{M}_{0,n}$, we would ideally like the compactification $\overline{\mathcal{H}}_{d,g}$ to be likewise finite over $\overline{\mathcal{M}}_{0,n}$.

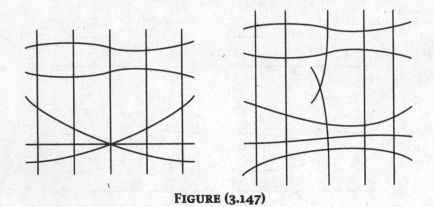

FIGURE (3.147)

Can we in fact implement this idea? The key question to answer is: what happens to a cover $\pi : C \to \mathbb{P}^1$ branched over b distinct points $p_i(t)$ as the smooth b-pointed curve $(\mathbb{P}^1; p_1(t), \ldots, p_b(t))$ approaches a general stable b-pointed rational curve $(B; p_1, \ldots, p_b)$? Happily, it isn't impossible to visualize this. To start, consider the simplest case in which some subset of the points $p_i(t)$ come together at comparable speed — i.e., the family of b-pointed curves is as shown on the left side of Figure (3.147). In this case, we can simply blow up once at the point where the $p_i(t)$ meet to obtain the stable limit B shown on the right of the figure. The curve B is simply a union of two copies of $\mathbb{P}^1$ meeting at one point, with the limits of those $p_i(t)$ that remained distinct on one component and the limits of those that came together on the other.

To see what is going on topologically, we may draw the general fiber, the fibers near $t = 0$, and, the special fiber as shown at the top, middle and bottom, respectively of Figure (3.148).

Suppose now we have a family of branched covers $C_t \to \mathbb{P}^1$ of the general fibers of this family. Can we fill it in with a branched cover $C_0 \to B$ of the special fiber? The answer is yes: just as the node of the special fiber B of the family of stable b-pointed rational curves is created by contracting the loop around the neck of the sphere in the middle picture above, so the inverse image of that loop in the

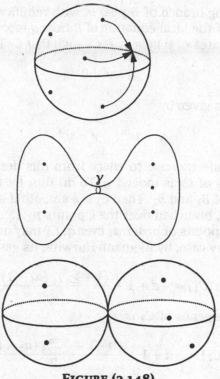

FIGURE (3.148)

covers C_t will consist of a collection of disjoint loops, each of which can likewise be contracted to form a node of C_0.

In fact, we can describe C_0 explicitly. Suppose q is the node of B. Then the inverse image $\pi^{-1}(B \setminus \{q\})$ will be smooth and simply branched over the points $p_i \in B$. This leaves only the question: what goes on at the points of $\pi^{-1}(q)$?

To answer this question, we need to fix a bit of notation. Suppose (after reordering if necessary) that the points $\{p_1(t), \ldots, p_k(t)\}$ remain distinct and that B_1 is the component of B on which their limits lie while the points $\{p_{k+1}(t), \ldots, p_b(t)\}$ come together and have limits on the other component B_2 of B. Let $\sigma_i \in \mathfrak{S}_d$ be the monodromy action associated to the branch point $p_i(t)$ on the nearby smooth fiber with respect to a base point on the "collar" γ shown in the middle of Figure (3.148). The monodromy around the loop γ is then the product

$$\sigma = \sigma_1 \cdot \ldots \cdot \sigma_k = (\sigma_{k+1} \cdot \ldots \cdot \sigma_b)^{-1},$$

and the connected components $\gamma_1, \ldots, \gamma_m$ of $\pi^{-1}(\gamma)$ are loops corresponding bijectively to the cycles of the permutation σ. If γ_i corresponds to an a_i-cycle in σ, then it'll be a cover of degree a_i of γ. Thus, each branch of the corresponding node r_i of C_0 will be ramified over the corresponding branch of B near q, with ramification of order a_i. In other words, if the local equation of B near q is $xy = 0$, then there are local coordinates u, v near $r_i \in C_0$ such that C_0 is given locally as

$$u \cdot v = 0$$

and the map π is given by

$$x = u^{a_i}; \quad y = v^{a_i}.$$

It's a worthwhile exercise to check from this description that the arithmetic genus of C_0 is indeed g. To do this, let C_1 and C_2 be the inverse images of B_1 and B_2. Then C_1 is a smooth d-sheeted branched cover of $B_1 \cong \mathbb{P}^1$, branched over the k points $p_1, \ldots, p_k$ and also having ramification points of order a_i over q. C_1 may or may not be connected, but in any case, by Riemann-Hurwitz, its genus is

$$g(C_1) = -d + 1 + \frac{(k + \sum_{i=1}^{m} (a_i - 1))}{2}$$

and similarly the genus of C_2 is

$$g(C_2) = -d + 1 + \frac{(b - k + \sum_{i=1}^{m} (a_i - 1))}{2}.$$

Finally, C_1 and C_2 meet at $d - \sum_{i=1}^{m} (a_i - 1)$ points, so that the genus of C is

$$g(C) = g(C_1) + g(C_2) + d - \sum_{i=1}^{m} (a_i - 1) - 1$$

$$= -d + 1 + \frac{b}{2}$$

$$= g.$$

Note that this could also serve to confirm that the singularities of C_0 are nodes.

The generalization of this description to the case of an arbitrary stable b-pointed curve B is immediate and yields the:

DEFINITION (3.149) *Let $(B; p_1, \ldots, p_b)$ be a stable, b-pointed curve of genus 0 and let $q_1, \ldots, q_k$ the nodes of the curve B. By an* admissible cover *of the curve B we'll mean a nodal curve C and regular map $\pi : C \longrightarrow B$ such that:*

1) $\pi^{-1}(B_{ns}) = C_{ns}$, *and the restriction of the map π to this open set is simply branched over the points p_i and otherwise unramified; and,*

2) $\pi^{-1}(B_{\text{sing}}) = C_{\text{sing}}$, and for every node q of B and every node r of C lying over it, the two branches of C near r map to the branches of B near q with the same ramification index. Equivalently, we can find local analytic coordinates x, y on B and u, v on C such that for some m,

$$xy = 0, \quad uv = 0, \quad \pi^* x = u^m, \quad \text{and} \quad \pi^* y = v^m.$$

The definition of a family of admissible covers of a family of stable b-pointed curves of genus 0 is analogous.

The main theorem about admissible covers says simply that they have all the desirable properties we hoped for at the start of this section:

THEOREM (3.150) (EXISTENCE OF $\overline{\mathcal{H}}_{d,g}$ [82]) *There exists a coarse moduli space $\overline{\mathcal{H}}_{d,g}$ for admissible covers which satisfies the conditions of (3.144).*

By definition, $\overline{\mathcal{H}}_{d,g}$ is modular. Its local geometry is also accessible: an analytic neighborhood of the point in $\overline{\mathcal{H}}_{d,g}$ corresponding to a degree d cover $\pi : C \rightarrow (B, p_1, \ldots, p_b)$ with C of genus G is a quotient of a $(2d + 2g - 5)$-dimensional polydisc by the (finite) automorphism group of the cover π. Likewise, it's straightforward to describe the boundary $\Delta = \overline{\mathcal{H}}_{d,g} - \mathcal{H}_{d,g}$: for example, the locus Δ_δ of covers $(C; (B, p_1, \ldots, p_b), \pi)$ in which B has $\delta + 1$ irreducible components (or, equivalently, δ nodes) has pure codimension δ in $\overline{\mathcal{H}}_{d,g}$ and lies in the closure of $\Delta_{\delta-1}$.

EXERCISE (3.151) How many irreducible components does the boundary Δ have?

What isn't so straightforward is the existence of the mapping to $\overline{\mathcal{M}}_g$ in (3.145) and we won't give a proof of the main theorem here. Instead, following our usual practice, we will show, in a few settings that arise commonly in applications, how to describe the admissible covers that arise as limits of families of smooth branched covers.

To begin with, suppose that $\pi : C_t \rightarrow \mathbb{P}^1$ is a family of branched covers with π_t branched over the points $p_1(t), \ldots, p_b(t)$. Consider the three cases illustrated in Figure (3.146) in which two of the branch points — say p_1 and p_2 — come together. We know, of course, what the limit of the family of stable b-pointed rational curves $\{(\mathbb{P}^1, p_1(t), \ldots, p_b(t))\}$ is: it's the curve of genus 0 in Figure (3.152) having two components B_1 and B_2, with p_1 and p_2 on B_2 and the rest (not shown in the figure) on B_1.

Suppose first that the transpositions giving the monodromies σ_1 and σ_2 around the two points p_1 and p_2 in the cover $C_t \rightarrow \mathbb{P}^1$ are

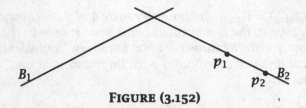

FIGURE (3.152)

equal. In this case the monodromy around the neck of the barbell
pictured in Figure (3.148) is trivial, so that the limiting admissible
cover $C_0 \to B$ will be unramified over the node q of B. Moreover, since
the group generated by σ_1 and σ_2 has one orbit of order 2 and the
rest of order 1, the inverse image of B_2 in C_0 will consist of $d - 1$
components. Of these, $d - 2$ will map one-to-one to B_2 and one will
be the degree 2 of B_2 branched over p_1 and p_2. (Note that all these
components must be rational.) We may thus draw the picture of the
cover $C_0 \to B$ schematically as in Figure (3.153).

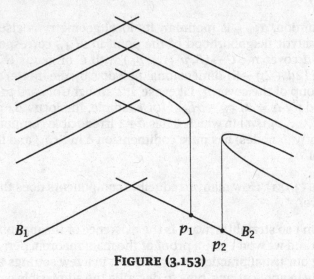

FIGURE (3.153)

In this case, the stable models $\overline{C}_0$ of the admissible cover C_0 display
a surprising variety. The "new" side of the cover (B_2 above) consists
of $(d - 2)$ rational curves meeting the "main" side (over B_1) in a single
node over the node of the base and one rational curve meeting the
main side in two nodes. In the stable model, all of the former curves
contract away completely and the latter contracts to a node r. It's
tempting to conclude that the stable model lies in Δ_0. This, in fact,
is what happens "most" of the time; but, depending on the combinatorics of the branch points that remain on the main side of the cover,
the stable model of $\overline{C}_0$ can lie in *any* of the Δ_i's or even outside the
boundary.

The key fact to recall is that the connected components of the curve C_t correspond bijectively to the orbits in a general fiber of C_t of the subgroup of the symmetric group generated by the simple transpositions associated to the monodromy around the full set of branch points $p_1, \ldots, p_b$. Since the general curve C_t is connected, this subgroup must be transitive. However, the transpositions associated to the branch points $p_3, \ldots, p_b$ that remain on the stable model need no longer generate a transitive subgroup: the group they do generate will, in some cases, have two orbits (which were formerly connected by $\sigma_1 = \sigma_2$). Correspondingly, the stable model will have two components meeting at the node r. Depending on the number of ramification points lying on each component and on the degrees with which each covers B_1, the genera of these components need only satisfy the restriction that their sum is g. If the smaller of the two genera is $i > 0$, then the stable model will lie in Δ_i. If the smaller genus is 0, this rational component will again contract away in the stable model leaving a smooth curve of genus g that covers the line with degree less than d.

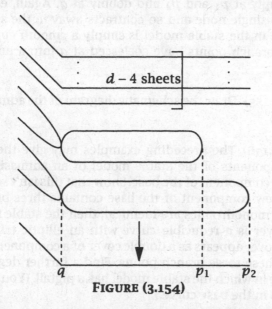

$d - 4$ sheets

q p_1 p_2

FIGURE (3.154)

When two branch points come together, there are also two other combinatorial possibilities for these monodromies. The analysis of these confirms our earlier conclusions, based on Figure (3.148), that in these cases the abstract curve C_t undergoes no visible degeneration as $t \to 0$.

One possibility, shown as the third in Figure (3.148), is that the monodromies around the two branch points p_1 and p_2 are given by a pair of disjoint transpositions. (If the degree of the covering is large, we expect this to happen most of the time.) In this case, the schematic picture of the admissible cover over the new component of the base (which we've shown meeting the main component of the base at a point q) must be as shown in Figure (3.154). Indeed, the sheets that ramify over each of p_1 and p_2 and no others must ramify over the node q since this monodromy is what occurs at the neck of the "barbell". Thus over the new component we have $(d - 2)$ rational curves each meeting the main part of the cover in a single node. In the stable model, these all contract away, leaving us with a smooth (!) curve in which two simple branch points on unrelated sheets happen to have become aligned over q.

The final possibility, represented by the middle picture in Figure (3.148), is that the transpositions corresponding to the monodromies around p_1 and p_2 have one index in common: their product is then a three-cycle that gives the monodromy around the point q. We thus have $(d - 3)$ rational components simply covering the new component of the base and one rational component that is a triple cover ramified simply at p_1 and p_2 and doubly at q. Again, each of these curves has a single node and so contracts away in the stable model. What we see in the stable model is simply a smooth curve in which two simple branch points have coalesced at q into a double branch point.

EXERCISE (3.155) Draw the schematic diagram of the admissible cover in this case.

EXERCISE (3.156) The preceding examples may give the impression that all components of the stable model of an admissible cover lie over a single component of the base. Show that this isn't so by showing that if the new component of the base contains three branch points all of whose monodromies are identical, then the stable model of the resulting cover is a reducible curve with an elliptic tail that in the admissible cover appears as a double cover of a component of the base containing these three branch points. Find a further degeneration of this example in which the stable model has a pigtail. (You'll need three components in the base curve.)

EXERCISE (3.157) List the distinct combinatorial possibilities for an admissible cover whose base curve has two components, one of which contains exactly three simple branch points.

As we indicated at the outset of this section, one of the main goals of the theory of admissible covers is to answer the question, "which stable curves lie in the closure of the locus of smooth k-gonal curves?". We will indicate now how this goes. To begin with, some terminology:

DEFINITION (3.158) *Let C be a stable curve. We say that a nodal curve C' is stably equivalent to C if C is obtained from C' by contracting to a point all smooth rational components of C' meeting the other components of C' in only one or two points (that is, containing only one or two nodes of C').*

DEFINITION (3.159) *Let C be a stable curve. We say that C is k-gonal (resp., hyperelliptic, trigonal) if it's a limit of smooth k-gonal (resp., hyperelliptic, trigonal) curves; that is, if [C] lies in the closure of the locus of smooth curves with a g_k^1 (resp., g_2^1, g_3^1).*

Now, since the moduli space of admissible covers is projective, we have the:

THEOREM (3.160) *A stable curve C is k-gonal if and only if there exists a k-sheeted admissible cover C' → B of a stable pointed curve of genus 0 whose domain C' is stably equivalent to C.*

With this theorem, we can answer completely the question of which stable curves are k-gonal. While the combinatorics can get complicated in general, we can usually arrive at a relatively simple answer if we restrict either to strata of low codimension in the boundary of $\overline{\mathcal{M}}_g$ or to subvarieties of low codimension in the closure of the locus of smooth k-gonal curves. The following exercises give some examples.

EXERCISE (3.161) Let C be a stable curve consisting of two smooth irreducible components C_1 and C_2 meeting at a single point p. Show that C is hyperelliptic if and only if $h^0(C_1, \mathcal{O}_{C_i}(2p)) = 2$ for $i = 1$ and 2; that is (in case both C_1 and C_2 have genus at least 2), if and only if each is hyperelliptic and if p is a Weierstrass point of each.

EXERCISE (3.162) Now let C be a stable curve consisting of two smooth irreducible components C_1 and C_2 meeting at two points p and q. Show that C is hyperelliptic if and only if $h^0(C_i, \mathcal{O}_{C_i}(p+q)) = 2$ for $i = 1$ and 2. That is (in case both C_1 and C_2 have genus at least 2), if and only if each is hyperelliptic and if the pair p and q is conjugate under the corresponding hyperelliptic involution. Note in particular that every stable curve of genus 3 consisting of two elliptic curves meeting at two points is hyperelliptic.

EXERCISE (3.163) Let $H \subset \mathcal{M}_g$ be the locus of hyperelliptic curves, and $\overline{H}$ its closure in $\overline{\mathcal{M}}_g$. Find the irreducible components of $\overline{H} \setminus H$ having codimension 1 in $\overline{H}$. Are these all the irreducible components of $\overline{H} \setminus H$?

EXERCISE (3.164) Which stable curves of the following types are trigonal?

1) Curves C consisting of two smooth irreducible components C_1 and C_2 meeting at a single point p.

2) Curves C consisting consisting of two smooth irreducible components C_1 and C_2 meeting at a two points p and q.

A final remark: since, in the definition of admissible cover, we never allow two of the smooth branch points to come together, or any one of them to approach a singular point, the definition and construction work just as well to compactify the space of covers of a curve B of arbitrary genus, with arbitrarily specified branching over b distinct points $p_1, \ldots, p_b$. Precisely, for any degree d, pair of genera h and g, and collection $\sigma_1, \ldots, \sigma_b$ of conjugacy classes in the symmetric group $\mathfrak{S}_d$, we define a *pseudo-admissible cover with branching* $(\sigma_1, \ldots, \sigma_b)$ to be a stable b-pointed curve $(B; p_1, \ldots, p_b)$, a nodal curve C, and a regular map $\pi : C \to B$ such that:

1) $\pi^{-1}(B_{\mathrm{ns}}) = C_{\mathrm{ns}}$; the monodromy of the map π around the point p_i is in the conjugacy class σ_i over the points p_i; and π is unramified over $C_{\mathrm{ns}} \setminus \{p_1, \ldots, p_b\}$; and

2) $\pi^{-1}(B_{\mathrm{sing}}) = C_{\mathrm{sing}}$, and over a neighborhood of each node q of B the map π is as described in the original definition (3.149) of admissible cover

The same argument as may be given for Theorem (3.150) shows simultaneously that there exists a coarse moduli space $\mathcal{H}_{d,g}$ for pseudo-admissible covers with given branching, and that this moduli space satisfies the conditions of (3.144). An example of the use of these covers (and their moduli) will be given in the proof of Diaz' theorem in Section 6.B.

H   The hyperelliptic locus in $\overline{\mathcal{M}}_3$ completed

To conclude this chapter, let's return to the problem of finding the class of the divisor $\overline{H}$ in the moduli space $\overline{\mathcal{M}}_g$ (or, equivalently, the class of the associated divisor class $\overline{h} \in \mathrm{Pic}_{\mathrm{fun}}(\overline{\mathcal{M}}_3) \otimes \mathbb{Q}$) and see how the description via admissible covers of the closure of the locus of hyperelliptic curves in $\mathcal{M}_3$ helps us to complete the calculation.

Consider the family in Exercise (3.136) obtained by attaching a fixed (general) elliptic curve E to a fixed (general) curve D of genus 2 at a variable point $q \in D$. We would like to know the degree of the divisor

$\overline{h}$ on this family, and clearly the place to start is to identify which fibers of this family belong to the closure $\overline{H}$ of $H \subset \mathcal{M}_3$.

The theory of admissible covers answers this readily. Let $C_q = D \cup E/p \sim q$ be the fiber of our family over $q \in D$. By what we've said, $[C_q]$ will lie in $\overline{H}$ if and only if there exists an admissible cover $\pi : C \to B$ whose domain curve C is a nodal curve stably equivalent to C_q; and we ask when this is the case. To answer this, observe that if C is such a curve, C will have components isomorphic to D and E, and that the map π will necessarily have degree 2 on each of them. They will thus comprise the inverse images of two components B_1 and B_2 of B. Next, note that if either D or E is unramified over a node r of B, then in the stable model of C the two points of D or E lying over r will be identified either to each other or to points of E or D respectively — in either case a contradiction. It follows in particular that the points $p \in E$ and $q \in D$ are ramification points of the map π. Of course, this imposes no restriction on $p \in E$ — it simply says that the map π restricted to E is the map associated to the linear series $|2p|$ — but it does imply that q *must be a Weierstrass point of* D. Conversely, if q is a Weierstrass point of D, the two maps of D and E to $\mathbb{P}^1$ associated to the pencils $|2q|$ and $|2p|$ together give an admissible cover $C_q \to \mathbb{P}^1 \cup \mathbb{P}^1$. We conclude, therefore, that the point $[C_q]$ will lie in the closure of H if and only if q is a Weierstrass point of D.

The next issue is whether the intersection of our curve $D \subset \overline{\mathcal{M}}_g$ with the divisor $\overline{H}$ is transverse or not. We can deal with this in two ways. One is to answer it directly, as follows. Since the curves of the form $\{C_q\}_{q \in D}$ give a fibration of an open subset of the boundary $\Delta_1 \subset \overline{\mathcal{M}}_3$, the intersection of a general such curve with $\overline{H}$ will be transverse if and only if the intersection of $\overline{H}$ with Δ_1 is. But the divisor $\overline{H}$ will restrict to a multiple divisor on Δ_1 only if the tangent space to $\overline{H}$ at a general point of $\overline{H} \cap \Delta_1$ is equal to the tangent space to Δ_1 — in other words, if there are no families of admissible covers $C_t \to B$ with $[C_0] \in \Delta_1$ and total space $\mathcal{C}$ smooth. However, we can write down such a family readily. To start, take one of the admissible covers $C_q \to \mathbb{P}^1 \cup \mathbb{P}^1$ described above, let x and y be local coordinates on B near r, and let u and v be local coordinates on C_q at the two identified points p and q respectively satisfying

$$xy = uv = 0, \quad x = u^2 \text{ and } y = v^2.$$

Then take the deformations of B and C_q given locally by $xy - t^2$ and $uv - t$.

We therefore conclude that the degree of the divisor $\overline{h}$ on our curve D is

$$\deg_D(\overline{h}) = 6.$$

Since we know that $\deg_D(\lambda) = \deg_D(\delta_0) = 0$ while $\deg_D(\delta_1) = -2$, we may plug into the general relation

$$\overline{H} = a \cdot \lambda + b \cdot \delta_0 + c \cdot \delta_1$$

to conclude that $c = -3$. Putting this together with the results of Section F, we arrive at the formula

$$\overline{h} = 9\lambda - \delta_0 - 3\delta_1.$$

Translating this into the Chow group of cycles on the moduli space $\overline{\mathcal{M}}_3$ using Proposition (3.92) and Proposition (3.93), we deduce that

(3.165) $[\overline{H}] = 18\lambda - 2\delta_0 - 3\delta_1.$

We said above that there were two ways to get around the problem of the transversality of intersection of a curve in $\overline{\mathcal{M}}_3$ with the divisor h. The second, which we should mention here if only as a check of the last equality, is simply to use a different family. For this purpose, we'll look at the third family introduced in the previous section: the family of stable curves obtained by attaching to a fixed general curve D of genus 2 a pencil $\{E_t\}$ of plane cubics by identifying a base point of the pencil with a fixed general point $q \in D$.

As before, we first ask which of the curves C_t in this family are hyperelliptic. The answer when E_t is smooth is already known: since $q \in D$ is general, and in particular not a Weierstrass point of D, these curves will never be hyperelliptic. On the other hand, the same argument as above can be used to show that the curves corresponding to singular E_t — that is, curves obtained by attaching to D an irreducible nodal cubic — are never hyperelliptic either: as before, any admissible cover $C \rightarrow B$ from a curve stably equivalent to C_t would have to have degree 2 on the components corresponding to the two components of C_t, and so would have to be ramified at $q \in D$. Thus, this family is disjoint from $\overline{H}$, and in particular $\deg_{\mathbb{P}^1}(h) = 0$. Plugging this in yields the relation

$$0 = a + 12b - c$$

and once more we can combine this with our previous calculations to recover (3.165).

You can test your mastery of these calculations with the following exercises. For the first, Exercise (3.166), which involves virtually all of the concepts introduced in this Chapter, an answer key has been provided in Table (3.167) so you can check your results. (You will need to recall that for a rational divisor class y on the moduli stack and a one-parameter family $\rho : \mathcal{X} \rightarrow B$ of curves whose general member is stable, by the *degree of y on B* we mean the degree of y on any family $\rho' : \mathcal{X}' \rightarrow B'$ obtained from $\rho : \mathcal{X} \rightarrow B$ by semistable reduction, divided by the degree of the base change morphism $B' \rightarrow B$ involved.)

EXERCISE (3.166) Find the degrees of the classes λ, δ_0, δ_1, κ and $\overline{h}$ on the following one-parameter families of curves of genus 3, and verify in each case the two relations $12\lambda = \delta_0 + \delta_1 + \kappa$ and $\overline{h} = 9\lambda - \delta_0 - 3\delta_1$ found above among these classes.

1) a general pencil of plane quartics including a cuspidal curve; that is, the family of curves $\{C_t\}$ given by polynomials $F(X,t) = t_0 G(X) + t_1 H(X)$ for G a general quartic and H a general quartic with a cusp;

2) a general pencil of plane quartics including a tacnodal curve;

3) a general pencil of plane quartics including a couble conic; and

4) a general pencil of plane sections of a general quartic surface $S \subset \mathbb{P}^3$.

	pencil #1	pencil #2	pencil #3	pencil #4
$\deg(\lambda)$	$2\frac{5}{6}$	$2\frac{3}{4}$	$\frac{3}{2}$	4
$\deg(\delta_0)$	25	$24\frac{1}{2}$	13	36
$\deg(\delta_1)$	$\frac{1}{6}$	0	0	0
$\deg(\kappa)$	$8\frac{5}{6}$	$8\frac{1}{2}$	5	12
$\deg(\overline{h})$	0	$\frac{1}{4}$	$\frac{1}{2}$	0

TABLE (3.167)

For a further discussion of the moduli of stable hyperelliptic curves (that is, the closure in $\overline{\mathcal{M}}_g$ of the locus of smooth hyperelliptic curves), see Section 6.C.

Here are some further exercises involving calculations of divisor classes in $\overline{\mathcal{M}}_g$:

EXERCISE (3.168) Find the class of the closure in $\overline{\mathcal{M}}_5$ of the locus of smooth trigonal curves.

EXERCISE (3.169) In terms of the generators λ, ω and σ_i of the Picard group of $\overline{\mathcal{C}}_3$ as described on page 62, find the class of the closure of the locus of pairs (C, p) where C is a smooth curve of genus 3 and p is a Weierstrass point of C. For extra credit: what is the branch divisor of this locus over $\overline{\mathcal{M}}_3$, and what does this have to do with the calculations in this subsection and in Exercise (3.170)?

EXERCISE (3.170) Find the class of the closure in $\overline{\mathcal{M}}_3$ of the locus of smooth plane quartics C with a *hyperflex*, that is, a point whose tangent line has contact of order 4 with C (equivalently, smooth curves C of genus 3 possessing a point $p \in C$ such that $\mathcal{O}_C(4p) \cong K_C$).

EXERCISE (3.171) Find the class of the closure in $\overline{\mathcal{M}}_4$ of the locus of smooth curves C possessing only one g_3^1 (equivalently, smooth curves C of genus 4 whose canonical models lie on singular quadrics).

The calculation of the classes of certain divisors in $\overline{\mathcal{M}}_g$ will be the main ingredient in the proof, in Section 6.F, that $\mathcal{M}_g$ is of general type for large g. In particular, the calculation carried out there subsumes (almost) Exercise (3.168). If you're interested, Exercises (3.169), (3.170) and (3.171) are generalized there, in Exercises (6.75), (6.76) and (6.78) respectively.

Chapter 4

Construction of $\overline{\mathcal{M}}_g$

This chapter is organized as follows. We review just enough of the basic notions of geometric invariant theory (G.I.T.) to indicate how it can be used to construct moduli of projective varieties: if you have some familiarity with G.I.T. you can safely skip this section. Then, we give a fairly complete discussion of the Hilbert-Mumford numerical criterion for the stability of Hilbert points, of Gieseker's criterion (which implies this numerical one), and of how it can be used to prove the stability of Hilbert points of smooth curves embedded by complete linear series of large degree. We omit only some arithmetic calculations.

At this point, we could but do not construct $\mathcal{M}_g$ as a coarse moduli space. Instead, we outline what else must be proved in order to construct the compactification $\overline{\mathcal{M}}_g$ in the approach of Gieseker and Mumford. The heart of the construction is the Potential Stability Theorem. We first try to motivate this result by showing how it leads to a construction of $\overline{\mathcal{M}}_g$ from which many basic geometric properties follow as easy corollaries. In fact, a complete proof of the Potential Stability Theorem would be too lengthy to include here so we have instead tried to indicate the main line of the argument leaving many technical lemmas and verifications to you in the exercises. Finally, we deduce the consequences of the theorem that are needed to construct $\overline{\mathcal{M}}_g$ as a coarse moduli space.

One important feature of the G.I.T. approach is that it's the only one that can be carried out in all characteristics (and even over more general base rings than fields). Therefore, while we will continue to work over $\mathbb{C}$, we'll deviate from the general approach of this book and make occasional remarks about how the complex case extends to positive characteristic.

A Background on geometric invariant theory

The G.I.T. strategy

Geometric invariant theory is a technique for forming quotient spaces in algebraic geometry that provides a fundamental method for the construction of moduli spaces of projective varieties. In this subsection, we'll recall the basic facts on which this method is based and outline the steps involved in verifying these. The remainder of this chapter deals with the details of carrying out this program for algebraic curves.

To begin with, let's review the basic problem, fixing the complex numbers as the ground field and taking a very naive point of view. We are given a *set* $\mathcal{M}$ of isomorphism classes of (complex) varieties — think of smooth curves of genus g. Our first task is to find a scheme structure on this set that is natural in the sense that given any flat proper family $\pi : \mathcal{C} \to B$ whose fibers are in $\mathcal{M}$, the natural set-theoretic moduli map φ below is actually a morphism of schemes.[1]

(4.1)
$$
\begin{array}{c}
\mathcal{C} \\
\downarrow{\scriptstyle \pi} \\
B \xrightarrow{\ \varphi\ } \mathcal{M}
\end{array}
$$

You should pause for a moment to reflect on how audacious an aspiration this really is by considering how hopeless it is in almost any other mathematical setting.

Next, we want to be able to work with fairly general complete families B and to be able to apply projective methods to the study of $\mathcal{M}$ itself. To do this, we need to find a set of varieties $\overline{\mathcal{M}}$ containing $\mathcal{M}$ on which we can construct a natural projective scheme structure with respect to which $\mathcal{M}$ is a dense open subscheme. No matter how nice the varieties in $\mathcal{M}$ itself are, any such $\overline{\mathcal{M}}$ is almost certain to contain singular elements. A simple example in genus 1 (easily generalized to any genus) is the family of curves

(4.2) $$ y^2 z = x(x - tz)(x - z). $$

---

[1]More properly, we should (as we did in Chapter 1 but here will not) ask for a scheme structure that makes $\mathcal{M}$ into a coarse moduli space for the associated moduli functor.

The curves C_t are, for small nonzero t, nonisomorphic smooth curves of genus 1. The *special* fiber C_0 is a nodal rational cubic curve. But, this family is its own semistable reduction so there is no hope of filling in the family with a smooth special fiber. On the other hand, recall the family (2.11) of plane cubics

$$y^2 z = x^3 - t^2 a x z^2 - t^3 b z^3$$

discussed in Section 2.C. For $t \neq 0$, the curves C_t are all isomorphic to the smooth elliptic curve C_1 while the curve C_0 with equation $y^2 z = x^3$ is rational and has a cusp at $(0, 0, 1)$. No reasonable algebraic structure on the set of isomorphism classes of plane cubics can coexist with a set-theoretic map that has one value for nonzero t and another for $t = 0$. This example shows that there are some degenerations that simply cannot be allowed into $\overline{\mathcal{M}}$ if we want a good scheme structure: let's call these, informally, *bad* degenerations. We thus face a subtle second problem, that of determining an $\overline{\mathcal{M}}$ that is large enough to complete $\mathcal{M}$ but small enough not to contain any bad elements.

To summarize, we have two basic problems:

Problem 1: Determine the set of "good" degenerations that can be allowed into $\overline{\mathcal{M}}$.

Problem 2: Construct natural scheme structures on $\mathcal{M}$ and $\overline{\mathcal{M}}$.

Geometric invariant theory (henceforth G.I.T.) provides a two-step strategy for answering both of these problems simultaneously.

Step 1: Parameterize the elements of a class $\mathcal{M}'$ *plus some extra structure* by the points of a projective subvariety $\mathcal{K}$ of $\mathbb{P}(W)$, where W is a linear representation of a reductive group G, so that the set of points of $\mathcal{K}$ parameterizing the extra structures on a fixed element C of $\mathcal{M}'$ forms a single G orbit in $\mathcal{K}$.

Step 2: Form a quotient $\mathcal{K}/G$ of $\mathcal{K}$ by the action of G.

Before going any further, we want to make a few simplifying remarks. The first is that, in our applications, G will always be the special linear group $\mathrm{SL}(V)$ of another vector space V and W will always be a representation constructed out of the basic representation V of G by multilinear algebra. In particular, the G-orbit of a point of $\mathbb{P}(W)$ will always parameterize different choices of a system of homogeneous coordinates on $\mathbb{P}(V)$.[2] We assume this henceforth. We also adopt the

---

[2]Most of what we have to say about G.I.T. applies in more general situations in which G is any reductive algebraic group that doesn't map onto $\mathbb{G}_m$ and $\mathcal{K}$ is any

convention that, if x is a nonzero point of a representation W, then $[x]$ denotes the corresponding point of $\mathbb{P}(W)$. Conversely, if $[x]$ is a point of $\mathbb{P}(W)$, then x denotes any lift to W and we'll often ascribe a property to $[x]$ — especially, stability — by reference to a property of a lift x that doesn't depend on which lift is chosen.

We're now ready to make a few comments about the G.I.T. strategy. The first step may seem to be merely a *more* difficult version of Problem 2 since constructing the parameter space $\mathcal{K}$ involves finding a natural scheme structure on the set of classes of "elements of $\mathcal{M}$' plus extra structure". In particular, the space $\mathcal{K}$ must have universal properties at least as strong as any we wish its quotient $\mathcal{M}$ to enjoy. In practice, adding a suitable extra structure often makes things much easier. The parameter spaces discussed in Chapter 1 illustrate this principle.

For example, the Hilbert scheme $\mathcal{H} = \mathcal{H}_{d,g,r}$ parameterizes "curves of arithmetic genus g plus a very ample linear system of degree d and $r + 1$ sections" or equivalently "subcurves of $\mathbb{P}^r$ of degree d and arithmetic genus g". If we let $V = \mathbb{C}^{r+1}$, the group $G = \mathrm{SL}(V)$ acts on $\mathcal{H}$ and a G-orbit consists of all elements of $\mathcal{H}$ corresponding to the same curve and linear system. Moreover, the Hilbert scheme is even a fine parameter space so has better universal properties than we're asking of our moduli spaces. In this example, there are many choices for the underlying representation W corresponding to the choice of large degree m used in the construction of $\mathcal{H}$ in Section 1.B. For a fixed m, W is the Plücker space $\Lambda^{P(m)} \mathrm{Sym}^m(V^\vee)$ of the Grassmannian of quotients of $\mathrm{Sym}^m(V^\vee)$ having with dimension $p(m) = dm - g + 1$.

We will eventually construct the spaces $\mathcal{M}_g$ and $\overline{\mathcal{M}}_g$ by carrying out Step 2 using the subschemes $\widetilde{\mathcal{K}}$ of $\mathcal{H}$ which were used in Section 3.B to produce universal deformations of stable curves: recall that these parameterize "n-canonically embedded stable curves of arithmetic genus g plus a basis of $H^0(C, \omega_C^{\otimes n})$". In essence, then Step 1 amounts to getting a start on the moduli problem 2 by solving a related but easier problem.

Finite generation of and separation by invariants

To take advantage of this simplification, we must form the quotient in Step 2 and it's here that the real difficulties in the G.I.T. approach surface. The basic idea, however, is very simple. If $X = \mathrm{Spec}(R)$ is a G-invariant affine subscheme of $\mathcal{K}$ and R^G is the subring of G-invariant

---

scheme on which G acts rationally, but to extend the conclusions we need requires the introduction of a number of technical concepts and considerable effort. If you're interested, you should consult Chapters 1–3 of [50].

elements of R, define $X/G = \mathrm{Spec}(R^G)$. Then the restriction of the quotient map q to X should be the rational map $X \dashrightarrow X/G$ dual to the inclusion of rings $R^G \longmapsto R$. We then patch these local characterizations together to the statement that if $\mathbb{C}[\mathcal{K}]$ is the homogeneous coordinate ring of $\mathcal{K}$ in $\mathbb{P}(W)$ and $\mathbb{C}[\mathcal{K}]^G$ is the subring of homogeneous G-invariants, then $\mathcal{K}/G = \mathrm{Proj}(\mathbb{C}[\mathcal{K}]^G)$.[3] In other words, the quotient map $q : \mathcal{K} \dashrightarrow \mathcal{K}/G$ is given by taking the values of homogeneous invariant polynomials on $\mathcal{K}$.

The first question we have to resolve is whether these definitions make sense: the implicit assertion that $\mathbb{C}[\mathcal{K}]^G$ is noetherian requires justification since it's only a sub-*ring* of $\mathbb{C}[\mathcal{K}]$. This and related questions are known as "Hilbert's 14th problem". We won't enter into this here except to remark that the finite generation for every $\mathcal{K}$ is equivalent to the reductivity of G: see Appendix A to [50]. There is even an explicit characteristic p counterexample due to Nagata [123] with $G = \mathbb{Z}/p\mathbb{Z}$. In other words, we must first check that *there aren't too many invariants*. This is the beautiful:

THEOREM (4.3) (HILBERT-WEYL-HABOUSH) *If a reductive algebraic group G acts rationally on a noetherian ring R, then R^G is finitely generated.*

The hypothesis of rationality means that for any $f \in R$, $\mathrm{span}\{g \cdot f \,|\, g \in G\}$ is finite-dimensional. This is automatic if the action is algebraic. Using it and the fact that any finite-dimensional representation of G is completely reducible, it's easy to construct a *canonical* $\mathbb{C}$-linear projection $\rho : \mathbb{C}[X] \to \mathbb{C}[X]^G$ called the Reynolds operator. The key property of this operator is expressed in the second part of the next exercise.

EXERCISE (4.4) 1) Verify the existence of the Reynolds operator asserted above. More precisely, suppose that G is a linearly reductive

---

[3] If you're familiar with G.I.T. you'll notice that we haven't introduced the notion of a linearization of the G-action on $\mathcal{K}$ needed to underpin such a patching argument. This omission is a deliberate one aimed at simplifying our treatment if you're seeing G.I.T. for the first time. If you want to see a discussion of the issues we're glossing over you can refer to Chapter 1, Section 3 of [50]. If you're familiar with linearizations you'll recognize that our omission is harmless in all the examples that will concern us here. These will always begin with the canonical linear action of $G = \mathrm{SL}(V)$ on V, pass to the induced G-action on a G-representation W functorially constructed from V by multilinear algebra and finally descend to the induced G-action on $\mathbb{P}(W)$. In this situation, there is a canonical choice of an ample line bundle — $\mathcal{O}_{\mathbb{P}(W)}(1)$ — and of a G-linearization with respect to this bundle. Moreover, with these canonical choices stability of a point $[x]$ of $\mathbb{P}(W)$ is equivalent to the stability of *any* lift x of $[x]$ to W. Likewise, when we consider the problem of quotienting a scheme $\mathcal{K}$, we'll always be dealing with a G-invariant subscheme of a $\mathbb{P}(W)$ constructed as above. Hence, the affine linear point of view we use amounts to a simplifying abuse of language.

group (i.e., every finite-dimensional linear representation V of G decomposes canonically as a direct sum of irreducible representations) and that R is a vector space — possibly infinite-dimensional — on which G acts rationally and show that there is a *canonical* G-linear projection $\rho : \mathbb{C}[X] \to \mathbb{C}[X]^G$.

2) Show that if R is both a noetherian ring and an S-algebra and $\varphi : R \to S$ is a surjective S-module homomorphism, then S is noetherian. Then prove, or simply apply, the following lemma and deduce Theorem (4.3).

LEMMA (4.5) (REYNOLDS LEMMA)   *The map* $\rho : \mathbb{C}[X] \to \mathbb{C}[X]^G$ *is a* $\mathbb{C}[X]^G$-*module homomorphism; that is, if* $f \in \mathbb{C}[X]^G$ *and* $h \in \mathbb{C}[X]$, *then* $\rho(fh) = f\rho(h)$.

Hilbert's proof of this theorem (for $G = \text{SL}(n, \mathbb{C})$), now translated into English in [86], marked the first nonconstructive use of the Hilbert basis theorem. Previous work had focused on calculating explicit finite bases of invariants in specific examples. The high point of this line of study was Gordan's proof of the finite generation of the invariants of $\text{SL}(2, \mathbb{C})$ acting on symmetric powers of $\mathbb{C}^2$. The length and complexity of this calculation should be remarked since it shows the necessity of finding some way of studying the quotient map without calculating explicit invariants. The proof sketched above is essentially due to Weyl [151].

In positive characteristic, reductive groups no longer act completely reducibly. (The simplest example, when $\text{char}(k) = 2$, is the action of $\text{SL}(2, k)$ on the space of 2×2 matrices over k, where the invariant line of scalar matrices has no invariant complement. Equivalently, there is no nontrivial linear invariant: in characteristic 0, this would be the trace which is identically 0 in this example.) Nonetheless, with somewhat more work, the conclusion that their rings of invariants are finitely generated can be extended to all characteristics. A group G is called *geometrically reductive* if for any representation of G, there is a nonconstant invariant homogeneous polynomial not identically 0 on the invariant subspace. (In the example above, the determinant is a suitable quadratic invariant.) Nagata [124] showed that this hypothesis could be used as a substitute for complete reducibility in proving the finite generation of rings of invariants and Haboush [69] proved that reductive groups in positive characteristics are geometrically reductive.

The next step in carrying out the G.I.T. program is to see that *there aren't too few invariants*. The next theorem states, in effect, that there are as many invariants as permitted by the obvious restriction that an invariant is constant on the closure of any G-orbit.

THEOREM (4.6) (HILBERT-NAGATA) *If G is geometrically reductive and W is a representation of G, then values of homogeneous invariant polynomials separate disjoint closed G-invariant subsets of $\mathbb{P}(W)$.*

In characteristic 0, this may be seen as follows. Suppose that X and Y are disjoint G-invariant subsets with ideals I and J respectively. Then, we can write $1 = f + g$ for some $f \in I$ and $g \in J$. Applying the Reynolds operator to this equation, we find that

$$1 = \rho(1) = \rho(f + g) = \rho(f) + \rho(g).$$

Since I and J are G-stable, $\rho(f) \in I^G$ and $\rho(g) \in J^G$. Thus, $\rho(f)$ is an invariant that is 0 on X and 1 on Y as required. An alternate proof is provided by the following exercise.

EXERCISE (4.7) (in char 0)[4] Show that if Y is a closed G-invariant subvariety of V there is a G-equivariant polynomial map from V to a representation V' of G such that Y is the inverse image of the origin in V'. Use this to deduce:

COROLLARY (4.8) *The quotient map $q : \mathbb{P}(W) \dashrightarrow \mathbb{P}(W)/G$ has base locus exactly the set of points $[x] \in \mathbb{P}(W)$ such that, for any (or equivalently, every) nonzero lifting x of $[x]$ to W, the origin 0 of W lies in the closure $\overline{G \cdot x}$ of the G-orbit of x.*

This also follows by standard arguments from the theorem as outlined in Exercise (4.10) below.

DEFINITION (4.9) *The base locus of q is called the* nonsemistable *locus of $\mathbb{P}(W)$ and is denoted $\mathbb{P}(W)_n$ and its complement $\mathbb{P}(W)_{ss}$ is called the* semistable *locus: thus, $[x]$ is semistable if 0 is not contained in the closure of $\overline{G \cdot x}$. A point $[x]$ is called* stable *if the orbit $G \cdot x$ of any lifting is closed and if the stabilizer $stab_G(x)$ is of minimal dimension amongst all stabilizers of points in $\mathbb{P}(W)$. The locus of such points is called the* stable locus *and is denoted $\mathbb{P}(W)_s$.*

In all the examples we'll consider (Exercise (4.12) excepted), the minimal dimension of a stabilizer will be 0. To simplify language, we assume henceforth that any $\mathcal{K} \subset \mathbb{P}(W)$ we consider contains points with finite stabilizers. Therefore, a point will be stable if the orbits of its liftings are closed and $stab_G(x)$ is finite. Given a subvariety $\mathcal{K}$ of $\mathbb{P}(W)$, we define analogous loci in $\mathcal{K}$ by restricting to $\mathcal{K}$ from $\mathbb{P}(W)$.[5]

---

[4]For extra credit, prove this in characteristic p also. A proof in this case may be found in [49] or in Appendix A to [50].

[5]All these properties have intrinsic analogues defined using only the G-action on $\mathcal{K}$ and not referring to a lift to W, but these yield exactly the same semistable and stable loci.

EXERCISE (4.10) This exercise characterizes semistable and stable points of $\mathbb{P}(W)$ in terms of the values of homogeneous invariant polynomials.

1) Show that if $[x]$ lies in $\mathcal{K}_{ss}$, then there is a *homogeneous G-invariant* polynomial f that does not vanish at $[x]$. *Hint*: Take $X = \overline{G \cdot x}$ and $Y = \{0\}$.

2) Show that if $[x]$ lies in $\mathcal{K}_{ss}$ and $[y]$ doesn't lie in the closure of $G \cdot [x]$, then there is a homogeneous G-invariant polynomial f that vanishes at $[y]$ but not at $[x]$. *Hint*: Take Y to be the cone over $G \cdot y$.

3) Show that the polynomials f in the preceding parts may be chosen to have degree bounded by a constant independent of the choice of $[x]$ and $[y]$.

COROLLARY (4.11)   1) *Two points $[x]$ and $[y]$ in $\mathcal{K}_s$ lie in the same G-orbit if and only if $q([x]) = q([y])$.*

2) *If $[x]$ and $[y]$ are in $\mathcal{K}_{ss}$, then $q([x]) = q([y])$ if and only if*

$$\overline{G \cdot x} \cap \overline{G \cdot y} \cap \mathcal{K}_{ss} \neq \emptyset.$$

Warnings. Two are in order here:

1) The standard term for the *nonsemistable* locus is the *unstable* locus. This choice is very unfortunate since the complement of this locus is *not* the stable locus but the semistable locus, as we remark above. The exercise below shows that the distinction is a real one; i.e., there are often points that are semistable but not stable. Such points are generally called *strictly semistable*. The standard terminology goes back to Mumford's foundational work [50]. We ask your indulgence of our doubtless quixotic attempt to rationalize custom.

2) There is no a priori connection between the notions of stable and semistable point introduced above and the notion of moduli stable or semistable curve. The main theorems of this chapter will show that nonetheless there is a very close connection between the PGL(V)-stability in this sense of the Hilbert points of suitable models of a curve C in $\mathbb{P}(V)$ and the moduli stability of the underlying abstract curve C. When necessary to avoid confusion, we'll refer to the abstract concepts defined in Chapter 2 as *moduli* stability and semistability.

EXERCISE (4.12) This exercise treats the action of SL($n, \mathbb{C}$) on the space $M_{n \times n}(\mathbb{C})$ by conjugation, with orbits the similarity classes of $n \times n$ complex matrices. It provides a rare case in which the ring of invariants can be explicitly computed. This is the one exception to our working assumption that points with finite stabilizers exist.

1) Show that the closure of the similarity class corresponding to a fixed Jordan normal form contains the diagonalizable similarity class

with the same characteristic polynomial. Determine the closure of each similarity class.

2) Show that any homogeneous invariant polynomial is constant on the set of matrices with given characteristic polynomial.

3) Show that the coefficients of the characteristic polynomial (viewed as homogeneous polynomial functions on $M_{n \times n}(\mathbb{C})$) generate the ring of invariants $\mathbb{C}[M_{n \times n}(\mathbb{C})]^{\mathrm{SL}(n,\mathbb{C})}$ and that these invariants are algebraically independent. In other words, the quotient map q in this case maps a matrix to its characteristic polynomial.

4) Determine the stable and unstable loci in $M_{n \times n}(\mathbb{C})$.

EXERCISE (4.13) Show that if x is a semistable point whose orbit isn't closed then the closure of the orbit $G \cdot x$ of x contains a unique closed orbit $G \cdot y$ and that $\dim \mathrm{stab}_G(y) > \dim \mathrm{stab}_G(x)$.

The numerical criterion

The corollaries above show that we can only hope to have a quotient whose closed points correspond to orbits in $\mathcal{K}$ (and hence to isomorphism classes in $\mathcal{M}'$) beneath the stable locus in $\mathcal{K}$. This, however, should be viewed as one of the key benefits of the G.I.T. approach. The theory simultaneously identifies those bad varieties that must be excluded to have a separated moduli space (these are identified as the ones whose Hilbert points are nonsemistable), while it constructs the scheme structure on the set of good or stable varieties. The fundamental problem of constructing moduli in this approach is then to determine the stable locus in the appropriate Hilbert schemes. For emphasis, we repeat that, Exercise (4.12) above notwithstanding, it's essentially never possible to determine the stable locus by explicit calculations with invariants or by directly determining which orbits in $\mathbb{P}(W)$ are closed. This difficulty prevented much use being made of G.I.T. for constructing moduli until Mumford found a means of circumventing it.

To see how Mumford's idea works, let's ask in what, if any, special cases it's easy to determine the stable and semistable loci. The answer is: when $G = \mathbb{G}_m$, or in our naive setting when $G = \mathbb{C}^*$.[6] As remarked above, we can and will forget about $\mathcal{K}$ for the purposes of this analysis. The decomposition

(4.14) $$\mathbb{C}[\mathbb{C}^*] = \mathbb{C}[t, t^{-1}] = \bigoplus_{i \in \mathbb{Z}} \mathbb{C}t^i$$

---

[6]This is the one point at which we need to think about representations W of a reductive group G that don't fall into the special case where $G = \mathrm{SL}(V)$ and W is derived from V by multilinear algebra.

shows that the characters of $\mathbb{C}^*$ — since $\mathbb{C}^*$ is abelian, all its irreducible representations are one-dimensional — correspond bijectively with the integers. Since $\mathbb{C}^*$ acts completely reducibly,[7] we can decompose any finite-dimensional representation $\lambda : \mathbb{C}^* \to GL(W)$ of $\mathbb{C}^*$ as

$$(4.15) \qquad\qquad W = \bigoplus_{i \in S_\lambda(W)} W_i$$

where W_i is the set of vectors $w \in W$ on which the element t of $\mathbb{C}^*$ acts by the rule $\lambda(t) \cdot w = t^i \cdot w$ and $S_\lambda(W)$ is the finite set of integers for which W_i is nonzero.

DEFINITION (4.16) *We call the set $S_\lambda(W)$ the λ-state of the representation W, the elements i of $S_\lambda(W)$ the λ-weights of W and the space W_i the ith λ-weight space of W. If x is any element of W, we define the λ-state $S_\lambda(x)$ of x to be the set of i in $S_\lambda(W)$ for which the component x_i of x in the subspace W_i is nonzero, call the elements of $S_\lambda(x)$ the λ-weights of x and define*

$$\mu_\lambda(x) = \min(S_\lambda(x)).$$

These invariants are also defined for points $[x]$ of $\mathbb{P}(W)$ by our usual convention of using their common value on any lifting x of $[x]$.

More naively, this amounts to using the fact that $\lambda(\mathbb{C}^*)$ is an abelian subgroup of $GL(W)$ to choose a basis $B = \{b_1, \ldots, b_N\}$ of W with respect to which, for each $t \in \mathbb{C}^*$, $\lambda(t)$ acts on W by the diagonal matrix $\text{diag}(t^{w_1}, \ldots, t^{w_N})$. The space W_i is then just the span of those b_j for which $w_j = i$, so $S_\lambda(W)$ is the set of distinct w_j's and $S_\lambda(x)$ is the set of distinct w_j's for which the jth coordinate of x in the basis B is nonzero.

The point of these definitions is that, since

$$\lambda(t) \cdot x = \sum_{i \in S_\lambda(x)} t^i \cdot x_i,$$

we have the equivalences

$$\mu_\lambda(x) \geq 0 \iff \lim_{t \to 0} \lambda(t) \cdot x \text{ exists} \iff x \text{ is not stable; and}$$

$$\mu_\lambda(x) > 0 \iff \lim_{t \to 0} \lambda(t) \cdot x = 0 \iff x \text{ is not semistable.}$$

The two left equivalences are clear. The limit in the middle condition on the first line is either x itself (in which case all of $\lambda(\mathbb{C}^*)$ lies in

---

[7]For once, the analogous statement is also true in characteristic p.

the stabilizer of x, which therefore has positive dimension) or a point of the closure of the λ-orbit of x not lying in that orbit. These possibilities correspond exactly to the two ways in which x can fail to be stable for the λ-action of $\mathbb{C}^*$. The right equivalences on the second line are an even more direct translation of the definition of nonsemistability to our special case. Our conventions immediately give the same equivalences for the (semistability) of $[x]$ as well.

Mumford's idea (building on some examples due to Hilbert) was that this easy case is actually the general one. Any algebraic group homomorphism $\lambda : \mathbb{C}^* \to G$ is called a *one-parameter subgroup* (often abbreviated one-parameter subgroup) of G and λ is called *nontrivial* if its image is. We have the:

THEOREM (4.17) (HILBERT-MUMFORD NUMERICAL CRITERION)

1) $[x]$ is G-nonsemistable $\iff$ For some one-parameter subgroup λ of G, $\mu_\lambda([x]) > 0$.

2) $[x]$ is G-semistable $\iff$ For every one-parameter subgroup λ of G, $\mu_\lambda([x]) \leq 0$.

3) $[x]$ is G-nonstable $\iff$ For some nontrivial one-parameter subgroup λ of G, $\mu_\lambda([x]) \geq 0$.

4) $[x]$ is G-stable $\iff$ For every nontrivial one-parameter subgroup λ of G, $\mu_\lambda([x]) < 0$.

This criterion is often rephrased as the statement that $[x]$ is stable [resp: semistable] if and only if every nontrivial one-parameter subgroup of G acts on it with negative [resp: nonpositive] weight(s). Since the weights of the one-parameter subgroup λ^{-1} (obtained by element-wise inversion of the image of λ) are minus those of λ, the criterion can also be stated more symmetrically as: $[x]$ is stable [resp: semistable] if and only if every nontrivial one-parameter subgroup of G acts on it with both positive and negative [resp: nonnegative and nonpositive] weight(s). This form best illustrates the motivation for using the term stable point.

If the orbit $G \cdot x$ isn't closed, then it's possible to find a disc Δ in $\mathbb{P}(V)$ such that $\Delta^* \subset G \cdot x$ but $\Delta \not\subset G \cdot x$ i.e., such that 0 lies in $\overline{G \cdot x}$ but not in $G \cdot x$. The content of the criterion is that we can actually make this disc the image of the unit circle in a one-parameter subgroup of G under the map $g \mapsto g \cdot x$. This, in turn, comes down to showing that any morphism of Δ^* into G can be taken to a one-parameter subgroup by pre- and post-multiplying by suitable regular maps of Δ^* into $SL(V)$. Proving this would take us too far afield, so we refer to [93], which also contains a beautiful quantitative study of the invariants μ_λ, for further details.

Stability of plane curves

We're now ready to carry out the G.I.T. program in the simple case when $\mathcal{M} = \mathcal{M}_1$, the set of smooth curves of genus 1. Our treatment of this example goes back to Hilbert. The first step of setting up the parameter space $\mathcal{K}$ is immediate. The vector space $\mathcal{K} = \mathbb{P}(\mathrm{Sym}^3(\mathbb{C}^3)^\vee)$ of homogeneous plane cubic curves parameterizes curves of arithmetic genus 1 and degree 3 in $\mathbb{P}^2(\mathbb{C})$. If C is a curve of genus 1, then C has a one-parameter family of g_3^2's, but these all yield projectively equivalent plane cubics. Thus the open subset $\mathcal{K} - \Delta$ of $\mathcal{K}$ obtained by removing the discriminant hypersurface Δ (the locus of singular cubic curves) parameterizes smooth curves of genus 1 plus a choice of homogeneous coordinates on $\mathbb{P}^2$. The basic vector space here is $V = \mathbb{C}^3$, the group G is thus SL(3), and the representation W is $\mathrm{Sym}^3(\mathbb{C}^3)^\vee$.

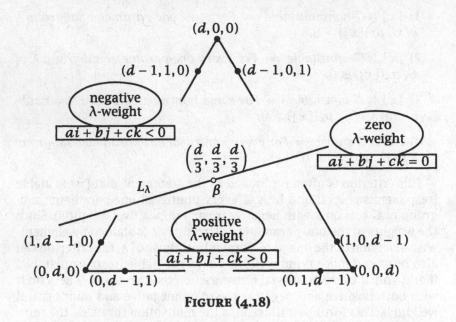

FIGURE (4.18)

Let's generalize for a moment to plane curves of arbitrary degree d. If we fix a one-parameter subgroup $\lambda : \mathbb{C}^* \to \mathrm{SL}(3)$, then there are homogeneous coordinates x, y and z on $\mathbb{C}^3$ with respect to which

$$\lambda(t) = \mathrm{diag}(t^a, t^b, t^c)$$

with $a + b + c = 0$ since $\det(\lambda(t)) = 1$. The basis B of $\mathrm{Sym}^d(\mathbb{C}^3)^\vee$ consisting of monomials of degree d in x, y and z also diagonalizes the action of λ:

$$\lambda(t) \cdot x^i y^j z^k = t^{(ai+bj+ck)} x^i y^j z^k.$$

If

$$f(x, y, z) = \sum_{i+j+k=d} c_{ijk} x^i y^j z^k = 0$$

is the equation of a curve C of degree d (i.e., f is the point of $\text{Sym}^d(\mathbb{C}^3)^\vee$ determined by C), then the B-coordinates of f are just its coefficients and the λ-state $S_\lambda(f)$ is simply the set of monomials whose coefficients in f are nonzero.

This setup can be represented in planar barycentric coordinates

$$i + j + k = d$$

by points of a plane triangle as shown in Figure (4.18). By linearity, we may speak of the weight of any point in this real plane and then the line L_λ with equation $ai + bj + ck = 0$ describes the locus of points of λ-weight 0. The numerical criterion can immediately be translated in these terms as follows. The curve C is λ-stable if and only if some monomial lying strictly on the negative side of the line L_λ appears with nonzero coefficient in the equation f; C is λ-semistable if and only if some monomial lying on or to the negative side of this line has nonzero coefficient.

Now we return to the case of cubics. If we fix $(a, b, c) = (-5, 1, 4)$ then we get the picture shown in Figure (4.19). We can now analyze

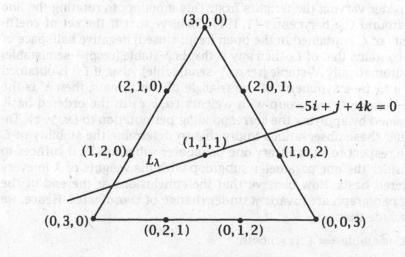

FIGURE (4.19)

what λ-stability means geometrically by supposing that successively greater numbers of the coefficients c_{ijk} of the equation f lying in the negative weight half-plane vanish ordering the coefficients by their distance from L_λ or equivalently by minus their λ-weights. At each stage, the equation f becomes more special and this is reflected in the increasingly singular character of the point $P = (1, 0, 0)$ on C.

In the summary below, we use an * to indicate a general nonzero form and $\langle\rangle$'s to indicate a general element of an ideal.

$c_{300} = 0$: $f = \langle y, z \rangle$, so P lies in C.

$c_{210} = 0$: $f = *x^2z + \langle y, z \rangle^2$, so $z = 0$ is tangent to C at P.

$c_{201} = 0$: $f = \langle y, z \rangle^2$, so P is a double point of C.

$c_{120} = 0$: $f = *yz + *z^2 + \langle y, z \rangle^3$, so $z = 0$ is tangent to a branch of C at P.

$c_{111} = 0$: $f = *z^2 + \langle y, z \rangle^3$, so P is a cusp of C with tangent cone twice the line $z = 0$.

$c_{102} = 0$: $f = \langle y, z \rangle^3$, so P is a triple point of C.

Comparing this summary with the diagram, we may conclude that:

$$C \text{ is smooth} \iff C \text{ is } \lambda\text{-stable; and}$$

$$C \text{ has at worst nodes} \iff C \text{ is } \lambda\text{-semistable.}$$

Next, what happens if we vary the weights $(a, b, c) = (-5, 1, 4)$ of λ and the coordinates (x, y, z) with respect to which it diagonalizes? Consider varying the weights first. This amounts to rotating the line L_λ around the barycenter $(1, 1, 1)$. Observe that if the set of coefficients of C contained in the open [resp: closed] negative halfspace of $L_{\lambda'}$ contains that of L_λ then any C that is λ-stable [resp: λ-semistable] is automatically λ'-stable [resp: λ'-semistable]. Also, if $L_{\lambda'}$ is obtained from L_λ by a symmetry of the triangle of coefficients, then λ' is the one-parameter subgroup with weights (a, b, c) in the ordered basis obtained by applying the corresponding permutation to (x, y, z). Together these observations imply that to determine the stability of C with respect to an arbitrary one-parameter subgroup λ', it suffices to consider the one-parameter subgroup with the weights of λ in every ordered basis. Now observe that the conclusions at the end of the last paragraph are invariant under change of coordinates. Hence, we conclude that:

C is stable $\iff$ C is smooth;

C is semistable $\iff$ C has at worst ordinary double points; and

C is nonsemistable $\iff$ C has a cusp or worse singularity.

(The third of these conclusions follows by negating both sides of the second.)

A few further comments about the respects in which this analysis is and is not typical are in order. The most typical feature is the very direct connection between the instability of f, "undesirable" geometric features of C, and the destabilizing one-parameter subgroup λ. Note, first, how these results confirm the intuition from the families discussed at the start of this section that we must avoid cuspidal cubics in order to have a nice moduli space but that we must include nodal cubics if the resulting space is to be complete. Also, note that the *reducible* cubics with nodes — a conic and a transversal line, or a triangle — are semistable. This phenomenon is again quite typical. Complete moduli spaces almost always include reducible varieties. Perhaps, the most subtle difficulty in the construction of $\overline{\mathcal{M}}_g$ is to control such curves.

Second, note that instability is due to a singularity of C and the destabilizing one-parameter subgroups are very closely tied to the character of the singularities: for a cusp, the singular point and its tangent cone are defined by the vanishing of coordinates that diagonalize λ and the weights have the property that the coefficients that must be nonzero to avoid a worse singularity are of equal λ-weight. It's a very general phenomenon that instability of the Hilbert point of a variety X is "caused" by a bad subscheme of X and that the destabilizing one-parameter subgroup is the one that most clearly "picks out" the subscheme. (A more precise statement of this principle is the main result of [93].)

The effect of this phenomenon is that it's generally very easy to show that a Hilbert point isn't stable. To show that such a point is stable, however, requires handling a general one-parameter subgroup λ, which may bear no relation to the geometry of X. Such proofs, as we'll see, involve using geometric estimates about sublinear series of the hyperplane bundle on X to obtain combinatorial estimates for the weights of a general λ. Unfortunately, the combinatorics involved are often extremely hard and lengthy. They are not even completely understood for Hilbert points of stable curves embedded by complete linear systems. This difficulty represents the main obstacle to using G.I.T. to construct moduli for higher-dimensional varieties. Because our Hilbert scheme is a projective space, this difficulty doesn't arise here.

What aspects of this example are not typical? First, the ring of invariants of this example is one of the few that has been computed explicitly. It's generated by two elements A of degree 4 and B of degree 6 that are essentially the coefficients of the Weierstrass normal form of the cubic. The discriminant is one geometrically meaningful invariant (since projectivities don't affect the smoothness of a curve): in fact, $\Delta = (27A^3 - 4B^2)$. The classical j-invariant is given by $j = A^3/\Delta$; it generates the invariant function field and hence gives a rational parameterization of the quotient $\mathcal{M}_1$. If C is a smooth cubic, then $\Delta(C)$ is nonzero and the moduli point of C is determined by its j-invariant. All three orbits of nodal cubics (corresponding to a rational nodal cubic, a conic and a transverse line, and a triangle) have $\Delta = 0$ but $A \neq 0$, hence $j = \infty$. Cubics with cusps or worse (amongst which we also find degenerate cases of the reducible semistable configurations such as a conic plus a tangent line) have $A = B = 0$ and hence undefined j-invariant.

We should also warn that as d increases, the singularities that can lie on a stable plane curve of degree d become more and more complex: see the exercises below. The case $d = 3$ is a convenient accident that confirms our philosophy. Conversely, certain smooth varieties are unstable — see [116] for examples.

EXERCISE (4.20) 1) Analyze the stability of quartic plane curves, showing that cusps are semistable on quartics.

2) Show that for every μ, there is a degree $d(\mu)$ such that ordinary plane curves of degree d with only ordinary μ-fold points have stable equations.

3) Use a tetrahedron of coefficients to analyze the stability of cubic and quartic surfaces in $\mathbb{P}^3$. *Hint*: The answer is given in Section 1 of [121].

PROBLEM (4.21) What cubic and quartic threefolds have stable equations?

B Stability of Hilbert points of smooth curves

In this section, we'll first interpret the numerical criterion for the stability of Hilbert points of general subvarieties of projective space. Then, we'll give several sufficient criteria and see how to combine one criterion due to Gieseker (4.30) with standard results about curves (Riemann-Roch and Clifford's theorem) to deduce the stability of Hilbert points of smooth curves embedded by complete linear systems of large degree [Theorem (4.34)].

The numerical criterion for Hilbert points

Our first task is to understand the meaning of the numerical criterion for the $SL(r + 1)$-action on the Hilbert scheme $\mathcal{H} = \mathcal{H}_{P(m),r}$ of subschemes of $\mathbb{P}^r$ with Hilbert polynomial $P(m)$. To do this, we fix a one-parameter subgroup $\lambda : \mathbb{C}^* \to \mathrm{SL}(r + 1)$ and homogeneous coordinates

$$B = B_\lambda = \{x_0, \ldots, x_i, \ldots, x_r\}$$

which we view as a basis of $(\mathbb{C}^{r+1})^\vee$ with respect to which

$$\lambda(t) = \mathrm{diag}\,(t^{w_0}, \ldots, t^{w_i}, \ldots, t^{w_r})$$

with $\sum_i w_i = 0$. The data of λ is thus equivalent to the data of B considered as a *weighted basis* (i.e., along with a set of integral weights w_i summing to 0) and we'll henceforth refer to B and λ interchangeably.

As for plane curves, the basis

$$B_m = \left\{ Y = \prod_i x_i^{m_i} \;\middle|\; \sum_i m_i = m \right\}$$

of $\mathrm{Sym}^m(\mathbb{C}^{r+1})^\vee$ consisting of monomials of degree m in x_i's also diagonalizes the action of λ: if we define the B-weight w_Y of Y by

$$w_Y = \sum_i w_i m_i,$$

then

$$\lambda(t) \cdot Y = t^{w_Y} Y.$$

For the rest of this section, fix a suitably large m and let

$$W = \Lambda^{P(m)}\left(\mathrm{Sym}^m(\mathbb{C}^{r+1})^\vee\right).$$

Then the construction of Section 1.B exhibits $\mathcal{H}$ as a subscheme of the Grassmannian G of $P(m)$-dimensional quotients of $\mathrm{Sym}^m(\mathbb{C}^{r+1})^\vee$, which in turn lies in $\mathbb{P}(W)$. The representation W of $\mathrm{SL}(r + 1)$ has a natural Plücker basis consisting of all unordered $P(m)$-element subsets

$$Z = \left\{ Y_{j_1}, \ldots, Y_{j_{P(m)}} \;\middle|\; Y_{j_k} \in B_m \right\}$$

of B_m. This basis diagonalizes the action of λ on W: if we set

$$w_Z = \sum_k w_{Y_{j_k}},$$

then

$$\lambda(t) \cdot Z = t^{w_Z} Z.$$

The Plücker coordinate Z is nonzero at the point $[Q]$ of the Grassmannian G corresponding to a quotient Q if and only if the images in Q of the $P(m)$ degree m monomials Y_{j_k} in Z form a basis of Q. Since the Hilbert point $[X]$ of a subscheme of X of $\mathbb{P}^r$ with Hilbert polynomial $P(m)$ corresponds to the quotient

$$\mathrm{Sym}^m(\mathbb{C}^{r+1})^\vee \xrightarrow{\ \mathrm{res}_X\ } H^0(X, \mathcal{O}_X(m)),$$

Z is nonzero at $[X]$ if and only if the restrictions $\text{res}_X(Y_{j_k})$ of the monomials in Z are a basis of $h^0(X, \mathcal{O}_X(m))$. We will call such a set of monomials a *B-monomial basis* of $H^0(X, \mathcal{O}_X(m))$.

The change of point of view from one-parameter subgroups λ to weighted bases B can be pushed a little further. First, note that, in the language of weighted bases, there is no need to maintain the requirements that the weights w_i be integral or sum to 0. Instead, we denote this sum by w_B. The second simplification involves the notion of a rational *weighted filtration F* of $V = \mathbb{C}^{r+1}$. This is just a collection of subspaces U_w of V, indexed by the rational numbers, with the property that $U_w \subset U_{w'}$ if and only if $w \geq w'$. Any weighted basis B determines a weighted filtration F_B by taking

$$U_w = \text{span}\{x_i | w_i \leq w\}.$$

We say that B is *compatible* with F if $F_B = F$. If so, then we define the weight w_F of F to be w_B: this clearly doesn't depend on which compatible B we choose.

Each F is determined by the subspaces associated to the finite number of weights w at which there is a jump in the dimension of U_w. It's convenient to use a notation that implicitly assumes that all these jumps in dimension are of size 1 and to view F as the collection of data:

$$F = F_1 : \mathbb{C}^{r+1} = V_0 \supsetneqq V_1 \supsetneqq \cdots \supsetneqq V_r \supsetneqq \{0\}$$

(4.22)

$$w_0 \geq w_1 \geq \cdots \geq w_r$$

Thus, $U_w = \cup_{w_i \leq w} V_i$ and an element x in V has weight $w(x) = w_i$ if and only if x lies in V_i but not in V_{i+1}. Of course, if $w_i = w_{i+1}$, then F has a larger jump and V_{i+1} is neither uniquely determined by F nor, indeed, needed to recover the filtration F. We ask you to accept this harmless ambiguity since it makes it possible to use the same indexing in discussing one-parameter subgroups, weighted bases and weighted filtrations.

By repeating the arguments above using any basis B compatible with F, we see that F determines weighted filtrations F_m of each $\text{Sym}^m(\mathbb{C}^{r+1})^\vee$. But anytime we have a weighted filtration on a space S and a *surjective* homomorphism $\varphi : S \rightarrow H$, we get a weighted filtration on H by the rule that the weight of an element h of H is the *minimum* of the weights of its preimages in S. Thus, F_m determines a weighted filtration, which we also denote by F_m, on $H^0(X, \mathcal{O}_X(m))$. We let $w_F(m)$ denote the weight of any basis of $H^0(X, \mathcal{O}_X(m))$ compatible with the filtration F_m: as the notation suggests, we'll shortly be viewing these weights as giving a function of m depending on F. With these preliminaries, we have:

PROPOSITION (4.23) (NUMERICAL CRITERION FOR HILBERT POINTS)
The mth Hilbert point $[X]_m$ of a subvariety X of $\mathbb{P}^r$ with Hilbert polynomial P is stable [resp: semistable] with respect to the natural $SL(r+1)$-action if and only if the equivalent conditions below hold:

1) *For every weighted basis B of $\mathbb{C}^{r+1}$, there is a B-monomial basis of $H^0(X, \mathcal{O}_X(m))$ whose B-weights have* negative *[resp: nonpositive] sum.*

2) *For every weighted filtration F of V whose weights w_i have average α,*

$$w_F(m) < m\alpha P(m) \qquad (resp: w_F(m) \le m\alpha P(m)).$$

PROOF. If we diagonalize the action of the one-parameter subgroup λ associated to B on W as above, then the B-monomial bases are just the nonzero Plücker coordinates of $[X]_m$ and their weights are the weights of $[X]_m$ with respect to λ. Thus 1) is an immediate translation of the numerical criterion (4.17). To see 2), observe that if B is any basis compatible with the filtration F and we set $w_i' = \beta(w_i - \alpha)$ where β is chosen so that all the weights w_i' are integral, then B becomes a weighted basis, and, moreover, every weighted basis B arises in this way from some F. The F-weight of any degree m monomial then differs from its B-weight by $m\alpha\beta$. Hence the weight of *any* B-monomial basis of $H^0(X, \mathcal{O}_X(m))$ will differ from $\beta w_F(m)$ by $\beta m\alpha h^0(X, \mathcal{O}_X(m)) = \beta m\alpha P(m)$. Therefore, the given inequality is equivalent to the negativity of the B-weights of such bases. ∎

EXERCISE (4.24) 1) Show that a collection of d distinct points in $\mathbb{P}^r$ has a stable Hilbert point if and only if, for every proper linear subspace L of projective dimension s the number of points lying in L is less than $d(\frac{r+1}{s+1})$. How should this criterion be modified to treat general zero-dimensional subschemes?

2) Formulate an analogous criterion for the stability of a cycle of k-linear subspaces of $\mathbb{P}^r$.

EXERCISE (4.25) This exercise gives an example of a *smooth* but Hilbert *unstable* variety: the Steiner surface. This is the surface S of degree 4 in $\mathbb{P}^4$ which is the closure of the image of $\mathbb{P}^2$ under the rational map given by all conics passing through a fixed point, say, $(0,0,1)$. Equivalently, it's the projection of the Veronese in $\mathbb{P}^5$ from a point lying on it.

1) Show that $H^0(S, \mathcal{O}_S(m))$ may be identified with the span of the monomials of degree $2m$ in the homogeneous coordinates (x, y, z) on $\mathbb{P}^2$ whose degree in z is at most m.

2) Use this, first, to define a filtration F on $H^0(S, \mathcal{O}_S(1))$ by assigning weight 0 to those sections *not* divisible by z and weight 1 to all others. Then, use it to show that the Hilbert point of S is unstable by showing the inequality $w_F(m) > m\alpha_F P(m)$ and applying Proposition (4.23).

3) Geometrically, S is an $\mathbb{F}_1$ rational ruled surface and the projection onto the weight 0 subspace of the filtration F has center the exceptional section of self-intersection -1 on S. Generalize this example to the projectivization of any unstable rank 2 bundle on a smooth curve (cf. [116]).

We will continue to write $\alpha := \alpha_F$ for the *average weight* of an element of a basis B of $\mathbb{C}^{r+1}$ compatible with F. We will also say simply that the variety X is *Hilbert stable with respect to F* if, for all large m, the inequalities of the proposition hold for F, and that X is *Hilbert stable* if for all large m, the mth Hilbert points of X are stable: i.e., the inequalities of the proposition hold for every nontrivial F. All the methods of verifying the stability of an mth Hilbert point that we will use apply to all sufficiently large m, the implicit lower bound depending only on the Hilbert polynomial P of X so this will not introduce any ambiguity. To see why this is so, we introduce an idea developed in [121]: the weights $w_F(m)$ are given for large m by a numerical *polynomial* in m of degree $(\dim(X) + 1)$. For our purposes, all we'll need is the:

LEMMA (4.26) (ASYMPTOTIC NUMERICAL CRITERION) *Let X be a subscheme of dimension n and degree d in $\mathbb{P}^r$.*

1) *There are constants C and M depending only on the Hilbert polynomial P of X, and, for each F, a constant e_F depending on F such that, for all $m \geq M$,*

$$\left| w_F(m) - e_F \frac{m^{n+1}}{(n+1)!} \right| < Cm^n.$$

2) *If $e_F < \alpha_F(n+1)d$, then X is Hilbert stable with respect to F; and if $e_F > \alpha_F(n+1)d$, then X is Hilbert nonsemistable with respect to F.*

3) *Fix a Hilbert polynomial P and a subscheme $\mathcal{S}$ of $\mathcal{H} = \mathrm{Hilb}_{P,r}$. Suppose that there is a $\delta > 0$ such that*

$$e_F < \alpha_F(n+1)d - \delta$$

for all weighted filtrations F associated to the Hilbert point of any X in $\mathcal{S}$. Then there is an M, depending only on $\mathcal{S}$, such that the mth Hilbert point $[X]_m$ of X is stable for all $m \geq M$ and all X in $\mathcal{S}$.

PROOF. For the first assertion, which follows by standard arguments, we'll simply refer to [121]. The second then follows by taking leading coefficients in the second form of the numerical criterion and using Riemann-Roch to provide the estimate, for large m,

$$P(m) = h^0(X, \mathcal{O}_X(m)) = \frac{dm^n}{n!} + O(m^{n-1}).$$

However, while this comparison of leading coefficients shows that $w_F(m)$ will be negative for m greater than some large M, exactly *how large* this M must be taken depends on the ratio of the constant C in part 1) to the difference $\alpha_F(n+1)d - e_F$ in part 2). To get the uniform assertion of part 3), we need both a uniform lower bound (given by δ) for this last difference and the uniform upper bound, provided by Mumford, for C. ∎

Remark. It's possible to carry out the construction of $\overline{\mathcal{M}}_g$ using the Chow scheme as a parameter space. The inequalities in 2) of the lemma are respectively equivalent to stability and instability of Chow points. Equality in 2) is equivalent to Chow semistability but gives no information on Hilbert stability. For further discussion, see [116] or [121].

Gieseker's criterion

We close these technical preliminaries with a fundamental estimate due to Gieseker [57] for e_F. While this is in no way essential, we'll now simplify by assuming that X is a *smooth* curve, which we denote C embedded in $\mathbb{P}^r = \mathbb{P}(V)$ by a linear series with a fixed Hilbert polynomial P. In order to state this criterion, we need to define an additional set of invariants of the filtration F given as in (4.22). These are the *degrees* d_j of the subsheaf generated by the sections in $|V_j|$. Equivalently, d_j is the degree of the image of C under the projection from $\mathbb{P}^r$ to $\mathbb{P}(V_j)$ multiplied by the degree of this projection. It's also convenient to define $e_j = d - d_j$ so that e_j is the *codegree*, or drop in degree, under this projection.

Gieseker first fixes a subsequence

$$0 = j_0 > j_1 > \cdots > j_h = r$$

of $(0, \ldots, r)$. He next introduces two auxiliary positive integers p and n to be fixed later and considers the filtration of $H^0(C, \mathcal{O}_C(n(p+1)))$ given by the images $U_{k,i}^n$ under restriction to X of the subspaces

$$W_{k,i}^n = \mathrm{Sym}^n(V_0) \cdot \mathrm{Sym}^{n(p-i)}(V_{j_k}) \cdot \mathrm{Sym}^{ni}(V_{j_{(k+1)}})$$

of $\mathrm{Sym}^{n(p+1)}(V)$. Here the index k runs from 0 to $h-1$ and, for each k, the index i ranges between 0 and p.

Gieseker's key claim is that, for any fixed choice of Hilbert polynomial P and integers n and p, there is an N depending only on these three choices but *not* on the Hilbert point $[C]$ or the weighted filtration F being considered, such that the dimension formula

(4.27) $\dim(U_{k,i}^n) = n\left(d + (p-1)d_{j_k} + id_{j_{k+1}}\right) - g + 1$

holds for every $n \geq N$ and for every k and i. Since every section in $U_{k,i}^n$ has weight at most $n(w_0 + (p-i)w_{j_k} + iw_{j_{k+1}})$, this claim leads, as we shall see more precisely in a moment, to upper bounds for $w_F(m)$, and eventually for e_F.

To see (4.27) pointwise, observe that, if L_j is the line bundle on C generated by the sections in V_j, then we can view $U_{k,i}^n$ as a sub-linear series of $H^0(C, (M_{k,i})^{\otimes n})$ where

$$M_{k,i} = L \otimes (L_{j_k})^{(p-i)} \otimes (L_{j_{k+1}})^i$$

is a line bundle of degree $d + (p-i)d_{j_k} + id_{j_{k+1}}$. Now, $|V_0|$ restricts to a very ample linear series on C since it's the linear series that realizes the embedding of C in $\mathbb{P}^r$. Therefore, the linear space $W_{k,i}^1$ restricts, on C, to a very ample base point free linear subseries of the complete linear series $H^0(C, M_{k,i})$. This, in turn, implies that, for large n, we have equalities

$$H^0(\mathbb{P}(W_{k,i}^1), \mathcal{O}(n)) = \mathrm{Sym}^n(W_{k,i}^1) = W_{k,i}^n$$

and, moreover, that the map

$$\varphi_{k,i}^n : W_{k,i}^n \rightarrow H^0(C, (M_{k,i})^{\otimes n})$$

given by restriction to C will be surjective. Taking $n \geq 2g$, the estimate of the claim follows from Riemann-Roch applied to $(M_{k,i})^{\otimes n}$.

EXERCISE (4.28) This exercise shows how to obtain the uniform version of (4.27) from the pointwise version. Fix the Hilbert polynomial P and integers n and p as above.

1) Use the Uniform m lemma (1.11) to show that there is an $N_{D,R}$ with the property that whenever the linear series $\varphi_{k,i}^1(W_{k,i}^1)$ has degree D and dimension R, then for all $n \geq N_{D,R}$, all $k \leq h-1$ and all $i \leq p$,

$$H^1(\mathbb{P}(W_{k,i}^1), \mathcal{I}_C(n)) = 0,$$

and hence $\varphi_{k,i}^n$ is surjective.

2) Show that both D and R above can be bounded in terms of P, n and p alone and hence deduce the uniform version of (4.27).

To prepare to extract estimates for $w_F(m)$ from (4.27), we abstract our setup for a moment. Suppose we're given a filtration of a vector space T by subspaces $T_{k,i}$ such that

$$
\begin{aligned}
T_{0,0} &\supset T_{0,1} \supset \cdots \supset T_{0,p-1} \supset T_{0,p} \\
&= T_{1,0} \supset T_{1,1} \supset \cdots \supset T_{1,p-1} \supset T_{1,p} \\
&= \cdots \\
&= T_{h-1,0} \supset T_{h-1,1} \supset \cdots \supset T_{h-1,p-1} \supset T_{h-1,p} \\
&= T_{h,0}.
\end{aligned}
$$

Suppose, further, that we know that each $T_{k,i}$ has dimension $D_{k,i}$ and that its elements have weight at most $R_{k,i}$. Then, any basis of T compatible with the filtration above would have weight at most

$$
\Big(\sum_{k=0}^{h-1} \sum_{i=0}^{p-1} (D_{k,i} - D_{k,i+1}) R_{k,i} \Big) + D_{h,0} R_{h,0}
$$

$$
= D_{0,0} R_{0,0} + \Big(\sum_{k=0}^{h-1} \sum_{i=1}^{p} D_{k,i} (R_{k,i} - R_{k,i-1}) \Big).
$$

We now apply this observation taking T to be $H^0(C, L^{n(p+1)})$ and $T_{k,i} = U_{k,i}^n$. We can thus take

$$
D_{k,i} = n(d + (p-i)d_{j_k} + (i)d_{j_{k+1}}) - g + 1
$$

and

$$
R_{k,i} = n\Big(w_0 + (p-i)w_{j_k} + i w_{j_{k+1}} \Big).
$$

Using $d = d_0$ gives an upper bound for $w_F(n(p+1))$ of

$$
\big(n(p+1)d - g + 1 \big)\big(n(p+1)w_0 \big)
$$

$$
+ \Big[\sum_{k=0}^{h-1} \Big(\sum_{i=1}^{p} \big(n(d + (p-i)d_{j_k} + (i)d_{j_{k+1}}) - g + 1 \big) \Big)
$$

$$
\cdot \big(n(w_{j_{k+1}} - w_{j_k}) \big) \Big]
$$

which, by applying standard formulae to the interior summation, we may rewrite as

$$\bigl(n(p+1)d - g + 1\bigr)\bigl(n(p+1)w_0\bigr)$$

$$+ \left[\sum_{k=0}^{h-1} \left(npd + n(p^2 - p)\frac{d_{j_k}}{2} + n(p^2 + p)\frac{d_{j_{k+1}}}{2} - p(g - 1) \right) \right.$$

$$\left. \cdot \bigl(n(w_{j_{k+1}} - w_{j_k})\bigr) \right].$$

Expanding this last expression in powers of n and p, we obtain

$$w_F(n(p+1))$$

$$\le n^2 p^2 \left[dw_0 + \frac{1}{2}\sum_{k=0}^{h-1} \bigl(d_{j_k} + d_{j_{k+1}}\bigr)\bigl(w_{j_{k+1}} - w_{j_k}\bigr) \right]$$

$$+ n^2 p \left[2dw_0 + \frac{1}{2}\sum_{k=0}^{h-1} \bigl(2d - d_{j_k} + d_{j_{k+1}}\bigr)\bigl(w_{j_{k+1}} - w_{j_k}\bigr) \right]$$

$$+ n^2 \left[dw_0 \right] + np \left[-(g-1)\bigl(w_0 + \frac{1}{2}\sum_{k=0}^{h-1}(w_{j_{k+1}} - w_{j_k})\bigr) \right]$$

$$+ n \left[-(g-1)w_0 \right].$$

Because we have these inequalities whenever $p \gg 0$ and $n \gg p$, they yield the estimate

$$e_F \le 2dw_0 + \sum_{k=0}^{h-1} \bigl(d_{j_k} + d_{j_{k+1}}\bigr)\bigl(w_{j_{k+1}} - w_{j_k}\bigr).$$

To see this, first note that on the right we just have the $n^2 p^2$ coefficient from the previous line, which we denote, for a moment, by A. This would also be the $n^2(p+1)^2$ coefficient were we to have expanded in powers of n and $(p + 1)$. Therefore, given any $\varepsilon > 0$, we may choose p so large that $w_F(n(p+1)) \le n^2(p+1)^2(A + \varepsilon)$ for all $n \gg p$. Taking leading coefficients (and recalling that our normalization of these introduced a factor of 2 on the right), this proves the desired inequality to within ε.

Next, observe that

$$d_{j_k} + d_{j_{k+1}} = \bigl(d - e_{j_k}\bigr) + \bigl(d - e_{j_{k+1}}\bigr) = -(e_{j_k} + e_{j_{k+1}} - 2d)$$

and that, if we suppose that $w_r = 0$, then

$$w_0 = \sum_{k=0}^{h-1} \bigl(w_{j_k} - w_{j_{k+1}}\bigr).$$

These allow us to rewrite the estimate for e_F above as

$$e_F \leq \sum_{k=0}^{h-1} \left(e_{j_k} + e_{j_{k+1}} \right) \left(w_{j_k} - w_{j_{k+1}} \right).$$

Finally, define ε_F to be the minimum of these estimates over all subsequences of $(0, \dots, r)$:

$$(4.29) \qquad \varepsilon_F = \min_{0=j_0<j_1\cdots<j_h=r} \left(\sum_{k=0}^{h-1} \left((e_{j_k} + e_{j_{k+1}})(w_{j_k} - w_{j_{k+1}}) \right) \right).$$

We then obtain:

LEMMA (4.30) (GIESEKER'S CRITERION)

1) *A smooth curve C is Hilbert stable with respect to a filtration F with $w_r = 0$ if $\varepsilon_F < 2d\alpha_F$.*

2) *Fix a smooth curve C of degree d and genus g in $\mathbb{P}^r$ as above, and numbers ε_i which are upper bounds for the codegree of every subspace V_i of codimension i in V. Define ε_C by*

$$(4.31) \quad \varepsilon_C = \max_{\substack{w_0\geq\cdots\geq w_r=0 \\ \sum_{i=0}^{r} w_i=1}} \left(\min_{0=j_0<\cdots<j_h=r} \left(\sum_{k=0}^{h-1} (\varepsilon_{j_k} + \varepsilon_{j_{k+1}})(w_{j_k} - w_{j_{k+1}}) \right) \right)$$

Then, C is Hilbert stable if

$$\varepsilon_C < \frac{2d}{(r+1)}.$$

3) *Fix integers d, g and r and a subscheme S of the Hilbert scheme of curves of arithmetic genus g and degree d in $\mathbb{P}^r$ such that every curve in S is smooth. If there is a $\delta > 0$ such that*

$$\varepsilon_C < \frac{2d}{(r+1)} - \delta$$

for every curve C in S, then there is an M such that the mth Hilbert point $[C]_m$ of C is stable for all $m \geq M$ and all curves C in S.

PROOF. The first assertion follows directly from the asymptotic numerical criterion by the argument given above. On the other hand, the quantity defining ε_F depends only on the codegrees e_i and weights w_i of F. We may translate the weights of F so that the smallest equals 0 and then rescale the weights of F so that they sum to 1 and hence have average

$$\alpha_F = \frac{1}{(r+1)}$$

without affecting the F-stability of C. We will call such filtrations *normalized*. The right side of (4.29) is increasing as a function of the e_i's so the maximum in the definition of ε_C is an upper bound for the ε_F of every normalized weighted filtration. Therefore, the inequality of part 2 implies Hilbert stability with respect to every F. Finally, the uniform assertion in part 3 follows from the uniform version of the asymptotic numerical criterion (4.26). ∎

Remark. This criterion, and its higher-dimensional analogues, are the best tools currently available for verifying the stability of Hilbert points. Unfortunately, there are many varieties with stable Hilbert points for which even the first inequality is violated for some filtrations F.

EXERCISE (4.32) 1) Show that, for a nodal plane cubic, $\varepsilon_C = 2 = (\frac{2d}{r+1})$.
2) If C is a nondegenerate moduli stable curve in $\mathbb{P}^r$ with n nodes and $r \geq 2n$, show that there are filtrations F with codegrees $e_i \geq 2i$ for $0 \leq i \leq n$.
3) Give an example of a complete linear system on an irreducible moduli stable curve for which $\varepsilon_F > (\frac{2d}{r+1})$.

The construction of $\overline{\mathcal{M}}_g$ will show that if $d \gg g$, then curves like those in the last part of the exercise do have stable points. In other words, Gieseker's criterion is far from being a sharp estimate of e_F. In dimensions greater than 1, the analogous inequality can even be false for smooth varieties: see [116] for an example. For a wide variety of applications, therefore, it would be of great interest to have good answers to the:

PROBLEM (4.33) Find better estimates for e_F.

Stability of smooth curves

We're now ready to state the fundamental:

THEOREM (4.34) (STABILITY OF SMOOTH CURVES OF HIGH DEGREE) *Suppose that C is a smooth curve of genus $g \geq 2$ embedded in $\mathbb{P}^r$ by a complete linear system L of degree $d \geq 2g$. Then C is Hilbert stable. Moreover, an M such that the mth Hilbert point $[C]_m$ is stable for all $m \geq M$ may be chosen uniformly for all such curves C.*

We will follow the argument given in [121] following ideas of Gieseker. First, imagine that both the ε_i's and the w_i's in Gieseker's criterion are fixed and plot the points (ε_i, w_i) as shown in Figure (4.35).

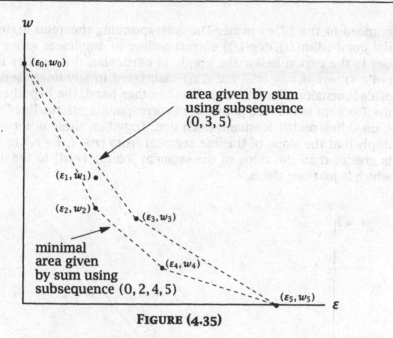

FIGURE (4.35)

The key observation is that the sum

$$\sum_{k=0}^{h-1} (\varepsilon_{j_k} + \varepsilon_{j_{k+1}})(w_{j_k} - w_{j_{k+1}})$$

associated to a subsequence $0 = j_0 > j_1 > \cdots > j_h = r + 1$ in the definition of ε_C in (4.31) represents twice the area in the first quadrant bounded by the axes and the "curve" obtained by joining the pairs of points $(\varepsilon_{j_k}, w_{j_k})$ and $(\varepsilon_{j_{k+1}}, w_{j_{k+1}})$ by straight line segments. Taking the minimum of these sums over all such subsequences amounts to computing twice the area under the lower convex envelope E of these points.

Now allow the w_i's to vary, keeping the ε_i's fixed. If any of the points (ε_i, w_i) does *not* lie on E, then moving it down onto E will leave the minimum in Gieseker's criterion unchanged while reducing the sum of the w_i's. Dually, this means that the maximum over sets of weights summing to 1 in (4.31) must occur when the weights are chosen so that *all* the points (ε_i, w_i) lie on E. For such weights, the sum associated to the full sequence — that is, $j_i = i$ for all i from 0 to r — realizes the minimum over all subsequences.

Next, we claim that we can take

$$\varepsilon_i = \frac{d}{(r+1)} i.$$

This is most easily seen from the graph in Figure (4.36) in which the Riemann-Roch line $d = r + g$ and the Clifford line $d = 2r$ are graphed in the (d, r)-plane. The corresponding theorems state that the point $(\dim(U), \deg(U))$ corresponding to *any* linear series on C lies in the region below the graph. In particular, this applies to the point $(r - i, d_i) = (r - i, d - e_i)$ associated to any linear series V_i of codimension i in $H^0(C, L)$. On the other hand, the hypothesis of the theorem is that the point (d, r) corresponding to the line bundle L on C lies *on* the Riemann-Roch line. Together, these observations imply that the slope of the line segment from $(r - i, d - e_i)$ to (r, d) is greater than the slope of the segment joining (r, d) to the origin, which is just our claim.

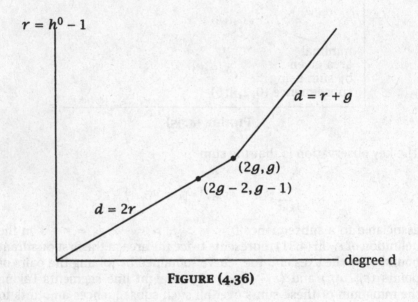

FIGURE (4.36)

The claim in turn shows that

$$\varepsilon_C \leq \sum_{j=0}^{r-1} \left(\varepsilon_j + \varepsilon_{j+1} \right) \left(w_j - w_{j+1} \right)$$

$$\leq \frac{d}{r+1} \sum_{j=0}^{r-1} (j + (j+1)) \left(w_j - w_{j+1} \right)$$

$$= \frac{d}{r+1} \sum_{j=0}^{r-1} \left(2w_j \right).$$

Using our assumption that the w_i's sum to 1 and are decreasing — and hence that $w_0 \geq \left(\frac{1}{r+1} \right)$ — this immediately implies

$$\varepsilon_C \leq \frac{2d}{r+1} (1 - 2w_0) \leq \frac{2d}{r+1} \left(1 - \frac{1}{r+1} \right).$$

The pointwise assertion of the theorem now follows immediately from the second part of (4.30) and the uniform assertion from the third part taking $\delta = (2d)/(r+1)^2$.

EXERCISE (4.37) This exercise gives a closed form for the expression that arises in Gieseker's criterion (4.30). This is useful in studying a variety of stability problems involving special curves, vector bundles on curves and $K3$ surfaces.

Fix an increasing sequence $(\varepsilon_0, \ldots, \varepsilon_{r+1})$ and define ε in terms of this sequence by (4.31).

1) Show that, up to a common rescaling, the weights w_i that maximize ε are those that minimize

$$e = \left(\min_{0=j_0 < j_1 < \cdots < j_h = r} \left(\sum_{k=0}^{h-1} (\varepsilon_{j_k} + \varepsilon_{j_{k+1}})(w_{j_k} - w_{j_{k+1}}) \right) \right)$$

subject to the constraint $e = 1$. Argue as in the proof of (4.34) that the w_j's that realize this minimum have the property that the points (ε_j, w_j) *all* lie on their own lower convex envelope, and hence, that, for such weights

$$(4.38) \qquad e = \sum_{j=0}^{r-1} (\varepsilon_j + \varepsilon_{j+1})(w_j - w_{j-1}) = \sum_{j=0}^{r-1} (\varepsilon_j + \varepsilon_{j+1}) x_j.$$

where, on the right, we've set $x_j := w_j - w_{j+1}$.

2) Show that the w_j's decrease and that the points (ε_i, w_i) all lie on their own lower convex envelope if and only if the sequence $(x_j/(\varepsilon_{j+1} - \varepsilon_0))$ is decreasing. Deduce that these conditions define an $(r-1)$-simplex in the hyperplane

$$\sum_{j=1}^{r} j x_j = \sum_{j=0}^{r} w_j$$

in x-space and hence that the linear function e in (4.38) achieves its maximum value at one of the vertices of this simplex.

3) Show that a set of w_j's corresponds to a vertex of this simplex if and only if, for some i between 1 and r, we have

$$w_j = \frac{(\varepsilon_i - \varepsilon_j)}{i\varepsilon_i - \varepsilon_1 - \cdots - \varepsilon_{i-1}}$$

for $j \le i$ and $w_j = 0$ for $j > i$, and that, for these weights, the sum defining e equals

$$\frac{(\varepsilon_i)^2}{i\varepsilon_i - \varepsilon_1 - \cdots - \varepsilon_{i-1}}.$$

4) Use Figure (4.36) to show that $\varepsilon_i \le i$ for $i \le d - 2g + 1$ and $\varepsilon_i \le i + l$ for $i = d - 2g + 1 + l$. Combine this with part 3) to reprove (4.34) by explicitly evaluating the ε_C of Gieseker's criterion (4.30).

C Construction of $\overline{\mathcal{M}}_g$ via the Potential Stability Theorem

The plan of the construction and a few corollaries

Our goal in this section is to outline the main ideas in the G.I.T. construction of $\overline{\mathcal{M}}_g$ when $g \ge 2$. Although the construction is direct and global in nature and has the projectivity of $\overline{\mathcal{M}}_g$ as an immediate corollary, the main technical result [Theorem (4.45)] involves performing many small instability calculations. To keep this section brief, we leave the details of some steps to you, generally by setting them in the form of exercises; and we work pointwise, indicating in parenthetical remarks when stronger uniform results are required and leaving you to supply the necessary arguments.

The basic technique is to show that suitable projective models of moduli stable curves have stable Hilbert points and apply G.I.T. However, no direct proof that Hilbert points of singular moduli stable curves verify the numerical criterion is known. In particular, as shown in Exercise (4.32), Gieseker's criterion may fail for such points. The paradoxical idea for verifying the stability of certain Hilbert points of stable curves, due to Gieseker and Mumford, has two parts.

First, consider curves embedded by linear systems of *degree sufficiently large relative to the arithmetic genus*. The Potential Stability Theorem (4.45) shows that if such a curve isn't moduli semistable then it has a nonsemistable Hilbert point, and if it's moduli semistable, then the linear system that embeds it must have a number of good properties. These results came as a big surprise when they were first discovered since stable curves in the plane and other low-dimensional projective spaces can have arbitrarily bad singularities: see part 2) of Exercise (4.20). The key idea is that imposing the degree hypothesis above on the embedding does away with these pathologies. The proof of the Potential Stability Theorem is the heart of the construction of $\overline{\mathcal{M}}_g$ and involves most of the work. [8]

The second part of the proof involves considering a one-parameter smoothing of a pluricanonically embedded stable curve C. By the valuative criterion, the pluricanonical Hilbert points of the smooth fibers

---

[8]However, the rest of the construction and its consequences are independent of this proof, so if you want you can simply accept this result and omit its proof.

in such a family have a Hilbert semistable limit in the corresponding Hilbert scheme. The Potential Stability Theorem is then used to deduce that this limit can only be the Hilbert point of the pluricanonical model of C. This approach has recently been extended by Caporaso [16], at the cost of considerably greater technical complications, to prove a converse to the Potential Stability Theorem that is then applied to construct modular compactifications of the universal Picard varieties $\mathcal{P}_{d,g}$ discussed in Section B.

The remainder of this section is organized as follows. First, we claim four properties for the pluricanonical locus constructed in C. Assuming these, we construct $\overline{\mathcal{M}}_g$ and deduce a few important properties. We then turn to the statement and proof of the Potential Stability Theorem. Finally, we deduce the claims from this theorem.

Definition of $\overline{\mathcal{M}}_g$ and verification of its properties

For the rest of this chapter, we fix a genus $g \geq 2$ and an integer $n \geq 5$ and define integers r, s and d in terms of these, as in (3.14), by

$$s = r + 1 = (2 \cdot n - 1)(g - 1) \quad \text{and}$$

$$d = 2 \cdot n(g - 1).$$

(In this section, our curves will live in a projective space of dimension r, but it'll simplify many of the formulas in the proof of the Potential Stability Theorem to express them in terms of the corresponding affine dimension $s = r + 1$.) We let $\mathcal{H} = \mathcal{H}_{d,g,r}$ and let $\widetilde{\mathcal{K}}$ be the locus in $\mathcal{H}$ of moduli stable curves C of genus g embedded in $\mathbb{P}^r$ by the nth power of their dualizing sheaves.

Finally, we define $\widetilde{\mathcal{K}}_{ss}$ to be the intersection of $\widetilde{\mathcal{K}}$ with the *semistable locus* in $\mathcal{H}$ for the natural action of $G = \mathrm{SL}(r + 1)$. We will continue to abuse language slightly and refer to all curves in $\widetilde{\mathcal{K}}$ as *n-canonically embedded* (even when they aren't smooth) and will likewise refer to $\widetilde{\mathcal{K}}_{ss}$ as the locus of *n-canonically embedded Hilbert-semistable curves* in $\mathcal{H}$.

CLAIM (4.39)   1) $\widetilde{\mathcal{K}}_{ss}$ *is smooth.*

2) $\widetilde{\mathcal{K}}_{ss}$ *is closed in the semistable locus $\mathcal{H}_{ss}$ in $\mathcal{H}$.*

3) $\widetilde{\mathcal{K}}_{ss} = \widetilde{\mathcal{K}}_{s}$; *i.e., every curve whose Hilbert point lies in $\widetilde{\mathcal{K}}_{ss}$ is Hilbert stable.*

4) $\widetilde{\mathcal{K}}_{ss}$ *contains the n-canonical models of every moduli stable curve of genus g.*

Of course, the last property simply states that $\widetilde{\mathcal{K}}_{ss} = \widetilde{\mathcal{K}}$. We will, however, only see this at the very end of the section; hence the need for the notation $\widetilde{\mathcal{K}}_{ss}$ here. The first statement follows immediately from Lemma (3.35) which proves the smoothness of $\widetilde{\mathcal{K}}$. The others will be proved at the end of the section once the Potential Stability Theorem (4.45) is established.

DEFINITION (4.40) *We define $\overline{\mathcal{M}}_g$ to be the G.I.T. quotient for the action of G on $\widetilde{\mathcal{K}}_{ss}$; that is, if we denote the quotient map by $q : \widetilde{\mathcal{K}}_{ss} \to \widetilde{\mathcal{K}}_{ss}/G$, then $\overline{\mathcal{M}}_g = \widetilde{\mathcal{K}}_{ss}/G$.*

Implicit in this definition, of course, is the assertion that this quotient is a coarse moduli space for stable curves of genus g. Let's check this now.

Let $\varphi : \mathcal{C} \to S$ be a family of stable curves of genus g. Each frame for the direct image $\varphi_*((\omega_{\mathcal{C}/S})^{\otimes n})$ of the nth power $(\omega_{\mathcal{C}/S})^{\otimes n}$ of the relative dualizing sheaf realizes $\mathcal{C}$ as a family of n-canonically embedded curves of genus g in $\mathbb{P}^r \times B$ and hence corresponds to a unique morphism $\alpha : B \to \mathcal{H}$. Claim 3 above implies that every such α factors through $\widetilde{\mathcal{K}}_{ss}$ to give a map $B \to \widetilde{\mathcal{K}}_{ss}$, which we continue to denote by α. By construction, the composition $\chi = q \circ \alpha : B \to \overline{\mathcal{M}}_g$ is independent of the initial choice of basis. The universal properties of the Hilbert scheme (i.e., the fact that the map α is supplied by an isomorphism of functors between the functor of points of $\widetilde{\mathcal{K}}_{ss}$ and the moduli functor of "isomorphism classes of stable curves of genus g plus a basis for the sections of the n-canonical sheaf") immediately imply that the maps χ paste together to yield a natural transformation Ψ from the moduli functor F of "isomorphism classes of stable curves of genus g" to the functor of points of $\overline{\mathcal{M}}_g$.

To see that $\overline{\mathcal{M}}_g$ is a coarse moduli space, it remains to check that Ψ satisfies properties 1) and 2) of Definition (1.3). Conditions 2) and 3) above immediately imply property 1): complex points of $\overline{\mathcal{M}}_g$ correspond bijectively to G-orbits in $\widetilde{\mathcal{K}}_{ss}$ and these in turn correspond bijectively to moduli stable curves of genus g over $\mathbb{C}$.

To check property 2, suppose we're given another scheme $\mathcal{M}'$ and a natural transformation Ψ' from F to $\text{Mor}_{\mathcal{M}'}$. Applying Ψ' to the restriction to $\widetilde{\mathcal{K}}_{ss}$ of the universal curve $\mathcal{C} \to \mathcal{H}$ over the Hilbert scheme $\mathcal{H}$ yields a morphism $\rho : \widetilde{\mathcal{K}}_{ss} \to \mathcal{M}'$. Since F is the functor of "isomorphism classes of stable curves of genus g", the map ρ must be constant on each G-orbit in $\widetilde{\mathcal{K}}_{ss}$ and we can therefore factor ρ uniquely through the quotient $\overline{\mathcal{M}}_g$ of $\widetilde{\mathcal{K}}_{ss}$ by G; that is, $\rho = \pi \circ q$ for a *unique* morphism $\pi : \overline{\mathcal{M}}_g \to \mathcal{M}'$. The universal properties of the Hilbert scheme likewise imply that the natural transformation Π induced by the maps π satisfies the relation $\Psi' = \Pi \circ \Psi$ required in property 2). Modulo the claims in (4.39), we've proved:

THEOREM (4.41) $\overline{\mathcal{M}}_g$ *is the course moduli space for stable curves of genus g.*

Our use of the definite article above is justified by Exercise (1.4), which shows that any two models of a fixed coarse moduli space must be canonically isomorphic.

Projectivity and irreducibility of $\overline{\mathcal{M}}_g$

Condition 2. of Claim (4.39) seems to be extraneous to the argument above. In fact, it's the key to proving conditions 3. and 4. and is the point at which the Potential Stability Theorem is crucial. The assertion that $\widetilde{\mathcal{K}}_{ss}$ is closed in the semistable locus has an immediate corollary of fundamental importance both in psychological and practical terms:

COROLLARY (4.42) $\overline{\mathcal{M}}_g$ *is projective.*

This was first proved by Knudsen [97] by a delicate study of the fibers of the Torelli map from $\overline{\mathcal{M}}_g$ to the Satake compactification of the moduli space of principally polarized abelian varieties.

The next corollary depends on the assertion of Lemma (3.15) that the stabilizer $\text{stab}_{\text{SL}(r+1)}([C])$ of any Hilbert point $[C]$ in $\widetilde{\mathcal{K}}_{ss}$ is finite and reduced. This, together with the smoothness of $\widetilde{\mathcal{K}}_{ss}$, immediately confirms the local description of $\overline{\mathcal{M}}_g$ suggested by the versal deformation theory of stable curves and announced on page 53. The only singularities are at curves with automorphisms, and all such curves correspond to singular points (with a few exceptions in small genera where there is a divisor or curves with automorphisms).

Everything we've said thus far can be proved over an algebraically closed field of any characteristic without substantially more work. With less ease, Seshadri has shown that it's possible to work over fairly general base rings [139]. In keeping with the philosophy of this book, we won't say too much about this here except to remark that the G.I.T. approach to the construction of $\overline{\mathcal{M}}_g$ is the only one that can be carried out in positive characteristics. We cannot, however, resist giving Gieseker's proof that the irreducibility of $\overline{\mathcal{M}}_g$ in positive characteristics as a consequence of the classical irreducibility of the complex moduli space of smooth curves. This beautifully simple proof also illustrates the usefulness in applications of the explicit G.I.T. construction of $\overline{\mathcal{M}}_g$ as a projective variety.

THEOREM (4.43) $\overline{\mathcal{M}}_g$ *is irreducible over any algebraically closed field.*

The proof depends on the classical analytic fact that this is true in characteristic 0 which will be given in Section 6.A. Now, if our ground field k has characteristic p, choose a discrete valuation ring R having quotient field F of characteristic zero and having k as its residue field: for example, R can be taken to be the ring of Witt vectors of k. We can carry out the constructions of $\overline{\mathcal{M}}_g$ as a quotient of $\widetilde{\mathcal{K}}_{ss}$ by PGL$(r + 1)$ over R. (We've been working with SL$(r + 1)$ for convenience but, of course, it is PGL$(r + 1)$ which acts effectively on $\mathcal{H}$ and hence on $\widetilde{\mathcal{K}}_{ss}$). Since $\overline{\mathcal{M}}_g(R)$ is projective and its generic fiber $\overline{\mathcal{M}}_g(F)$ is connected, Zariski's connectedness theorem implies that the special fiber $\overline{\mathcal{M}}_g(k)$ is also connected. Since PGL$(r + 1, k)$ is irreducible, $\widetilde{\mathcal{K}}_{ss}(k)$ is also connected. Since $\widetilde{\mathcal{K}}_{ss}(k)$ is smooth and connected, it's irreducible. Thus, $\overline{\mathcal{M}}_g(k)$ is also irreducible.

This last corollary brings us back to the origins of stable curves. It was in order to show the irreducibility of $\mathcal{M}_g$ in positive characteristics by an induction on g that Deligne and Mumford first worked out the theory of the moduli stack of stable curves in [29]. It's interesting to compare the proof above with the somewhat intricate inductive proof that, lacking a concrete projective construction of $\overline{\mathcal{M}}_g$, Deligne and Mumford were obliged to give.

The Potential Stability Theorem

Statement and preliminaries

DEFINITION (4.44) *We call a connected curve C of genus g and degree d in $\mathbb{P}^r = \mathbb{P}(V^s)$ where $s = r + 1 = d - g + 1$ potentially stable if:*

1) *The embedded curve C is nondegenerate (i.e., spans $\mathbb{P}^r$).*

2) *The abstract curve C is moduli semistable.*

3) *The linear series embedding C is complete and nonspecial: i.e., $h^0(C, \mathcal{O}_C(1)) = s$ and $h^1(C, \mathcal{O}_C(1)) = 0$.*

4) *Any chain of smooth rational components of C meeting the rest of C in exactly two points has length 1.*

5) *If R is a smooth rational component meeting the rest of C in exactly two points, then $\deg_R(\mathcal{O}_C(1)) = 1$; that is, R is embedded as a line.*

6) *If Y is a complete subcurve of C of arithmetic genus g_Y meeting the rest of C in k_Y points, then*

$$\left| \deg_Y(\mathcal{O}_C(1)) - \frac{d}{g-1}\left(g_r - 1 + \frac{k_Y}{2}\right) \right| \le \frac{k_Y}{2}.$$

Conditions 4) and 5) are actually consequences of 6) — we'll see this shortly — but we've stated them separately because they indicate how close to moduli stable the abstract curve C underlying a potentially stable curve C in $\mathbb{P}^r(\mathbb{C})$ must be. We will continue, as usual, to abuse language and speak of a "potentially stable curve C" when the implied embedding is clear from the context. The justification for this somewhat baroque definition lies in the following theorem, which with Theorem (4.34) forms the heart of the whole G.I.T. construction of $\overline{\mathcal{M}}_g$.

THEOREM (4.45) (POTENTIAL STABILITY THEOREM) *Fix integers g and d with $g \geq 2$ and $d > 9(g - 1)$. Then there is an M depending only on d and g such that if $m \geq M$ and C in $\mathbb{P}^r(\mathbb{C})$ is a connected curve with semistable m^{th} Hilbert point, then C is potentially stable.*

Here we'll only prove a pointwise version of the theorem (with m allowed to depend on C), footnoting points at which our argument must be refined to get the uniform assertion of the theorem.

The answer to the natural question — is potential stability also sufficient for Hilbert semistability? — is, essentially, yes. With slightly weaker numerical hypotheses, this converse is proved by Caporasoin [16] and shown to yield modular compactifications of the universal Picard varieties. In particular, the moduli strictly semistable curves not ruled out by the theorem will have models of large degree — those satisfying conditions 3 and 6 — with semistable Hilbert points. We will only prove the converse for pluricanonical curves in the next subsection. Caporaso uses a generalization of the indirect approach taken there, but her proof requires an order of magnitude more combinatorial and geometric effort.

Even the proof of the necessity is somewhat lengthy: it occupies pages 35 to 87 of [58]. A condensed proof of a slightly weaker result, which suffices for the construction of $\overline{\mathcal{M}}_g$, is given in [60]. Our argument is most closely modelled on this one.

Despite the complications that ensue, the essential strategy is very simple: if C fails to have some property covered by (4.44), find the filtration F of V that highlights this failure most clearly and check that F is destabilizing by showing some form of the numerical criterion (4.23) is violated. Usually, we'll see that $e_F > 2d\alpha_F$ and apply the asymptotic version of Lemma (4.26). A certain care is needed in the order in which the properties are established since it is often necessary to assume some of these properties to check that the failure of others is destabilizing. We will indicate the correct order of these steps, giving the filtration F in each case. In a few steps, we check that F is destabilizing; in others, these checks are left as exercises. A few technical lemmas needed on the way have also been left as exercises.

We begin with a few notational preliminaries. If Y is a complete subcurve of C, we'll let

$$Y_{\text{red}} = \text{the underlying reduced subcurve}$$

$$n_Y : Y_{ns} \longrightarrow Y_{\text{red}} = \text{the normalization of } Y_{\text{red}}$$

$$\mathcal{L}_Y = \mathcal{O}_C(1)|_Y$$

$$d_Y = \deg_Y(\mathcal{L}_Y)$$

$$s_Y = h^0(Y, \mathcal{L}_Y)$$

$$g_Y = \text{arithmetic genus of } Y.$$

In the definitions given in Section B of the invariants $w_F(m)$ and e_F of a weighted filtration F on V, we implicitly had a fixed curve C and its mth Hilbert points in mind: $w_F(m)$ was the least F-weight of a basis of $H^0(C, \mathcal{O}_C(m))$ and e_F was the leading coefficient of the numerical polynomial representing $w_F(m)$. Here, we'll often wish to consider these invariants for subcurves Y of C: when we do, we'll write $w_F(Y, m)$ for $H^0(Y, \mathcal{O}_Y(m))$ and $e_F(Y)$ for the leading coefficient of this polynomial. If $Y = C$, we usually suppress the Y's in all these notations and we'll also generally replace $\mathcal{L}_C$ by $\mathcal{O}_C(1)$.

We also let $F(Y)$ denote the filtration (possibly trivial) of V that gives weight 0 to the kernel U_Y of the canonical restriction map

$$\varphi_Y : H^0(C, \mathcal{L}_C) \longrightarrow H^0(Y, \mathcal{L}_Y)$$

and weight 1 to all other sections. We note, for future reference, the obvious estimate

$$(4.46) \qquad\qquad \alpha_{F(Y)} \leq \frac{s_Y}{s}$$

with equality if and only if φ_Y is surjective.[9]

If Z is another complete subcurve of C having no common component with Y, we'll let $K_{Y,Z} = Y \cap Z$ and $k_{Y,Z} = |\{Y \cap Z\}|$. We will denote by $\tilde{Y}$ the closure of the complement of Y in C and write K_Y and k_Y for $K_{Y,\tilde{Y}}$ and $k_{Y,\tilde{Y}}$. We will refer to the nodes in K_Y as *boundary* nodes of Y. The other nodes in Y will be called *internal* to Y while the other nodes of $\tilde{Y}$ will be called *external* to Y. Finally, we define the *quasigenus*

$$h_Y = g_Y - 1 + \frac{k_Y}{2}.$$

The first virtue of the somewhat strange looking expression h_Y is that it provides an *additive* form of the genus in the sense that, if Y and Z have no common components, then

$$h_{Y \cup Z} = h_Y + h_Z.$$

---

[9]This is a typical example of a formula more naturally expressed in terms of the affine dimension s rather than the projective one r.

This follows directly from the definition by combining the ordinary arithmetic genus formula

$$g_{Y \cup Z} = g_Y + g_Z + k_{Y,Z} - 1$$

with the fact that, as sets,

$$K_{Y,Z} \cup K_{Y \cup Z} = K_Y \cup K_Z$$

with both unions disjoint. In effect, we view the boundary nodes as lying half on Y and half on $\tilde{Y}$ and so contributing $\frac{1}{2}$ to the genus of each.

We will mainly use these notations to help extract consequences of property 6) of (4.44). To get a first feel for this condition, suppose that Y is any chain of smooth rational curves in the potentially stable curve C with $k_Y \leq 2$. Then $g_Y = 0$ so $g_y - 1 + \frac{k_Y}{2} \leq 0$ and property 6) implies $d_Y \leq \frac{k_Y}{2}$. Since $d_Y > 0$, this immediately rules out $k_Y = 1$ and shows that any nodal, potentially semistable C is moduli semistable. Moreover, if $k_Y = 2$, then $d_Y = 1$ so that Y must be a single rational curve embedded as a line. The next exercise gives two restatements of property 6) of (4.44).

EXERCISE (4.47) Let C in $\mathbb{P}^r$ be a connected, reduced, nodal curve and let Y be a complete subcurve of C. Assume that $H^1(C, \mathcal{O}_C(1)) = 0$ (and hence also that $H^1(Y, \mathcal{L}_Y) = 0$).

1) Show that property 6) of (4.44) for *both* Y and $\tilde{Y}$ is equivalent to either *pair* of inequalities below:

i) $\left(\dfrac{d}{h}\right) h_Y \leq d_Y + \dfrac{k_Y}{2}$ and $\left(\dfrac{d}{h}\right) h_{\tilde{Y}} \leq d_{\tilde{Y}} + \dfrac{k_{\tilde{Y}}}{2}$,

or,

ii) $\left(\dfrac{s_Y}{s}\right) \geq \dfrac{\left(d_Y + \frac{k_Y}{2}\right)}{d}$ and $\left(\dfrac{s_{\tilde{Y}}}{s}\right) \geq \dfrac{\left(d_{\tilde{Y}} + \frac{k_{\tilde{Y}}}{2}\right)}{d}$.

2) Use the description of the dualizing sheaf of a nodal curve given in (3.3) to show that $\deg_Y(\omega_C|_Y) = \deg_Y(\omega_Y) + k_Y = 2h_Y$. Conclude that property 6) of (4.44) for Y is also equivalent to

$$\left| d_Y - \left(\frac{d}{\deg_C(\omega_C)}\right) \deg_Y(\omega_C|_Y) \right| \leq \frac{k_Y}{2}.$$

By part 1 of this exercise, the Potential Stability Theorem will follow if we show that the curves C of the theorem are reduced, nodal and nondegenerate, that the linear series embedding such a C is complete and nonspecial, and that

(4.48) $$\frac{s_Y}{s} \geq \frac{\left(d_Y + \frac{k_Y}{2}\right)}{d}$$

for *every* complete subcurve Y.

We leave as exercises two geometric estimates that will be our main tools for doing this.

EXERCISE (4.49) [FIRST BASIC ESTIMATE] Fix a weighted filtration F on V. If $S = \{Y_i | i \in I\}$ is a collection of subcurves of C such that the natural map

$$\varphi : \mathcal{O}_C \longrightarrow \bigoplus_{i \in I} \mathcal{O}_{Y_i}$$

has finite kernel and cokernel, then

$$e_F \geq \sum_{i \in I} e_F(Y_i).$$

If the weights of F are nonnegative, then the same conclusion holds assuming only that φ has finite cokernel.

EXERCISE (4.50) [SECOND BASIC ESTIMATE] Fix a weighted filtration $F = \{(V_i, w_i)\}$ on V with integer weights as in (4.22). Assume that:

1. Y is an irreducible subcurve of C with generic multiplicity μ.

2. For some j, V_j maps to 0 in $H^0(Y_{\mathrm{red}}, \mathcal{L}_{\mathrm{red}})$.

3. There is an effective divisor K on Y_{red} such that, for each $i < j$, V_i maps to
$$H^0(Y_{\mathrm{red}}, \mathcal{L}_{Y_{\mathrm{red}}}(-(w_0 - w_i)K)).$$

Then, $e_F \geq e_F(Y) \geq \left(w_0 - w_{j-1}\right)^2 \deg(K) + 2\mu w_{j-1} d_{Y_{\mathrm{red}}}$.

Remark. Proving the uniform version of the Potential Stability Theorem requires slightly stronger forms of the Basic Estimates. The assertion that F is destabilizing is by (4.23) a strict inequality for the polynomial $w_F(m)$. Even pointwise, no such comparison can be deduced from a nonstrict inequality on the leading coefficient e_F of $w_F(m)$ such as the estimates give. At first, it might seem that what we really need are versions of the Estimates with each e_F replaced by the corresponding $w_F(m)$ for some (pointwise or uniform) m. However, the conclusion that F is destabilizing will always involve combining one of the Basic Estimates with some form of the inequality $d > 9(g-1)$ in the hypothesis of Theorem (4.45). This introduces a "margin of error" that tautologically depends only on d and g. This margin is enough to allow us to apply the Estimates as stated above to obtain pointwise results. It's even enough to get uniform results if we strengthen the Estimates by adding some control over the nonleading terms of the polynomials $w_F(m)$. More precisely, we need to show that there are constants N, N' and M, depending only on d and g, such that, for $m \geq M$, we have

$$w_F(m) + Nm \geq \sum_{i \in I} w_F(Y_i, m)$$

and

$$w_F(m) + (N + N')m \geq w_F(Y) + N'm$$

$$\geq \left((w_0 - w_{j-1})^2 \deg(K) + 2\mu w_{j-1} d_{Y_{\text{red}}}\right)\left(\frac{m^2}{2}\right)$$

in the situations of the first and second estimate respectively.

The proofs of the Basic Estimates outlined in the following hints also give this refinement pointwise. Uniform versions can then be obtained by variations of the arguments used to get the uniform bounds in the Uniform m lemma (1.11) and Gieseker's criterion (4.30). We leave you to supply these refinements if you're interested.

Hints and sketch of proof: Both proofs are variations on the ideas used in the proof of Gieseker's criterion (4.30), the main difference being that lower bounds for e_F are derived from upper bounds for the dimensions of spaces of sections of small weight, rather than the reverse.

For example, the First Basic Estimate follows directly from the fact that the restriction maps

$$H^0(C, \mathcal{L}_C^{\otimes m}) \longrightarrow \bigoplus_{i \in I} H^0(C_i, \mathcal{L}_i^{\otimes m})$$

have kernel and cokernel of dimensions bounded by those of φ.

The Second Basic Estimate requires a bit more work. First, observe that replacing Y by the curve defined by the μth power of the ideal of Y_{red} doesn't affect either the hypotheses or any quantities in the inequality for e_F. Then, use the fact that

$$h^0(Y, \mathcal{O}(m)) = \mu m d_{Y_{\text{red}}} + O(1)$$

to see that it suffices to consider the case where $w_{j-1} = w_{s-1} = 0$. Next, use the fact that $w_F(m) \geq w_F(Y_{\text{red}}, m)$ — since the w_i's are now positive — to reduce to the case where Y is reduced.

To treat this case, the key observation is that any monomial of weight at most w restricts on Y_{red} to a section of

$$H^0\left(Y_{\text{red}}, (\mathcal{L}_{Y_{\text{red}}})^{\otimes m}(-(mw_0 - w)K)\right)$$

and that the dimension of every such space differs from that of its preimage in $H^0(Y, \mathcal{O}_Y(m))$ by a uniform constant. The estimate

$$e_F(Y) \geq (w_0 - w_{j-1})^2 \deg(K) + 2\mu w_{j-1} d_{Y_{\text{red}}}$$

then follows by applying Riemann-Roch, summing over w and taking leading coefficients. Since the First Basic Estimate applies with $S = \{Y\}$ we get the claimed estimate for e_F as well. We will mainly use this estimate in the extreme cases where either $j = 1$ or $j = s$.

Outline of the proof

We're now ready to turn to the proof of Theorem (4.45), which we present as a series of steps.

Step 1: C_{red} *is nondegenerate.*

If not, use the filtration F that gives weight -1 to the sections that vanish on C_{red}, and weight $w > 0$ to the others where w is chosen so that the average α_F of the weights of F is 0. Choose an integer q so that the qth power of any nilpotent in the ideal sheaf of C is 0.[10] Then no monomial that contains more than q factors of weight -1 can even be nonzero modulo this ideal. Hence if $(m - q)w > q$, every element of a monomial basis of $H^0(C, \mathcal{O}_C(m))$ has strictly positive weight and $w_F(m)$ is a fortiori positive. By Proposition (4.23), C is Hilbert nonsemistable.

Henceforth we assume that $d > 9(g-1)$. Combining this with Step 1 and the Riemann-Roch estimate $s \leq d - g + 1$, we get an estimate that we'll often use without comment in the sequel:

$$(4.51) \qquad\qquad \frac{d}{s} < \frac{9}{8}.$$

We will also often use, when Y is irreducible, the estimate $r_Y \leq d_Y + 1$ with equality if and only if Y is a rational normal curve.

Step 2: *Every component of C is generically reduced.*

Suppose that Y is a multiple component of C of multiplicity μ. We claim that there is a nonzero section of $\mathcal{O}_C(1)$ vanishing on Y_{red}; that is, the filtration $F = F(Y_{\mathrm{red}})$ is nontrivial. If not, then since C_{red} is nondegenerate, we would have, by Riemann-Roch,

$$s_{Y_{\mathrm{red}}} \geq s \geq \left(\frac{8}{9}\right) d.$$

The trivial estimates

$$s_{Y_{\mathrm{red}}} \leq d_{Y_{\mathrm{red}}} + 1 \leq \left(\frac{1}{2}\right) d + 1$$

---

[10]Again, proving the uniform version of Theorem (4.45) requires a bound for q depending only on d and g that we leave to you if you're interested.

(the last, because $\mu \geq 2$), show that this cannot happen. The same estimate shows that

$$\alpha_F \leq \frac{d_{Y_{red}} + 1}{s},$$

and hence

$$2\alpha_F d < 2\left(\frac{9}{8}\right)(d_{Y_{red}} + 1).$$

On the other hand, the Second Basic Estimate (4.50) applies to F if we there take Y as the curve, the kernel W of the restriction map φ_Y as the subspace V_j, and the empty divisor as K. Since $w_0 = \cdots = w_{j-1} = 1$, we find that $e_F \geq 2\mu d_{Y_{red}}$. Since $\alpha_F < 1$, these show F is destabilizing unless

$$2\mu d_{Y_{red}} < 2\left(\frac{9}{8}\right)(d_{Y_{red}} + 1),$$

which can only happen if $\mu = 1$ as desired or $\mu = 2$ and $d_{Y_{red}} = 1$.

If we're in the last case, then $2\alpha_F d < 2\left(\frac{9}{8}\right)(d_{Y_{red}} + 1) < \frac{9}{2}$. Since $d_Y = 2$, there must be another component Z of C meeting Y in some point P. We can apply the Second Basic Estimate again to Z, this time with $V_j = \{0\}$ and $K = \{P\}$ — so $w_0 = 1$ and $w_{j-1} = 0$ — and conclude that $e_F(Z) \geq 1$. Finally, we may combine our estimates for $e_F(Y)$ and $e_F(Z)$ using the First Basic Estimate (4.49) with $\mathcal{S} = \{Y, Z\}$ to find $e_F \geq e_F(Y) + e_F(Z) \geq 2 \cdot 2 \cdot 1 + 1 = 5 > 2\alpha_F d$, which, by the Asymptotic Stability Criterion (4.26), shows that F is destabilizing.

Step 3: *If an irreducible subcurve Y of C is not a rational normal curve, then $d_Y \geq 4$.*

If Y is not rational normal, then $s_{Y_{red}} \leq d_Y$. If $d_Y < 4$ as well, the filtration $F = F(Y_{red})$ is nontrivial and $Y \neq C$. Therefore, there is a component Z of $\hat{Y}$ that meets Y in a point P. Arguing exactly as in the preceding step (except that now we know that $\mu = 1$), we find that $e_F(Y) \geq 2d_Y$, $e_F(Z) \geq 1$, and $e_F \geq e_F(Y) + e_F(Z) \geq 2d_Y + 1$. On the other hand we also have

$$2d\alpha_F \leq 2d\left(\frac{s_{Y_{red}}}{s}\right) < \left(\frac{9}{8}\right)s_{Y_{red}} \leq \left(\frac{9}{4}\right)d_Y.$$

This shows that F is destabilizing unless $d_Y \geq 4$.

Step 4: *If Y is a reduced irreducible subcurve of C, then $n_Y : Y_{ns} \to Y$ is unramified.*

Assume that n_Y ramifies at P. Consider the three-stage weighted filtration F that gives weight 0 to the space V_2 of sections whose image under restriction to Y and pullback via n_Y to Y_{ns} lie in $H^0(Y_{ns}, \mathcal{L}_{ns}(-3P))$, weight 1 to the space V_1 of sections with images in $H^0(Y_{ns}, \mathcal{L}_{ns}(-2P))$ and weight 3 to all others. Since n_Y ramifies, Y itself must be singular. Hence, $d_Y \geq 4$ by Step 3 and the hypotheses of the Second Basic Estimate hold with $\mathcal{S} = \{Y\}$, $K = P$, and $j = 2$.

Since $w_0 - w_j = 3$, we conclude that $e_F \geq 9$. Since P is ramified $\dim(V_0/V_1) \leq 1$, and in any case, $\dim(V_1/V_2) \leq 1$ so $\alpha_F \leq \frac{4}{5}$. Using (4.51), this means that $2d\alpha_F < 9$ and hence that F is destabilizing.

Step 5: *Every singular point of C_{red} has multiplicity 2.*

If P is a point of multiplicity 3 or greater, we claim that the two-stage filtration F that gives weight 0 to the space V_1 of sections vanishing at P and weight 1 to all others satisfies $e_F \geq 3$. Since α_F is clearly $\frac{1}{s}$ and $2d\alpha_F \leq 2\left(\frac{9}{8}\right)$, this will show that F is destabilizing. Suppose first that (at least) three distinct components Y_1, Y_2 and Y_3 meet at P. Pick points P_i on $(Y_i)_{ns}$ mapping to P. The Second Basic Estimate applies to each Y_i separately, taking $K = P_i$ and $j = 1$ to yield $e_F(Y_i) \geq 1$. Now, applying the First Basic Estimate, we conclude that $e_F \geq e_F(Y_1) + e_F(Y_2) + e_F(Y_3) \geq 3$ as desired. If only one component Y passes through P, then by Step 4, there are at least three distinct points Q, R and S lying over P on Y_{ns}. Moreover, Y is singular at P, so by Step 3, $d_Y \geq 4$. The Second Basic Estimate can then be applied to Y taking $K = Q + R + S$ and $j = 1$ to yield $e_F \geq e_F(Y) \geq 3$.

EXERCISE (4.52) Show that $e_F \geq 3$ in the case where exactly two components pass through P and complete the verification of Step 5.

Step 6: *Every double point of C_{red} is a node.*

In view of Steps 4 and 5, this follows from the following exercise.

EXERCISE (4.53) Suppose that two distinct points Q and R of the normalization C_{ns} of C map to a point P on C and that the corresponding branches of C have a common tangent line L at P. Consider the three-stage filtration F that gives weight 0 to the space V_2 of sections vanishing on L, weight 1 to the space V_1 of those vanishing at P but not along L, and weight 2 to all others.

1) Suppose both Q and R lie on a component Y of C and $K = Q + R$. Show that $d_Y \geq 4$, that the filtration induced on $H^0(Y_{\mathrm{red}}, \mathcal{L}_{Y_{\mathrm{red}}})$ by F is

$$H^0(Y_{\mathrm{red}}, \mathcal{L}_{Y_{\mathrm{red}}}) \supset H^0(Y_{\mathrm{red}}, \mathcal{L}_{Y_{\mathrm{red}}}(-K)) \supset H^0(Y_{\mathrm{red}}, \mathcal{L}_{Y_{\mathrm{red}}}(-2K)),$$

and that $e_F(Y) \geq 8$.

2) Suppose Q_1 and Q_2 lie on different components Y and Z of C. Show that if $d_Y > 1$ then the filtration induced on $H^0(Y_{\mathrm{red}}, \mathcal{L}_{Y_{\mathrm{red}}})$ by F is

$$H^0(Y_{\mathrm{red}}, \mathcal{L}_{Y_{\mathrm{red}}}) \supset H^0(Y_{\mathrm{red}}, \mathcal{L}_{Y_{\mathrm{red}}}(-Q)) \supset H^0(Y_{\mathrm{red}}, \mathcal{L}_{Y_{\mathrm{red}}}(-2Q))$$

and that $e_F(Y) \geq 4$. Show that if $d_Y = 1$ then $e_F(Y) \geq 3$.

3) Conclude that $e_F \geq 7$ and hence is always destabilizing.

Hint: The first three parts follow by various applications of the Second Basic Estimate. Given these, the last part is immediate, except in the case where Q and R lie on different components when it follows from the First Basic Estimate by eliminating the possibility that both Y and Z have degree 1.

Step 7: $H^1(C_{\text{red}}, \mathcal{O}_C(1)) = \{0\}$.

First, an exercise based on Clifford's theorem (cf. [62, Lemma 9.1]), which will be used in the proof.

EXERCISE (4.54) Suppose that C is a connected nodal curve and that $\mathcal{L}$ is a line bundle on C generated by global sections such that $H^1(C, \mathcal{L}) \neq \{0\}$. Show that there is a subcurve Y of C for which $s_Y \leq (d_Y/2) + 1$.

If $H^1(C_{\text{red}}, \mathcal{O}_C(1)) \neq \{0\}$ and Y is chosen as in the exercise, we claim that $F = F_Y$ will be destabilizing. By applying the Second Basic Estimate separately to each component Z of Y with K empty and V_j the space of weight-zero sections on that component, we see that, for each Z, $e_F(Z) \geq 2d_Z$. Using the First Basic Estimate, we can sum these estimates over the components of Y to get $e_F \geq 2d_Y$. The exercise then gives us the estimate $2d\alpha_F < \frac{9}{8}(d_Y + 2)$, which is less than e_F unless $d_Y = 2$. In this last case, since $\mathcal{L}_Y$ is ample, Y must be a line, a smooth plane conic or a pair of lines meeting in one point, all of which violate the assumption $s_Y \leq (d_Y/2) + 1$.

Step 8: *C is reduced, so* $H^1(C, \mathcal{O}_C(1)) = \{0\}$ *and* $V = H^0(C, \mathcal{O}_C(1))$.

This is perhaps the prettiest point in the argument. Let $\mathcal{I}$ be the ideal sheaf of nilpotents in $\mathcal{O}_C$. Then we have an exact sequence

$$0 \longrightarrow \mathcal{I} \otimes \mathcal{O}_C(1) \longrightarrow \mathcal{O}_C(1) \longrightarrow \mathcal{O}_C(1)|_{C_{\text{red}}} \longrightarrow 0.$$

In cohomology, this gives,

$$H^1(C, \mathcal{I} \otimes \mathcal{O}_C(1)) \longrightarrow H^1(C, \mathcal{O}_C(1)) \longrightarrow H^1(C, \mathcal{O}_C(1)|_{C_{\text{red}}}) \longrightarrow 0.$$

Since we now know $\mathcal{I}$ has finite support, the first term is 0. Step 7 shows that the third is also 0, hence so is the second. But the map $H^0(C, \mathcal{O}_C(1)) \to H^0(C, \mathcal{O}_C(1)|_{C_{\text{red}}})$ is injective by Step 1, so

$$h^0(C, \mathcal{O}_C(1)) \leq h^0(C, \mathcal{O}_C(1)|_{C_{\text{red}}}) = h^0(C, \mathcal{O}_C(1)) - h^0(C, \mathcal{I} \otimes \mathcal{O}_C(1)).$$

Hence $h^0(C, \mathcal{I} \otimes \mathcal{O}_C(1)) = 0$. Again, since $\mathcal{I}$ has finite support, this implies that $\mathcal{I} = 0$.

We have now established parts 1 to 3 of (4.44). By Exercise (4.47) and the remarks following it, all that remains is to show that (4.48) holds for *every* complete subcurve Y. To prepare to show this as our final Step 10, we first sharpen the result of Step 3.

Step 9: *For every subcurve Y of C and every component E of $\tilde{Y}$, either*
   1) $d_E \geq k_{E,Y}$, *or*,
   2) E *is a rational normal curve for which* $d_E = k_{E,Y} - 1$.

Consider the subcurve $Z = Y \cup E$. We have $s_Z \geq s_Y$ since span$(Z) \supset$ span(Y), so $d_Z - g_Z + 1 \geq d_Y - g_Y + 1$ by using Riemann-Roch and Step 8 twice. Substituting $d_Z = d_Y + d_E$ and $g_Z = g_Y + g_E - 1 + k_{E,Y}$ yields $d_E - g_E + 1 \geq k_{E,Y}$, which in turn gives case 1) unless E is a smooth, rational curve as in case 2).

Step 10: *Inequality (4.48) holds for every subcurve Y of C.*

We will show that if the desired inequality doesn't hold, then the filtration $F = F_Y$ must be destabilizing. Since $\alpha_F \leq (s_Y/s)$, this will follow if we show that $e_F \geq 2d_Y + k_Y$. We will deduce this inequality from the following two claims:

   1. The Second Basic Estimate for Y itself yields $e_F(Y) \geq 2d_Y$;

   2. The Second Basic Estimate for each component E of $\tilde{Y}$ implies $e_F(E) \geq k_{E,Y}$.

Summing these using the First Basic Estimate immediately gives $e_F \geq 2d_Y + k_Y$. Claim 1. is immediate, taking the subspace V_j of the Second Basic Estimate to be the kernel U_Y of the restriction map from C to Y (so $w_0 = w_{j-1} = 1$). To prove claim 2. let's first suppose that $d_E \geq k_{E,Y}$. Then the hypotheses of the Second Basic Estimate hold for E with $V_j = \{0\}$ and $K = K_{E,Y}$. Since $w_0 = 1$ and $w_{j-1} = 0$, we obtain $e_F(E) \geq k_{E,Y}$ directly. If $d_E < k_{E,Y}$, then we're in Case 2) of Step 9 so $(k_{E,Y}/2) \leq d_E$. But then, since every section in U_Y vanishes on $K_{E,Y}$, U_Y maps to zero in $H^0(E, \mathcal{L}_E)$. Applying the Second Basic Estimate again with $V_j = U_Y$ gives $e_F \geq 2d_E$, which, by hypothesis, is at least $k_{E,Y}$.

Completion of the construction

We now return to the proof of the last three properties of $\widetilde{\mathcal{K}}_{ss}$ listed in Claim (4.39), thereby completing the construction of $\overline{\mathcal{M}}_g$. Because we've fixed $n \geq 5$, the hypothesis — $d > 9(g - 1)$ — of the Potential Stability Theorem (4.45), holds for the curves in $\widetilde{\mathcal{K}}_{ss}$. Thus, we know that every curve whose Hilbert point lies in the semistable locus $\widetilde{\mathcal{K}}_{ss}$ of $\mathcal{H}$ is potentially stable. We first prove Claim 2).

PROPOSITION (4.55) $\widetilde{\mathcal{K}}_{ss}$ *is closed in* $\mathcal{H}_{ss}$.

Since, by Lemma (3.34), we know $\widetilde{\mathcal{K}}$ is open in the full Hilbert scheme, it's at least locally closed in $\mathcal{H}_{ss}$. Write $\widetilde{\mathcal{K}}_{ss} = \mathcal{Y}_1 \cup \cdots \cup \mathcal{Y}_p$ with each $\mathcal{Y}_i$ irreducible and locally closed in $\mathcal{H}_{ss}$ and let $g_i : \mathcal{Y}_i \to \widetilde{\mathcal{K}}_{ss}$ be the corresponding inclusions. To show that $\widetilde{\mathcal{K}}_{ss}$ is closed, we must show that each g_i is proper. Applying the valuative criterion for properness, we must therefore show that given a discrete valuation ring R with residue field $\mathbb{C}$ and quotient field F, any map $\alpha : \mathrm{Spec}(R) \to \mathcal{H}_{ss}$ that takes the generic point $\eta = \mathrm{Spec}(F)$ of $\mathrm{Spec}(R)$ into $\widetilde{\mathcal{K}}_{ss}$ also takes the closed point $0 = \mathrm{Spec}(\mathbb{C})$ of $\mathrm{Spec}(R)$ into $\widetilde{\mathcal{K}}_{ss}$.[11] We first use α to pull back the universal curve $\mathcal{C} \to \mathcal{H}_{ss}$ to a curve $\rho : \mathcal{D} \to \mathrm{Spec}(R)$ and let $\omega = \omega_{\mathcal{D}/\mathrm{Spec}(R)}$ denote the relative dualizing sheaf of this family. It follows from the definition of $\widetilde{\mathcal{K}}_{ss}$ and the universal property of $\mathcal{H}$, first that

$$(4.56) \qquad \mathcal{O}_{\mathcal{D}}(1)|_{\mathcal{D}_\eta} \cong \omega^{\otimes n}|_{\mathcal{D}_\eta}$$

and further that $\alpha(0)$ will lie in $\widetilde{\mathcal{K}}_{ss}$ if and only if we can extend this isomorphism over the closed point 0.

If we decompose the special fiber C_0 of $\mathcal{D}$ into irreducible components

$$C_0 = \bigcup_{i=1}^{l} C_i,$$

then (4.56) implies that

$$\mathcal{O}_{\mathcal{D}}(1)|_{\mathcal{D}_\eta} \cong \omega^{\otimes n}\left(-\sum_{i=1}^{l} a_i C_i\right)$$

with the multiplicities a_i determined up to a common integer translation. (Since $\mathrm{Spec}(R)$ is affine, $\mathcal{O}_{\mathcal{D}}(-C) \cong \mathcal{O}_{\mathcal{D}}$.) We normalize the a_i's so that all are nonnegative and at least one equals 0.

What we must show, then, is that *all* the a_i's are 0. Note that this is automatic if C_0 is irreducible. To take care of reducible C_0's, we'll use property 6) of (4.44). Let Y be the subcurve of C_0 consisting of all C_i for which a_i is zero, and let $\widetilde{Y}$ be the remainder of C_0 — i.e., those components for which a_i is positive. Then a local equation for $\mathcal{O}_{\mathcal{D}}(-\sum_{i=1}^{l} a_i C_i)$ is identically zero on every component of $\widetilde{Y}$ and on no component of Y. In particular, such an equation is zero at each of the k_Y points of $K_Y = K_{Y,\widetilde{Y}} = Y \cap \widetilde{Y}$. Therefore, we find that

---

[11]We use terms chosen to emphasize that our argument is independent of the characteristic, but if you prefer complex analytic language you may replace, as usual, $\mathrm{Spec}(R)$ by a disc Δ, $\mathrm{Spec}(F)$ by Δ^* and 0 by the origin in Δ.

$$k_Y \leq \deg_Y\left(\mathcal{O}_D\left(-\sum_{i=1}^{l} a_i C_i\right)\right)$$

$$= \deg_Y\left(\mathcal{O}_D(1)|_{C_0}\right) - n \, \deg_y\left(\omega|_{C_0}\right)$$

$$= \deg_Y\left(\mathcal{O}_D(1)|_{C_0}\right) - \left(\frac{\deg_{C_0}\left(\mathcal{O}_D(1)|_{C_0}\right)}{\deg_{C_0}\left(\omega|_{C_0}\right)}\right) \deg_Y\left(\omega|_{C_0}\right)$$

$$\leq \frac{k_Y}{2}$$

by part 2) of Exercise (4.47). Therefore $k_Y = 0$ and since C_0 is connected, $a_i = 0$ for all i.

PROPOSITION (4.57) *Every curve C in $\mathbb{P}^r$ whose Hilbert point lies in $\widetilde{\mathcal{K}}_{ss}$ is moduli stable.*

Since every curve C in $\widetilde{\mathcal{K}}_{ss}$ is potentially stable, the only problem is that C may contain some smooth rational components meeting the rest of the curve in only two points. This cannot in fact occur since, on the one hand, the degree of the dualizing sheaf ω_C of C on such a component is zero while, on the other, $\omega_C^{\otimes n}$ is very ample on C because the Hilbert point of C lies in $\widetilde{\mathcal{K}}_{ss}$. ●

PROPOSITION (4.58) *Every moduli stable curve of genus g has a model in $\widetilde{\mathcal{K}}_{ss}$.*

For any moduli stable curve and any $n \geq 5$, $\omega_C^{\otimes n}$ is very ample on C, and thus embeds C as a curve in $\mathbb{P}^r$ whose Hilbert point $[C]$ lies in $\mathcal{H}$. To see that $[C]$ lies in $\widetilde{\mathcal{K}}_{ss}$ or, equivalently, in $\mathcal{H}_{ss}$, choose a one-parameter deformation $\mathcal{C} \to \mathrm{Spec}(R)$ of C to a smooth connected curve over a discrete valuation ring R; that is, the generic fiber C_η of $\mathcal{C}$ is a smooth curve of genus g and the special fiber is C. Then $\mathcal{C}$ is again a stable curve over $\mathrm{Spec}(R)$, so its n-canonical embedding realizes it as a family of curves in $\mathbb{P}^r(\mathbb{C})$ and hence corresponds to a unique morphism $\alpha : \mathrm{Spec}(R) \to \mathcal{H}$. Since the generic fiber C_η is smooth of degree $2n(g-1)$, its Hilbert point $[C_\eta]$ lies in $\mathcal{H}_{ss}$ by Theorem (4.34). (This is the only — but essential — point at which this theorem is used in the whole construction.)

Since the quotient of $\mathcal{H}_{ss}$ by $\mathrm{SL}(r+1)$ is projective, we can, after possibly making a finite change of base, find a map $\beta : \mathrm{Spec}(R) \to \mathcal{H}_{ss}$ that agrees with α at η. By pulling back the universal curve over $\mathcal{H}$ by β, we obtain a second curve $\mathcal{C}' \to \mathrm{Spec}(R)$ whose generic fiber is also C_η. By the uniqueness of the semistable reduction of a family of stable curves, the stable models of the special fibers C and C' are isomorphic. We cannot immediately assert that the curves themselves are isomorphic since the Potential Stability Theorem only asserts that C' is moduli semistable. However, $\widetilde{\mathcal{K}}_{ss}$ is closed in $\mathcal{H}_{ss}$. Therefore, since $\beta(\eta)$ lies in $\widetilde{\mathcal{K}}_{ss}$ and $\beta(0)$ lies in $\mathcal{H}_{ss}$, we conclude that $\beta(0)$ also lies in $\widetilde{\mathcal{K}}$. In other words, C' is also n-canonically embedded and hence must be moduli-*stable*. Thus C and C' are both abstractly isomorphic and projectively equivalent in $\mathbb{P}^r(\mathbb{C})$. But the Hilbert point $[C']$ is in $\mathcal{H}_{ss}$ by construction, hence so is $[C]$.

PROPOSITION (4.59) $\widetilde{\mathcal{K}}_{ss} = \widetilde{\mathcal{K}}_s$: *every curve whose Hilbert point lies in $\widetilde{\mathcal{K}}_{ss}$ is Hilbert stable.*

Every curve C whose Hilbert point lies in $\widetilde{\mathcal{K}}_{ss}$ is, by definition, Hilbert semistable. If the Hilbert point of such a curve did not have a closed $SL(r + 1)$-orbit, then its closure would contain a semistable orbit with stabilizer of positive dimension by Exercise (4.13). Since every curve whose Hilbert point lies in $\widetilde{\mathcal{K}}_{ss}$ is nondegenerate, this orbit would correspond to a curve C' with infinitely many automorphisms, and since $\widetilde{\mathcal{K}}_{ss}$ is closed in $\mathcal{H}_{ss}$, the Hilbert point of C' would lie in $\widetilde{\mathcal{K}}_{ss}$. Since every such C' is moduli stable and hence has only finitely many automorphisms this leads to a contradiction. Thus every point of $\widetilde{\mathcal{K}}_{ss}$ has closed orbit and finite stabilizer, which means that $\widetilde{\mathcal{K}}_{ss} = \widetilde{\mathcal{K}}_s$ as desired.

We have therefore completed the construction of $\overline{\mathcal{M}}_g$. This proof is based on a yoga due to Gieseker, which ought, morally, to be more widely applicable. We wish to conclude by laying out the main steps in his approach to constructing a compactification of a moduli space $\mathcal{M}$ for a set of smooth varieties. Gieseker's idea is to use geometric invariant theory, not merely as a technical tool to construct the right scheme structure on the moduli space but also as a guide to understanding what class of degenerations should appear at the boundary.

The first step in his plan is to show directly that suitable projective models of the varieties in $\mathcal{M}$ have stable Hilbert points. In our example, this is the stability of smooth curves. Next, we try to eliminate possible candidates for the boundary points by showing that they have unstable Hilbert points. Typically, this amounts to finding restrictions both on the intrinsic geometry of varieties with semistable Hilbert points and on how they are projectively embedded: here this role is played by the Potential Stability Theorem. In one sense, this step is easier than the first step. As we've seen above, a destabilizing filtration F is generally closely tied to some more-or-less pathological geometric property of the unstable variety and this makes the estimation of the weight function $w_F(m)$ fairly easy. Proving a variety stable, on the other hand, requires dealing with an arbitrary F; some idea of just how difficult this can be in higher dimensions, even for smooth varieties, can be obtained by examining Gieseker's tour-de-force verification of the stability of sufficiently pluricanonical models of surfaces of general type [57]. In another sense, however, the second step can be harder, since the number of nonsemistable cases to be ruled out can be dauntingly large. Now comes Gieseker's key idea: whatever class of varieties we cannot by our best efforts eliminate in the second step ought to form the boundary of the compactified moduli space $\overline{\mathcal{M}}$. The final step is to try to make this concrete by combining valuative methods with a study of degenerations of elements of $\mathcal{M}$ to these "potential" boundary varieties to yield an indirect proof of their Hilbert stability. In the construction of $\overline{\mathcal{M}}_g$, this involves the deformation theoretic results and the tricks with the valuative criterion used in this section. A bonus of this approach is that, when it works, the completion $\overline{\mathcal{M}}$ is automatically projective.

All three steps pose substantial difficulties and represent a challenge for further study. Other than $\overline{\mathcal{M}}_g$, the only case in which this program has been completely worked out is that of degenerations of vector bundles on curves: see [127], [62] and [61]. It would be more satisfying, but seems even harder, to modify the last step by verifying directly the stability of the Hilbert points of varieties on the boundary. As we've remarked earlier, this has not even been done for stable curves.

Somewhat less precise results have been obtained for higher-dimensional varieties of general type by rather different methods. We close with a few brief pointers to the main results. For more details and definitions of unfamiliar terms, see the cited references.

Viehweg has constructed a quasiprojective coarse moduli space for smooth varieties of arbitrary dimension having ample canonical divisor and fixed pluricanonical Hilbert polynomial: his approach, described in [148], is via G.I.T. but uses a strategy different from Gieseker's and yields slightly weaker existence results for curves and surfaces than those obtained by Gieseker.

By combining this with ideas of Kollár which essentially replace the further use of G.I.T. by methods from minimal model theory, Karu [90] has recently shown that the minimal model program in dimension $n + 1$ implies the existence of a projective coarse moduli space for almost smoothable n-folds with semi-log-canonical singularities, ample canonical divisor and fixed pluricanonical Hilbert polynomial. The keystone of his proof is a theorem of Kollár [99] which states that the corresponding moduli functor is represented by a projective scheme if it has various other expected properties — more precisely, if it is bounded, locally closed, separated, complete, has tame automorphisms and has semi-positive canonical polarization.

Karu's key innovation is the deduction of the boundedness of the moduli functor — the existence of a family over a projective base in which every such variety is a fiber — from the minimal model program. His strategy, based on the notion of weak semistable reduction introduced in Abramovich-Karu [1], avoids any need for G.I.T. but remains close in spirit to the proof of of the projectivity of $\overline{\mathcal{M}}_g$ given here. It simplifies earlier work of Alexeev in the surface case — see Karu's paper for details and all further references — and also relies on an earlier boundedness theorem of Matsusaka and work of Kawamata and Siu on invariance of plurigenera. The remaining properties in Kollár's criterion were known or follow fairly directly by combining boundedness with work of other authors including Iitaka, Kleiman, Kollár, Shepherd-Barron and Viehweg.

Exciting as this result is, it is only a step towards being able to deal with moduli of higher-dimensional varieties as we do with those of curves, even for surfaces where the minimal model hypothesis is known. Much work remains before we have the explicit description of the boundary so essential to nearly all the applications which follow in the rest of this book. For example, no one has yet enumerated the boundary divisors in the compactification of the space of quintic surfaces in $\mathbb{P}^3$.

Chapter 5

Limit Linear Series and Brill-Noether theory

In this chapter, we want to illustrate how the moduli space of stable curves can be used as a tool to prove theorems that deal with a single curve. In most such applications, the role of the moduli space is to allow us to deduce facts about certain smooth curves by studying what happens when these curves undergo suitable degenerations.

As our example, we've chosen the theory of special linear series. We will develop a theory of limits of linear series on some singular curves, and use this to give proofs of the basic results of Brill-Noether theory.

A Introductory remarks on degenerations

Before getting to work, a few words are in order concerning the nature of the theorems we will discuss, their history and their various proofs. We first recall the statements of the two most important results. The Kempf/Kleiman-Laksov/Griffiths-Harris/Brill-Noether theorem (which, following custom, we'll henceforth refer to as simply "Brill-Noether") says that a general curve C of genus g carries a g_d^r if and only if the *Brill-Noether* number $\rho = \rho(g, r, d)$ defined by

$$(5.1) \qquad \rho = (r+1)(d-r) - rg = g - (r+1)(g-d+r)$$

is nonnegative; and if so, then ρ is the dimension of the locus $W_d^r(C)$ of such linear series in $\operatorname{Pic}^d(C)$. The Gieseker-Petri theorem says that the multiplication map

$$(5.2) \qquad \mu_0 : H^0(L) \otimes H^0(K \otimes L^{-1}) \to H^0(K)$$

is injective for any line bundle L on a general curve C.

EXERCISE (5.3) Show that a g_d^r with negative ρ corresponds to a line bundle L for which the domain of μ_0 has dimension at least $g + 1$ and conclude that Gieseker-Petri implies the nonexistence statement in Brill-Noether. Give an example of a line bundle L on a curve C for which μ_0 fails to be injective where this isn't forced by dimensional considerations (the first such example occurs in genus 4).

What distinguishes these theorems from more elementary results on linear series like Riemann-Roch and Clifford? The obvious answer is that they aren't true on every curve C of genus g: they apply only to an open dense subset in, but not to *all* of, $\mathcal{M}_g$. (This isn't true of the existence half of the Brill-Noether theorem ($\rho \geq 0 \Rightarrow W_d^r(C) \neq \varnothing$), which was indeed the first of the statements proved, independently by Kempf and by Kleiman and Laksov: see [92], [95], [96] and [7] for an overview. But it's certainly true of the nonexistence half ($\rho < 0 \Rightarrow W_d^r(C) = \varnothing$) and of the Gieseker-Petri theorem.) Moreover, we have no independent way to characterize the loci over which they do hold. It follows that any proof of such results must be fundamentally different from proofs of the more elementary results, which take place on an arbitrary, fixed curve.

The most direct approach would be to work on a curve that is no longer arbitrary but merely sufficiently general. Indeed, the fact that these theorems concern conditions that are open on proper, smooth families of curves and that we're only required to prove them on an open subset of $\mathcal{M}_g$ means that we could prove any one by exhibiting a *single* smooth curve satisfying it. This doesn't, however, seem to help: as of this writing, no one yet knows how to write down for large g (at least, say, 16), a single complete, smooth curve satisfying any of these theorems. The curves we can write down for large g, such as hyperelliptic and trigonal curves, complete intersections, and the like, are invariably special with respect to all the properties that these theorems assert to be general. (See, in this connection, the discussion at the end of Section 6.F.)

One resolution of this problem is to work not on fixed curves but in families, so as to incorporate variational elements into the proofs. This idea, as we shall see in a moment, goes back to the classical Italian geometers. However, their approach amounted to replacing the search for general curves with one for general sets of Schubert cycles. Once again, it turned out that all examples that can be analyzed explicitly are special. Although this path did eventually lead to a proof of the Brill-Noether statement itself, it gradually became clear that more refined results like the Gieseker-Petri theorem would be possible only with better control of both the families used and the linear series on them. With hindsight, we can now see that what was needed was to find degenerations to curves that are sufficiently special that we can carry out the necessary analysis explicitly but that remain general in the senses of the theorems. Such curves do exist, but all known examples are not merely singular but highly reducible. This explains why the approaches to these problems via the theory of limit linear series remained unexplored until the theory of stable curves had been developed and, incidentally, why, although their statements deal with a single curve, they are naturally treated in a book on moduli of curves.

Let's now examine this history a bit more closely. To begin with, Brill and Noether asserted the truth of the theorem based, apparently, on a naive dimension count bolstered by the calculation of examples in low genus. Exactly how the desired variational element might enter into the proofs was first suggested by Castelnuovo. His goal was not to establish any of the present theorems. Rather, he assumed the statement of the Brill-Noether theorem and applied it to compute the *number* of g_d^r's on a general curve in the case $\rho = 0$, when we expect it to be finite. To do this, Castelnuovo looked not at any smooth curve of genus g, but at a g-nodal curve C_0: that is, a rational curve with g nodes $r_1, \ldots, r_g$ obtained by identifying g pairs of points (p_i, q_i) on $\mathbb{P}^1$. Any g_d^r on C_0, Castelnuovo reasoned, would pull back to a g_d^r on $\mathbb{P}^1$, which could then be represented as the linear series cut out on a rational normal curve $C \cong \mathbb{P}^1 \longrightarrow \mathbb{P}^d$ of degree d by those hyperplanes containing a fixed $(d - r - 1)$-plane $\Lambda \subset \mathbb{P}^d$. The condition that the g_d^r on $\mathbb{P}^1$ be the pullback of one on C_0 was simply that every divisor of the g_d^r containing p_i should contain q_i as well; in other words, Λ should meet each of the chords $\overline{p_i q_i}$ to C in $\mathbb{P}^d$. The number of g_d^r's on a general curve of genus g was thus, according to Castelnuovo, the number of $(d - r - 1)$-planes in $\mathbb{P}^d$ meeting each of g lines, a problem in Schubert calculus that Castelnuovo went on to solve (to obtain the correct value for the number of g_d^r's on a general curve in [20]).

It was Severi who first pointed out, some twenty years later, that Castelnuovo's computation might serve as the basis of a proof of the Brill-Noether statement. Severi's idea was to argue in two steps: first, if one had a family of curves $\{C_t\}$ with C_0 as above and C_t a complete, smooth curve for $t \neq 0$, then by upper semicontinuity of dimension we should have $\dim(W_d^r(C_t)) \leq \dim(W_d^r(C_0))$ for general t; and second, since the requirement that a $(d - r - 1)$-plane $\Lambda \subset \mathbb{P}^d$ meet a line in $\mathbb{P}^d$ is a codimension r condition, we should have

$$\dim(W_d^r(C_0)) = \dim(\mathbb{G}(d - r - 1, d)) - rg$$

$$= (r + 1)(d - r) - rg$$

$$= \rho$$

Problems abound with both halves of this argument. First of all, if one defines $W_d^r(C_0)$ to be line bundles of degree d on C_0 with $r + 1$

or more sections, there one runs into the difficulty that the family $\{W_d^r(C_t)\}$ may not be proper, simply because the family $\{\text{Pic}^d(C_t)\}$ isn't; i.e., the limit of linear series need not be a linear series. (The classic example of this is the g_2^1 associated to the meromorphic function $(\frac{y}{x})$ on the cubic plane curve C_t given by $y^2 = x(x+1)(x-t)$ for t different from 0 and (-1), or, equivalently, to the pencil cut out on C_t by lines through the origin.) This difficulty may be overcome, as it was first by Kleiman [94], by using the fact that the varieties parameterizing *torsion-free sheaves of rank 1* on C_t do form a proper family, and that for each subset $I = \{i_1, \dots, i_k\} \subset \{1, \dots, g\}$ those torsion-free sheaves on C_0 that fail to be locally free exactly at $\{r_{i_1}, \dots, r_{i_k}\}$ are direct images of invertible sheaves of degree $d - k$ on the partial normalization C_I of C_0 at $\{r_{i_1}, \dots, r_{i_k}\}$. Since $\rho(d-k, g-k, r) \le \rho(d, g, r)$, we need only verify the Brill-Noether statement $\dim(W_d^r) = \rho$ on all the partial normalizations of C_0.

As for the second half of Severi's intended argument, it's not the case that the g sets of conditions on a $(d - r - 1)$-plane $\Lambda \subset \mathbb{P}^d$ that it meet the chords $\overline{p_i q_i}$ to C are algebraically independent — i.e., that the corresponding Schubert cycles intersect in the expected codimension rg in $\mathbb{G}(d - r - 1, d)$ — for all choices of $p_i, q_i \in C$. The simplest example of this is the hyperelliptic case $d = 2, r = 1$. Here the g_2^1's on the curve C_0 correspond to points in $\mathbb{P}^2$ lying on each of the g chords $\overline{p_i q_i}$ to the conic $C \subset \mathbb{P}^2$. If $g = 2$, of course there is always a unique such point, corresponding to the fact that every genus 2 has a unique g_2^1. If $g = 3$, however, there may or may not exist such a point, depending on the choice of the points p_i, q_i, corresponding to the fact that while the general curve of genus 3 doesn't possess a g_2^1, some do.

Once again, we might hope to overcome this problem by introducing a further variational element into the argument: that is, to consider a further specialization of the points $p_i, q_i \in \mathbb{P}^1$. As before, however, nobody knows, for large g, even a single choice of such points for which the corresponding Schubert cycles are dimensionally transverse. Hence, this approach must also involve a further degeneration of the underlying curve. What worked in the end was to let the points $q_1, p_2, q_2, p_3, q_3, \dots$ tend to p_1, one at a time, in that order. This was done by Griffiths and Harris in [65], who were able to conclude the Brill-Noether statement.

Griffiths and Harris did not consider, in their paper, what happened in the limit to the curve C_0 as they carried out their degeneration. There, the Brill-Noether problem had been transposed into a question of whether certain Schubert cycles in $\mathbb{G}(d - r - 1, d)$ intersected transversely, and they were only concerned with the behavior, in the limit, of those Schubert cycles. It was Gieseker, in his proof of Petri's conjecture in [59], who first thought to follow the curve C_0 into the limit.

Gieseker worked with a family of curves obtained by taking $\mathbb{P}^1 \times \Delta_t$ and identifying sections $p_i(t)$ with $q_i(t)$ over Δ, with $p_i(t)$ and $q_i(t)$ all distinct for $t \neq 0$ and coming together with different orders of contact with p_1 at $t = 0$, as in Figure (5.4).

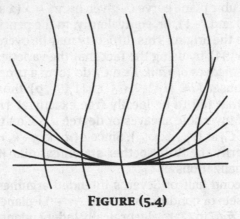

FIGURE (5.4)

In order to make sense out of "the curve C_0 obtained by identifying $p_i(t)$ with $q_i(t)$ on $\mathbb{P}^1 \times \{t\}$" when $t = 0$, it's natural before making the identifications to blow up the point $p_1(0)$ until the sections become disjoint. This results in a family shown in Figure (5.5). After

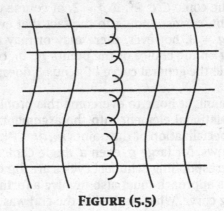

FIGURE (5.5)

making the appropriate identifications, the resulting C_0 looks like the one pictured in Figure (5.6).

One benefit of Gieseker's approach is that if one applies a base change to a family of curves degenerating to a curve of the type pictured in Figure (5.6), and then minimally resolves the resulting singularities of the total space, we again get such a family. We can also blow up a node and make a double base change with the same effect. It follows, then, that any family of line bundles away from the central fiber of such a family of curves may be assumed to be single-valued over

FIGURE (5.6)

the base, and then to extend over the central fiber. Thus, the need to deal with torsion-free sheaves as in Kleiman's reworking of the first part of Severi's argument is avoided. This benefit comes at a price of a new technical complication, however: we're now dealing with linear series on a reducible curve and have to develop a formalism for doing this. Setting this up will occupy most of the remainder of this chapter.

In another, more or less orthogonal development, it was noticed by Eisenbud and Harris ([37], [38]) that the proof of the Brill-Noether statement could be substantially simplified by specializing to *cuspidal*, rather than nodal curves. In its simplest case, cited above, of g_2^1's on curves of genus 3, this amounts to the observation that, while three *chords* to a conic in the plane may or may not be concurrent (in fact, they are so in codimension 1), three *tangent lines* can never be. Thus, while the Brill-Noether condition doesn't hold for all nodal curves of genus 3, it apparently does for all cuspidal rational curves. This phenomenon is in fact general: as it turns out, the Plücker formulas applied to a linear series on $\mathbb{P}^1$ directly imply the Brill-Noether property for an arbitrary g-cuspidal curve. (In the case of g_2^1's on curves of genus 3, the relevant Plücker formula is just the Riemann-Hurwitz formula, that a pencil of degree 2 on $\mathbb{P}^1$ cannot have more than two branch points.)

In these circumstances, it's natural to ask if it's possible to prove the Gieseker-Petri theorem via a specialization to a g-cuspidal curve. This turns out to be not as easy as Brill-Noether: for one thing, it isn't true that every g-cuspidal curve satisfies Petri's condition, and consequently further degeneration is required. The obvious thing to do is to let the cusps come together one at a time. On the other hand, in order to analyze linear series on the limiting curve à la Gieseker, we'll need to find a well-behaved model for this limit by considering the semistable reduction of such a degeneration. Since cuspidal curves aren't themselves semistable, the first thing to do is to apply a semistable reduction to the family of curves specializing to a g-

cuspidal one. This yields a family of smooth curves specializing to a rational curve with g elliptic tails attached (shown as the "S"-curves in Figure (5.7) and those that follow). Next, we can bring the points q_i at

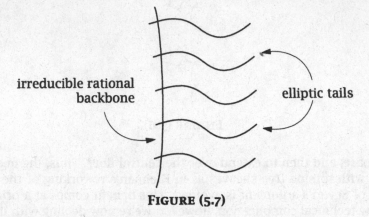

irreducible rational backbone

elliptic tails

FIGURE (5.7)

which the tails are attached together one at a time (in effect, making the cusps of the original family come together) to arrive at the curve pictured in Figure (5.8). Finally, to allow for the possibility of further

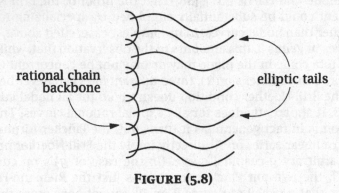

rational chain backbone

elliptic tails

FIGURE (5.8)

base change and/or blowing up of nodes and subsequent semistable reduction, we generalize this to the curve X_0 pictured in Figure (5.10). In this diagram, all the components are rational except the g elliptic tails on the at the right end of each chain.

These are the curves that we'll use in our proofs of the Brill-Noether and Gieseker-Petri theorems; we'll call them *flag curves*. Specifically, we will show that:

THEOREM (5.9) *If $\pi : X \to B$ is any flat projective family of curves with smooth general fiber and special fiber $X_0 = \pi^{-1}(b_0)$ isomorphic to a flag curve, as pictured in Figure (5.10), then the general fiber of π satisfies the Brill-Noether and Petri conditions (5.1) and (5.2).*

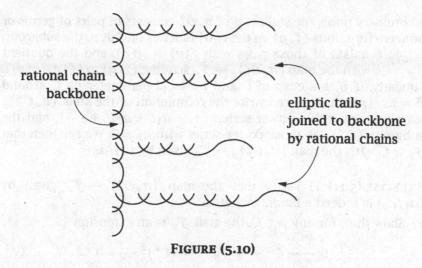

rational chain backbone

elliptic tails joined to backbone by rational chains

FIGURE (5.10)

In fact, the Brill-Noether statement will be shown for a larger class of curves: it will be true for the general member of any family of curves that includes a curve composed of a tree of rational curves with g elliptic tails attached. The Petri statement, however, will only be shown for the specific configuration of Figure (5.10).

We close this section by mentioning that there is another approach to these questions due to Lazarsfeld [107], which we won't go into here but which is perhaps the most elegant. Lazarsfeld uses a beautiful vector bundle argument to show that special linear series on certain curves lying on $K3$-surfaces must be cut out by linear series on the surface and is able to deduce that a curve whose class generates the Neron-Severi group of such a surface must be general in the senses of Brill-Noether and Gieseker-Petri.

B Limits of line bundles

We begin by assembling some basic facts about line bundles on individual stable curves and families of them. Let's consider, to begin with, the group $\mathrm{Pic}(C)$ of line bundles on a connected but possibly singular and/or reducible curve C. We will assume C is reduced, although a description of the group of line bundles on a nonreduced curve is a question of some interest as well. If we let $\pi : \tilde{C} \to C$ be the normalization of C, then an essentially complete description of $\mathrm{Pic}(C)$ is obtained by comparing the sheaf $\mathcal{O}_C^*$ of nonzero functions on C with the pushforward $\pi_* \mathcal{O}_{\tilde{C}}^*$ of that of $\tilde{C}$. Specifically, the map π gives an inclusion $\mathcal{O}_C^* \longmapsto \pi_* \mathcal{O}_{\tilde{C}}^*$. The quotient sheaf $\mathcal{F}$ is a skyscraper sheaf supported at the singular points of C, whose stalks are fairly easy to describe in terms of these singularities. Thus, for example, if $p \in C$ is an ordinary node, the stalk at p of $\pi_* \mathcal{O}_{\tilde{C}}^*$ consists of pairs of germs of nonzero functions (f, g) on the two branches of C at p, the subgroup $(\mathcal{O}_C^*)_p$ consists of those pairs with $f(p) = g(p)$, and the quotient $\mathcal{F}_p \cong \mathbb{C}^*$ with the map $(\pi_* \mathcal{O}_{\tilde{C}}^*)_p \to \mathcal{F}_p$ sending (f, g) to $(f(p)/g(p))$. Similarly, if p is a cusp of C and t a local parameter on $\tilde{C}$ around $\tilde{p} = \pi^{-1}(p)$, then we can write the completion of the stalk $(\pi_* \mathcal{O}_{\tilde{C}}^*)_p$ simply as invertible power series $\{a_0 + a_1 t + a_2 t^2 + \cdots\}$, and the subgroup $(\mathcal{O}_C^*)_p$ as those power series with $a_1 = 0$. We see then that $\mathcal{F}_p \cong \mathbb{C}$, with the map $(\pi_* \mathcal{O}_{\tilde{C}}^*)_p \to \mathcal{F}_p$ given by (a_1/a_0).

EXERCISE (5.11) 1) Verify that the map $(\pi_* \mathcal{O}_{\tilde{C}}^*)_p \to \mathcal{F}_p$ given by (a_1/a_0) is indeed a homomorphism.

2) Show that, for any $p \in C$, the stalk $\mathcal{F}_p$ is an extension

$$0 \longrightarrow \mathbb{C}^a \longrightarrow \mathcal{F}_p \longrightarrow (\mathbb{C}^*)^b \longrightarrow 0$$

where $(b + 1)$ is the number of branches of C at p (that is, $\#\pi^{-1}(p)$) and $a + b = \dim(\pi_* \mathcal{O}_{\tilde{C}}/\mathcal{O}_C)_p$ is the drop in genus associated to the singularity, that is, the contribution at p to the difference in arithmetic genus between C and $\tilde{C}$.

Having described $\mathcal{F}$, we turn now to the exact sequence

$$0 \longrightarrow \mathcal{O}_C^* \longrightarrow \pi_* \mathcal{O}_{\tilde{C}}^* \longrightarrow \mathcal{F} \longrightarrow 0.$$

Since C is connected, we have identifications

$$H^0(\mathcal{O}_C^*) = \mathbb{C}^*$$
$$H^0\left(\pi_* \mathcal{O}_{\tilde{C}}^*\right) = (\mathbb{C}^*)^\nu$$

where ν is the number of irreducible components of C. Using, for example, the Leray spectral sequence, we may also identify

$$H^1(\mathcal{O}_C^*) := \mathrm{Pic}(C)$$
$$H^1\left(\pi_* \mathcal{O}_{\tilde{C}}^*\right) \cong H^1\left(\mathcal{O}_{\tilde{C}}^*\right) = \mathrm{Pic}(\tilde{C}).$$

Putting these together, we obtain an exact sequence

(5.12) $0 \longrightarrow \mathbb{C}^* \longrightarrow (\mathbb{C}^*)^\nu \longrightarrow \Gamma(\mathcal{F}) \longrightarrow \mathrm{Pic}(C) \longrightarrow \mathrm{Pic}(\tilde{C}) \longrightarrow 0$

in which the first and last maps are just those induced by the pullback of the normalization map π^*. This sequence may thus be interpreted as saying:

1. To specify a line bundle L on C, we have to specify its pullback $\tilde{L} = \pi^*L$ to $\tilde{C}$, plus give "descent data", that is, specify when a section of $\tilde{L}$ is the pullback of a section of L. For example, if $p \in C$ is a node with $\pi^{-1}(p) = \{q, r\} \subset \tilde{C}$, we have to give an identification $\varphi_p : \tilde{L}_q \to \tilde{L}_r$ of the fibers of $\tilde{L}$ over p as one-dimensional complex vector spaces.

2. When $\tilde{L}$ is trivial, this descent data corresponds to giving a coset of $\mathcal{O}_C^*$ in $\pi_* \mathcal{O}_{\tilde{C}}^*$. For example, if p is a node as above and we fix a trivialization of $\tilde{L}$ near p, then φ_p is simply an element of $\mathbb{C}^*$ as above.

3. The descent data are only determined up to our choice of trivialization of $\tilde{L}$ over the nodes, which we may alter by composing a given trivialization with multiplication by a nonzero scalar on each connected component of $\tilde{C}$.

4. However, altering the trivialization in this way by multiplication by the *same* nonzero scalar on each component of $\tilde{C}$ does not, of course, change the descent data.

Finally, we see that in the analytic topology the coboundary map in the exponential sheaf sequence

$$c_1 : H^1(C, \mathcal{O}_C^*) \to H^2(C, \mathbb{Z}) \cong \mathbb{Z}^\nu$$

carries a line bundle L on C to its degrees on each of the irreducible components of C. We define $\operatorname{Pic}^0(C)$, or $J(C)$ called the *Jacobian* of C to be the connected component of the identity in the Picard group $\operatorname{Pic}(C)$, that is, the group of line bundles of degree 0 on every component. We then have the sequence

$$(5.13) \qquad 0 \longrightarrow \mathbb{C}^* \longrightarrow (\mathbb{C}^*)^\nu \longrightarrow \Gamma(\mathcal{F}) \longrightarrow J(C) \longrightarrow J(\tilde{C}) \longrightarrow 0.$$

From this it's usually straightforward to describe $J(C)$ in terms of the singularities of C. Here are two first examples.

EXERCISE (5.14) 1) If C is a g-nodal curve (that is, $\mathbb{P}^1$ with g pairs of points identified), show that $J(C) = (\mathbb{C}^*)^g$.

2) If C is a g-cuspidal curve, show that $J(C) = \mathbb{C}^g$.

More generally, suppose that C is any connected, reduced curve whose only singularities are nodes. We may associate to C what is called its *dual graph* $\Gamma(C)$, a one-dimensional cell complex defined as follows. Take one zero-cell or vertex for each irreducible component of C. Then, for each node of C attach a one-cell or edge by glueing the two ends of the edge to the vertices (not necessarily distinct) associated

FIGURE (5.15)

to the components containing the branches at that node. Thus, if C is a g-nodal curve, its dual graph is a bouquet of g loops (shown in Figure (5.15) for $g = 4$).

The dual graph lets us succinctly summarize what the sequence (5.13) says about the Jacobian of C by noting that we have a sequence

$$0 \longrightarrow (\mathbb{C}^*)^b \longrightarrow J(C) \longrightarrow J(\tilde{C}) \longrightarrow 0$$

where b is the first Betti number of $\Gamma(C)$. The g-nodal curve represents, for this sequence, one extreme, in which $J(\tilde{C}) = \{0\}$ and $J(C) \cong (\mathbb{C}^*)^g$. At the other extreme, we have nodal curves C satisfying the equivalent conditions:

1. $J(C)$ is compact;

2. The sum of the geometric genera of the components of C is g;

3. The dual graph of C is a tree.

We will say such a curve is of *compact type*. In particular, note that if C is of compact type, each irreducible component of C will be smooth, and no two components will meet in more than one point. Note also that the curve of Figure (5.10), which will be our main object of study, is of compact type, with Jacobian

$$J = \prod_{i=1}^{g} J(E_i) = \prod_{i=1}^{g} E_i$$

where the E_i are the elliptic components of the curve.

Remark. While we've defined the Jacobian of C as the group of line bundles of degree 0 on C, it may also be defined, in analogy with the smooth case, as linear functionals on the space of global sections of the dualizing sheaf ω_C modulo those linear functionals arising from integration over closed loops in $C - C_{\text{sing}}$. This is immediate if we recall from (3.5) that the sections of ω_C over an open set $U \subset C$ are given by meromorphic one-forms ω on $\pi^{-1}(U) \subset \tilde{C}$ such that for any $p \in U$ and $f \in \mathcal{O}_{C,p}$,

$$\sum_{q \in \pi^{-1}(p)} \text{Res}_q(\omega \cdot \pi^* f) = 0,$$

so that integration of sections of ω over cycles avoiding the singular points of C makes sense. As usual, we can write

$$J(C) = \left(H^0(C, \omega_C)^\vee\right) \Big/ \left(H_1(C - C_{\text{sing}}, \mathbb{Z})\right).$$

In particular, an Abel-type theorem holds: two divisors, $D = \sum p_i$ and $E = \sum q_i$, supported on the smooth locus of C will be linearly equivalent if and only they have the same degree on each irreducible component of C, and if, after reordering so that p_i and q_i lie on the same component, we can choose paths of integration γ_i from p_i to q_i for which

$$\sum_i \left(\int_{\gamma_i} \omega\right) = 0$$

for all $\omega \in H^0(C, \omega_C)$.

Having described the group of line bundles on a singular curve, we turn our attention now to families of line bundles on families of curves acquiring a singularity. Specifically, for the remainder of this chapter, we'll be concerned with a projective, flat family

$$\pi : X \to B$$

over a smooth curve B. We also fix a local parameter t at a point $0 \in B$ and assume that the fiber $X_t = \pi^{-1}(t)$ is smooth for $t \neq 0$ and that the special fiber X_0, while possibly singular and/or reducible, is always reduced. We will, in addition, work with families whose total spaces X are smooth, though this is really a luxury in which we indulge mainly to keep our statements as simple as possible.

What we want to develop are methods for obtaining information about the behavior of linear series in the general fiber X_t of such a family by looking at their "limits" on the central fiber X_0. This raises a second question, that of choosing a central fiber X_0 so that information about such limits is easy to obtain and work with. For example, we might wish, as Castelnuovo suggested, to take X_0 to be an irreducible, g-nodal curve. One difficulty with this, and many other choices, is that the limit of a family of line bundles on X_t may no longer be a line bundle, reflecting the fact that the Jacobian of X_0 need not be compact. Put another way, when we take the closure of the variety $\mathcal{W}_d^r = \{(t, L) : L \in W_d^r(X_t), t \neq 0\}$ over $B - \{0\}$ in the family of Picard varieties over B, the resulting scheme need not be proper over B.

One example of this is a family $X \to B$ of general curves X_t of genus 4 specializing to a curve X_0 with a node obtained by taking a hyperelliptic curve of genus 3 and identifying two points *not* conjugate under the hyperelliptic involution. (In this example, the family is general if X_t is nonhyperelliptic and if its canonical image lies on a *smooth* quadric Q_t.) What will happen then as t goes to zero is that the quadric Q_t will specialize to a quadric cone Q_0, and the canonical image of X_t to the intersection of Q_0 with a cubic passing through the vertex P of Q_0: in particular, X_0 will have a node at P. Then, since the g_3^1's on X_t are cut out by the rulings of Q_t, $\mathcal{W}_3^1$ will be a nice, connected two-sheeted cover of $B - \{0\}$ ramified only over 0. Moreover, since the lines on Q_0 aren't Cartier divisors and don't restrict to Cartier divisors on X_0, $\mathcal{W}_3^1$ will already be closed in the family of Picard varieties: that is, the limit of the linear series determined by L_t in $W_3^1(X_t)$ will not be a line bundle.

This difficulty will *not* arise if the line bundles L_t on each fiber X_t for $t \neq 0$ are restrictions of a single line bundle $\mathcal{L}$ on $\mathcal{X} \setminus X_0$. In this case, since we're assuming that the total space $\mathcal{X}$ is smooth, $\mathcal{L}$ will extend to a line bundle on all of $\mathcal{X}$ and hence the limit as $t \to 0$ of the L_t will be a line bundle. The problem here is that, in general, the *single-valued* section of $\mathcal{W}_d^r$ needed to define $\mathcal{L}$ may not exist, even if $W_d^r(X_t)$ is nonempty for each $t \neq 0$. For instance, in the example above, because $\mathcal{W}_3^1$ is a connected covering of $B - \{0\}$, the two g_3^1's on X_t cannot be distinguished. Moreover, if we apply a base change to make $\{L_t\}$ single-valued, then the total space of X will become singular, and line bundles on $\mathcal{X} - X_0$ no longer extend over all of X. Finally, if we resolve the singularities introduced into X by a base change, then the central fiber X_0 becomes a reducible curve. This, as we shall see in the sections that follow, introduces its own set of complications.

What can one do in this situation? The most natural thing to do might be to try and describe the limits of line bundles and/or linear series in families in which the total space is smooth and the central fiber is irreducible. The answers aren't too bad: the limit of a line bundle is always a rank 1 torsion-free sheaf, and in the limit a linear series may acquire a non-Cartier base divisor but will otherwise remain a bona fide linear series on a partial normalization of X_0. This approach was, as discussed in Section A, the basis of the first proofs of the Brill-Noether theorem. Carrying out this program requires a fairly large amount of machinery that doesn't seem to give the Gieseker-Petri theorem. The reason for this is that the Brill-Noether theorem deals only with a *general* g_d^r on a general curve: thus, by a suitable induction, we may avoid considering those L_t's on X_t that fail to specialize to line bundles. Petri's statement deals with *all* g_d^r's on a general curve, which forces us to handle all limiting g_d^r's.

Our approach here will be the opposite one: we'll allow ourselves to make base changes and to resolve singularities that this introduces and, correspondingly, will allow X_0 to be reducible. Having opened the door to the barn of reducible special fibers, there is no reason why we shouldn't go to the other extreme from Castelnuovo and avoid, as far as possible, the failures of properness discussed above by taking X_0 to be of compact type. In the following section, then, we'll develop the theory of *limit linear series*, which describes limits of linear series in families tending to such curves; and in the subsequent two sections we'll use this theory to give proofs of the Brill-Noether and Petri theorems.

C · Limits of linear series: motivation and examples

As we indicated in the last section, we want to consider here limits of linear series on a family of curves $\{X_t\}$ specializing to a reducible curve X_0 with the restriction that X_0 is of compact type. Our analysis here will allow us to give a proof of the Brill-Noether theorem (5.1) at the end of this section. In the final two sections of this chapter, we'll use it to prove the harder Gieseker-Petri theorem [Theorem (5.78)].

We begin by considering the simplest possible case. Fix a one-parameter family $\pi : X \to B$ of curves with smooth total space X, smooth fibers X_t for $t \neq 0$, and central fiber $X_0 = Y \cup Z$ the union of two smooth curves meeting at a single point p. B itself we'll take to be small enough to have trivial Picard group, e.g. a disc or the spectrum of a discrete valuation ring. We saw in the last section that, in general, a family of line bundles on the smooth fibers of such a degenerating family need not extend to a line bundle on the singular one. By contrast, what seems to be the difficulty in this case isn't the absence of an extension of a given line bundle $\tilde{\mathcal{L}}$ on $\tilde{X} = X \setminus X_0$ to X, but rather the presence of too many, no one of which really captures all the geometry of the linear series on X_t. Precisely, if $\tilde{\mathcal{L}}$ is a line bundle on $\tilde{X}$ there will always exist a line bundle $\mathcal{L}$ on X extending $\tilde{\mathcal{L}}$; but if $\mathcal{L}$ is any such bundle then so is the line bundle $\mathcal{L}(Y) = \mathcal{L} \otimes \mathcal{O}_X(Y)$.

To compare $\mathcal{L}$ and $\mathcal{L}(Y)$, note that the line bundle $\mathcal{O}_X(Y)$ clearly restricts to the line bundle $\mathcal{O}_Z(p)$ on Z. On the other hand, $\mathcal{O}_X(X_0)$ is trivial, so $\mathcal{O}_X(Y) = \mathcal{O}_X(-Z)$ and hence $\mathcal{O}_X(Y)$ must restrict to $\mathcal{O}_Y(-p)$ on Y. Thus, if $\mathcal{L}$ has degree α on Y and $d - \alpha$ on Z, the line bundle $\mathcal{L}(Y)$ will have degree $\alpha - 1$ on Y and $d - \alpha + 1$ on Z. In other words, we see that for given $\tilde{\mathcal{L}}$ on $\tilde{X}$ of relative degree d, there exists, for every $\alpha \in \mathbb{Z}$, a unique extension $\mathcal{L}_\alpha$ of $\mathcal{L}$ to X having degree α on Y and $d - \alpha$ on Z.

More generally now, suppose we assume only that X_0 is a nodal curve whose dual graph is a tree, and that Y and Z are components of X_0 meeting at a point p. By hypothesis, $X_0 - \{p\}$ has two connected components; let E be the union of the components of X_0 lying in the connected component of $X_0 \setminus \{p\}$ containing Y. Then the line bundle $\mathcal{O}_X(E)$ will have restrictions

$$\mathcal{O}_X(E) \otimes \mathcal{O}_Z = \mathcal{O}_Z(p),$$

(5.16) $$\mathcal{O}_X(E) \otimes \mathcal{O}_Y = \mathcal{O}_Y(-p) \quad \text{and}$$

$$\mathcal{O}_X(E) \otimes \mathcal{O}_W = \mathcal{O}_W$$

for any component W of X_0 other than Y and Z.

Let A_d be the set of integer-valued functions on the set of irreducible components of X_0 whose values sum to d. It follows that: *if $\widetilde{L}$ is any line bundle of relative degree d on $\widetilde{X}$ and if α is any element of the set A_d, then there exists a unique extension $\mathcal{L}_\alpha$ of $\widetilde{L}$ to X such that*

(5.17) $$\deg(\mathcal{L}_\alpha \otimes \mathcal{O}_Y) = \alpha(Y)$$

for every component Y of X_0. If, moreover, Y and Z are two components of X_0 meeting at the point p, and β is obtained from α by adding 1 to $\alpha(Y)$ and subtracting 1 from $\alpha(Z)$, then

(5.18) $$\mathcal{L}_\beta \otimes \mathcal{O}_Y = \mathcal{L}_\alpha \otimes \mathcal{O}_Y(p) \quad \text{and}$$
$$\mathcal{L}_\beta \otimes \mathcal{O}_Z = \mathcal{L}_\alpha \otimes \mathcal{O}_Z(-p).$$

This completely answers the question of what data we get in the limit from a family of line bundles of degree d on the family of curves X_t. We get a collection of line bundles, indexed by A_d, satisfying relations (5.17) and (5.18) above.

Observe that such a collection of data depends only on the curve X_0 itself and not on the particular family X_t specializing to it. This phenomenon is special to limiting fibers of compact type; for a family specializing to a general nodal curve, what constitutes a limit of line bundles will depend on the family.

We come now to the main question of interest: suppose that we have not just a line bundle L_t on the general fiber X_t of our family, but a linear series $V_t \subset H^0(X_t, L_t)$ — in other words, we are given a line bundle $\widetilde{L}$ on $\widetilde{X}$ together with a locally free subsheaf $\widetilde{V} \subset \pi_*(\widetilde{L})$ of rank $r+1$. What data on X_0 can we associate to such a family that will provide information about the limiting geometry of the linear series V_t as $t \to 0$?

One natural answer to this question would be to look at *all* possible limits of the family $\widetilde{L}$ of line bundles and at the corresponding linear system in each. For each $\alpha \in A_d$, we have an extension $\mathcal{L}_\alpha$; the subbundle $\widetilde{V} \subset \pi_*(\widetilde{L})$ will extend to a subbundle $\mathcal{V}_\alpha \subset \pi_*(\mathcal{L}_\alpha)$; and hence the linear series V_t have a limit $V_\alpha \subset H^0(X_0, \mathcal{L}_\alpha \otimes \mathcal{O}_{X_0})$ (that is, the fiber of the sheaf $\mathcal{V}_\alpha$ over 0). The upshot is that we again have a collection of limiting linear series V_α indexed by A_d.

The problem with this approach is that this data is both redundant and inconvenient to manage. Fortunately, as we'll see, it's not necessary to look at all these linear series. For the most part, it's sufficient to focus on a much smaller subset of the set of linear series V_α. For each component Y of X_0, we denote by $\mathcal{L}_Y$ the unique extension of $\mathcal{L}$ to X with degree d on Y and degree 0 on all other components of X_0 and we let

$$V_Y = \lim_{t \to 0}(V_t) \subset H^0(X_0, \mathcal{L}_Y \otimes \mathcal{O}_{X_0}).$$

These extensions have the immediate advantage (over those given by general α) that a section of $\mathcal{L}_Y$ vanishing on Y vanishes on all of X_0, so that we have an inclusion

$$V_Y \subset H^0(X_0, \mathcal{L}_Y \otimes \mathcal{O}_{X_0}) \subset H^0(Y, \mathcal{L}_Y \otimes \mathcal{O}_Y).$$

Thus we can view V_Y as a g_d^r on the smooth curve Y, rather than on all of X_0. Anticipating the formal definitions of the next section, we'll refer to V_Y as the limiting *aspect* of $\widetilde{\mathcal{V}}$ on Y. To sum up: *Associated to a linear series $g_d^r(t)$ on the general fiber X_t of X is a collection of limiting aspects V_Y that are g_d^r's on the various components Y of X_0.*

The logical next question now becomes, what is the relationship among the various linear series V_Y arising in this way? To answer this, let's again look first at the case in which X_0 has just two components Y and Z meeting at a point p; let $\mathcal{L}_\alpha$ be the extension of a line bundle $\widetilde{\mathcal{L}}$ to X having degree α on Y and $d - \alpha$ on Z. Recall that the various extensions $\mathcal{L}_\alpha$ are related by $\mathcal{L}_\alpha = \mathcal{L}_{\alpha+1}(-Z)$ on Y. In other words, the linear series $|\mathcal{L}_\alpha|$ on the total space X is just the subseries of $|\mathcal{L}_{\alpha+1}|$ containing the component Z with this fixed component removed. Thus, if D is any divisor in $|\mathcal{L}_\alpha|$, then

$$D + Z \in |\mathcal{L}_{\alpha+1}|,$$

and if D doesn't contain Y, then

$$(D + Z)|_Y = D|_Y + p.$$

This immediately gives the containments of linear series:

$$(V_{\alpha+1})|_Y \supset (V_\alpha)|_Y + p$$

or, dually,

$$(V_\alpha)|_Y \subset (V_{\alpha+1})|_Y - p.$$

(Here and in the sequel we denote by $V(-D)$, or simply $V - D$, the linear subseries of V obtained by imposing the base point condition given by the effective divisor D, and then subtracting D; that is, $V(-D) = \{E - D : E \in V \text{ and } E - D \geq 0\}$.)

These containments explain why, for most purposes, it suffices to look at the extreme linear series $V_Y = (V_d)|_Y$ and $V_Z = (V_0)|_Z$ introduced above, since they show us that

$$(V_\alpha)|_Y \subset V_Y(-(d-\alpha)p) \quad \text{and} \quad (V_\alpha)|_Z \subset V_Z(-\alpha p).$$

Further, these inclusions give rise to the basic relations between V_Y and V_Z. Since $\dim(V_\alpha) = r$, we must have

$$\dim(V_\alpha|_Y) + \dim(V_\alpha|_Z)$$

(5.19)

$$= \dim(V_Y(-(d-\alpha)p)) + \dim(V_Z(-\alpha p))$$

$$\geq \begin{cases} r, & \text{if } p \text{ is not a base point of } V_\alpha \\ r-1, & \text{if } p \text{ is a base point of } V_\alpha. \end{cases}$$

Happily, there is a more convenient way to express such relations. For any linear series V on a curve Y and any smooth point $p \in Y$, we introduce the *vanishing sequence*

$$0 \leq a_0(V,p) < a_1(V,p) < \cdots < a_r(V,p),$$

which is defined by the condition that, as sets, the $a_j(V,p)$'s are just the orders with which nonzero elements of V vanish at p:

$$\{a_j(V,p) \mid i = 0,\ldots,r\} = \{\operatorname{ord}_p(\sigma) \mid \sigma \neq 0 \in V\}.$$

Vanishing sequences encode information about how the map $\varphi_V : Y \to \mathbb{P}^r$ associated to V is inflected at p. For example, $a_0(V,p)$ is just the multiplicity with which p occurs as a base point of V. More generally, we have

(5.20) $\dim(V(-bp)) = r - i \Longleftrightarrow a_{i-1}(V,p) < b \leq a_i(V,p).$

Thus, to say that V has no base point at p means that $a_0(V,p) = 0$; assuming this is the case and that φ_V is birational onto its image, to say that $\varphi_V(Y)$ doesn't have a cusp at $P = \varphi_V(p)$ means that $a_1(V,p) = 1$; and in this case, if $r = 2$, to say that P isn't a flex of $\varphi_V(Y)$ means that $a_2(V,p) = 2$. For a general p, we expect no inflectionary behavior in V and hence that

$$a_i(V,p) = i, \quad i = 0,1,\ldots,r.$$

This suggests defining the *ramification sequence* $b_i(V,p)$ of the point p as the difference between the vanishing sequence at p and the generic one

$$b_i(V,p) = a_i(V,p) - i, \quad i = 0,1,\ldots,r,$$

and the *ramification index* $b_i(V, p)$ of p as the sum

$$\beta(V, p) = \sum_{i=0}^{r} b_i(V, p).$$

The classical Plücker formula (cf. Exercise 1.C.13 of [7]) then gives a global measure of the ramification of V at all points:

LEMMA (5.21) (PLÜCKER FORMULA) *If V is any g_d^r on a smooth curve Y of genus g,*

$$\sum_{p \in Y} \beta(V, p) = (r + 1)d + (r + 1)r(g - 1).$$

In the context of limit linear series, vanishing sequences provide a convenient way to encode the relation between the aspects V_Y and V_Z expressed in (5.19), which in view of (5.20), immediately translate to the inequalities:

(5.22) $a_i(V_Y, p) + a_{r-i}(V_Z, p) \geq d, \quad i = 0, 1, \ldots, r.$

This is a strong condition on the pair of linear series V_Y, V_Z. For instance, if $a_0(V_Y, p) = 0$ (that is, p is not a base point of V_Y), then this implies $V_Z(-dp) \neq \emptyset$, so that $a_r(V_Z, p) = d$; this implies that p is a highly inflected point for the linear series V_Z on Z. In fact, (5.22) is often a sufficient, as well as a necessary, condition for a pair of linear series V_Y and V_Z to arise as limits of linear series on smooth curves tending to X_0.

We remark here that, as we'll verify shortly, (5.22) holds more generally for the collection $\{V_Y\}$ of aspects we obtain from a family of linear systems on curves tending to a nodal curve X_0 of compact type: if Y and Z are any two components of X_0 meeting at a point p, then we must have

$$a_i(V_Y, p) + a_{r-i}(V_Z, p) \geq d$$

for all i.

To make these notions more concrete, let's look at an example that sheds light on the contrast between irreducible and reducible limit curves. Suppose we have a family $\pi : \mathcal{X} \to B$ of curves specializing to a g-cuspidal curve X_0, a family $\tilde{\mathcal{L}}$ of line bundles on $\mathcal{X} - X_0$, and a linear series

$$V_t \subset H^0(X_t, L_t)$$

on X_t for $t \neq 0$. Let $\mathbb{P}^1 = \tilde{X}_0 \to X_0$ be the normalization of X_0. We consider the pullback to $\mathbb{P}^1$ of divisors $D \in |\tilde{\mathcal{L}}|$. These do in fact form a linear system $\tilde{V}_0$ on $\mathbb{P}^1$, which may be described as follows. For each cusp p_i of X_0, the limit as $t \to 0$ of the line bundles L_t, which is a priori just a torsion-free sheaf on X_0, may or may not, in fact, be locally free at p_i. If it is, then the series $\tilde{V}_0$ will factor through the map $\pi : \mathbb{P}^1 \rightarrow X_0$ near p_i. Informally, the linear series $\tilde{V}_0$ "has a cusp" at the point $\tilde{p}_i$ over p_i. If this limiting line bundle fails to be locally free at p_i, then the limiting linear series need no longer factor through π near p_i but, if it doesn't, it must have a base point at $\tilde{p}_i$. We can cover both these possibilities by the statement

$$a_1(\tilde{V}_0, \tilde{p}_i) \geq 2 \,.$$

Equivalently, $2\tilde{p}_i$ imposes only one condition on $\tilde{V}_0$, which we rephrase informally as "$\tilde{V}_0$ has *at least* a cusp at $\tilde{p}_i$".

Let's now compare this to what happens after we make a base change of order 6 and minimally resolve the singularities of the resulting surface. The central fiber X_0 of the resulting family consists of a copy of $\mathbb{P}^1$, which we'll call Y and which is the normalization of the original g-cuspidal curve, plus g elliptic curves $E_1, \ldots, E_g$ attached to Y at the g points p_i which were formerly cusps. Now, the limit of the line bundles L_t is always locally free. We will analyze it along the lines of the preceding discussion but without the results we've obtained in general.

Let L_Y be the limiting line bundle having degree d on Y and degree 0 on each E_i, and let V_Y be the restriction to Y of the linear series

$$V_0 = \lim_{t \rightarrow 0} V_t \subset H^0(X_0, L_Y)$$

on X_0. Thus V_Y is a linear series of degree d and dimension r on $Y \cong \mathbb{P}^1$, and to describe it we may use the analysis above, as follows. For each i, there are the two possibilities for $L_Y|_{E_i}$. If $L_Y|_{E_i} \neq \mathcal{O}_{E_i}$, i.e., is nontrivial, then $H^0(E_i, L_Y \otimes \mathcal{O}_{E_i}) = 0$, so that every section of L_Y on X_0 vanishes on E_i. Correspondingly, V_Y *has a base point at* p_i.

The case $L_Y|_{E_i} = \mathcal{O}_{E_i}$ is the more interesting possibility. In this case, we want to consider also the limiting line bundles L_i having degrees $d-1$ on Y, 1 on E_i, and 0 on E_j for $j \neq i$, and the corresponding limiting linear series $V_{0,i}$ in $H^0(X_0, L_i)$. The point is that the line bundle

$$L_i|_{E_i} = \mathcal{O}_{E_i}(p_i)$$

still has only one global section that vanishes at p_i. On the other hand, the linear series $V_{0,i}$ restricted to Y has dimension at least $r-1$ and is contained in $V_0(-p_i)$. Since $\dim(V_{0,i}) = r$, we conclude that $V_{0,i}$ must have a base point at p_i; i.e., $2p_i$ *imposes only one condition on* V_Y.

Our conclusion, then, about the limiting series V_Y remains as it was for the linear series $\tilde{V}_0$ on $\mathbb{P}^1$ in the first analysis: the series V_Y must have at least a cusp at each p_i.

Observe, finally, that this conclusion about V_Y may be obtained directly from the relation (5.22): if V_{E_i} is the limiting g_d^r on E_i, we cannot have two sections of V_{E_i} vanishing to order $d-1$ at p_i, since otherwise the pencil they spanned would be a g_1^1 plus the fixed divisor $(d-1)p_i$. Thus

$$a_{r-1}(V_{E_i}, p_i) \le d - 2,$$

and by (5.22), we conclude

$$a_1(V_Y, p_i) \ge 2;$$

so V_Y has at least a cusp at p_i.

With this as motivation, we now want to go back and rederive the basic relation (5.22), this time algebraically and for arbitrary X_0 of compact type. As before, we'll let Y and Z be two components of X_0 meeting at a point p; again, we'll denote by E and F the divisors consisting of the sum of the curves in the connected components of $X_0 \setminus \{p\}$ containing Z and Y respectively. Then by (5.16), for any line bundle $\tilde{\mathcal{L}}$ on $X \setminus X_0$, we have

(5.23) $\mathcal{L}_Y \cong \mathcal{L}_Z(dE)$ and $\mathcal{L}_Z \cong \mathcal{L}_Y(-dE).$

Now, suppose we're given a family of divisors $D_t \in |L_t|$ for $t \ne 0$ — i.e., a divisor $\tilde{D} \in |\tilde{\mathcal{L}}|$ on $X \setminus X_0$ — and we're asked to find the limit as t approaches 0 of (D_t) as a divisor in the linear series $|\mathcal{L}_Z \otimes \mathcal{O}_Z|$. To do this, we simply write $\tilde{D} = (\sigma)$ and $D_t = (\sigma)|_{X_t}$ for some section σ of $\tilde{L}$ over $X \setminus X_0$, and then multiply σ by the (unique) correct power of t so that it extends to a holomorphic section σ_Z of $\mathcal{L}_Z$ on all of X, not vanishing identically on X_0. If $\sigma_Z|_{X_0} \ne 0$ we must have $\sigma_Z|_Z \ne 0$, and then, of course, the limit of the D_t is the divisor $(\sigma_Z|_Z)$.

In terms of this prescription, it's easy to relate the limit of the D_t as a divisor in $|\mathcal{L}_Y \otimes \mathcal{O}_Y|$ to the limit in Z: in view of (5.23), the section σ_Y of $\mathcal{L}_Y$ extending σ is obtained by multiplying σ_Z by the section of $\mathcal{O}_X(dE)$ vanishing d times on E, then dividing by the highest power t^α of t that divides the product. Equivalently,

(5.24)
$$\begin{aligned} (\sigma_Y) &= (\sigma_Z) + dE - \alpha X_0 \\ &= (\sigma_Z) + (d - \sigma)E - \alpha F \end{aligned}$$

where α is determined by the requirements that (σ_Y) be effective, but that $(\sigma_Y) - X_0$ not be.

To go in the other direction is even easier: by (5.23), we may view $\mathcal{L}_Z$ as a subsheaf of $\mathcal{L}_Y$. Given σ_Y, σ_Z will simply be $t^\alpha \sigma_Y$ where α is the smallest integer such that $t^\alpha \sigma_Y \in \Gamma(\mathcal{L}_Z)$; that is, the smallest integer such that $(\sigma_Y) + \alpha X_0 \ge dE$. Note that this requirement means that $(\sigma_Y) \ge (d - \alpha)E$, so that in particular the order of vanishing

$$\operatorname{ord}_p \left(\sigma_Y|_Y \right) \ge d - \alpha.$$

Similarly, by (5.24) we see that $(\sigma_Z) \geq \alpha F$, and hence

$$\mathrm{ord}_p\left(\sigma_Z|_Z\right) \geq \alpha.$$

We combine these in the:

LEMMA (5.25) *If $\sigma_Y \in \Gamma(\mathcal{L}_Y)$ and $\sigma_Z \in \Gamma(\mathcal{L}_Z)$ with neither vanishing identically on X_0 and $(\sigma_Y) = (\sigma_Z)$ on $X \backslash X_0$, then $\sigma_Z = t^\alpha \sigma_Y$ and*

$$d - \mathrm{ord}_p(\sigma_Y|_Y) \leq \alpha \leq \mathrm{ord}_p(\sigma_Z|_Z).$$

We remark for future use that as a consequence we have, trivially

$$(5.26) \qquad \mathrm{ord}_{p'}(\sigma_Y|_Y) \leq \mathrm{ord}_p(\sigma_Z|_Z)$$

where p' is any other point of Y, since $\mathrm{ord}_{p'}(\sigma_Y|_Y) + \mathrm{ord}_p(\sigma_Y|_Y) \leq d$.

To see how the linear series V_Y and V_Z relate to one another, we use a second lemma.

LEMMA (5.27) *There exist sections $\sigma_0, \ldots, \sigma_r$ of $\mathcal{L}_Y$ and $\tau_0, \ldots, \tau_r$ of $\mathcal{L}_Z$ generating $\pi_* \mathcal{L}_Y$ and $\pi_* \mathcal{L}_Z$ such that:*

1) *In terms of the inclusion $\mathcal{L}_Z \subset \mathcal{L}_Y$, $\tau_i = t^{\alpha_i} \sigma_i$.*

2) *The orders $\mathrm{ord}_p(\sigma_i|_Y)$ are all distinct.*

PROOF. The matrix expressing the inclusion of free $\mathcal{O}_{B,0}$-modules $\pi_* \mathcal{L}_Z \to \pi_* \mathcal{L}_Y$ can be diagonalized over $\mathcal{O}_{B,0}$ by Gaussian elimination; this yields bases σ_i and τ_i satisfying condition 1). Moreover, if $g(t)$ is holomorphic on B and $\alpha_i \geq \alpha_j$, we can replace σ_i by $\sigma_i + g\sigma_j$, and τ_i by $t^{\alpha_i}(\sigma_i + g\sigma_j)$ without affecting 1); such transformations allow us to achieve 2) as well. ∎

We can express our conclusions in terms of the vanishing sequences of V_Y and V_Z by applying (5.25) and (5.26) to the bases for V_Y and V_Z found in Lemma (5.27). The result is:

THEOREM (5.28) *Let $\pi : X \to B$ be a family of curves over a smooth, one-dimensional base, with general fiber smooth and special fiber X_0 a curve of compact type. Let Y and Z be irreducible components of X_0 meeting at a point p. Let $\widetilde{\mathcal{L}}$ be a line bundle on $X \backslash X_0$, $\widetilde{V} \subset \pi_* \widetilde{\mathcal{L}}$ a family of g_d^r's on $X \backslash X_0$, $\mathcal{L}_Y$ [resp. $\mathcal{L}_Z$] the unique extension of $\widetilde{\mathcal{L}}$ to X having degree d on Y [resp. Z] and degree 0 on all other components of X_0, and V_Y [resp. V_Z] the limit of $\widetilde{V}$ in $H^0(Y, \mathcal{L}_Y)$ [resp. $H^0(Z, \mathcal{L}_Z)$]. Then*

$$a_i(V_Y, p) + a_{r-i}(V_Z, p) \geq d.$$

PROOF. The previous two lemmas imply that, for some permutation ρ of $\{0,\ldots,r\}$

$$a_{r-i}(V_Z,p) \geq d - a_{\rho(i)}(V_Y,p).$$

The theorem then follows by applying the following lemma. ∎

LEMMA (5.29) *If $a_0 < a_1 < \cdots < a_r$ and $b_0 < b_1 < \cdots < b_r$ are sequences and for some permutation π of $\{0,\ldots,r\}$, we have $a_i \geq b_{\pi(i)}$ for all i, then $a_i \geq b_i$ for all i.*

This analysis actually yields a slight refinement for which we'll have use later.

LEMMA (5.30) *Let Y, Z and p be as above, and let p' be any point of Y other than p. Then,*

1) $a_i(V_Z,p) \geq a_i(V_Y,p')$;

2) *Equality holds in 1) for at most as many values of i as there are independent sections of V_Y vanishing only at p and p'.*

We also note that, by summing the inequalities of (5.28) over i and expressing the result in terms of ramification indices, we obtain:

COROLLARY (5.31) *Let Y, Z and p be as above. Then,*

$$\beta(V_Y,p) + \beta(V_Z,p) \geq (r+1)(d-r).$$

To see how this sort of analysis is applied in practice, let's use it to prove the Brill-Noether theorem, redeeming our earlier promise to do so. We start by taking X_0 to be any semistable curve of genus g consisting of a tree of N rational curves Y_i, to which g elliptic curves E_i are attached, with each E_i attached at one point p_i and the points p_i distinct: for example, the curve in Figure (5.10) will do. Let $\pi : X \to B$ be any smoothing of the curve X_0. We claim that *for general $t \in B$, the fiber X_t satisfies the Brill-Noether condition* $\dim W_d^r(X_t) \leq \rho$.

To prove this, suppose we have a family of g_d^r's on the smooth fibers of X, given as above, and consider the limit linear series associated to it. (Note that, if the general fiber X_t has a g_d^r, we can always assume we have such a family after making a base change and blowing up: the new central fiber will have more components but will still meet the conditions above.)

By the Plücker formula (5.21), the total ramification of each of the aspects $|V_{Y_i}|$ of our limiting g_d^r on any of the rational components of X_0 is $(r+1)(d-r)$. On the other hand, since the elliptic curves E_i don't have any rational functions with only one pole, the vanishing sequence of the aspect $|V_{E_i}|$ at the point p_i will be term-by-term bounded by the sequence

(5.32) $(d - r - 1, d - r, \ldots, d - 3, d - 2, d)$

and the ramification index of $|V_{E_i}|$ at p_i is therefore at most $(r + 1)(d - r) - r$. Thus the sum of the ramification indices of the aspects of the limiting g_d^r at the nodes of X_0 is at most

$$(N + g)(r + 1)(d - r) - rg.$$

But the curve X_0 has $N + g - 1$ nodes, and at each of those nodes the sum of the ramification indices of the relevant aspects must be at least $(r + 1)(d - r)$ by Corollary (5.31). We must therefore have

$$(N + g)(r + 1)(d - r) - rg \geq (N + g - 1)(r + 1)(d - r);$$

i.e., $(r + 1)(d - r) \geq rg$; or, equivalently, $\rho \geq 0$. This proves that *the general curve X_t possesses no g_d^r's with $\rho < 0$.*

To complete the proof, observe that equality can hold in the inequality (5.32) above only if the line bundle $\mathcal{L}_{E_i}$ restricted to E_i is isomorphic to $\mathcal{O}_{E_i}(d \cdot p_i)$, or, equivalently, if for any rational component Y of X_0 the line bundle $\mathcal{L}_Y$ restricted to E_i is trivial. It follows that if our family of g_d^r's has Brill-Noether number ρ, then *the line bundle $\mathcal{L}_Y$ will have to be trivial on at least $g - \rho$ of the curves E_i.* In terms of the identification of the Jacobian of X_0 with the product of the elliptic curves E_i, this says that the limit in $\text{Pic}_d(X_0)$ of the varieties $W_d^r(X_t)$ will be supported on the union of the ρ-dimensional coordinate planes, and hence that

$$\dim(W_d^r(X_t)) \leq \rho.$$

We remark that the same method can be used to prove the existence of g_d^r's on a general (and hence on every) curve of genus g whenever $\rho(g, r, d) \geq 0$. We won't do this here since the methods of Kempf, Kleiman-Laksov and Fulton-Lazarsfeld (as described in [7]) yield more information; but there are times when we may want to use this approach to prove the existence of linear series satisfying some additional conditions. We will illustrate this in a series of exercises proving the existence of curves with certain special Weierstrass points [Exercise (5.48)]. To do this, however, we must ask a basic question: when, conversely, does a collection of linear series $\{V_Y\}$ on the components of a curve X_0 of compact type, satisfying the inequalities of Theorem (5.28), actually arise as the limit of a family of g_d^r's on smooth curves X_t tending to X_0? We will give the best known answer to this [Theorem (5.41)] after first introducing the necessary formalism in the following section.

D Limit linear series: definitions and applications

In this section, the results of the previous section are formalized in the theory of limit linear series. While we won't give a complete account of this theory here, we indicate the basic statements and a few applications; in particular we'll reinterpret and give a stronger form of the Brill-Noether theorem.

Limit linear series

The first thing to do is to make a precise definition:

DEFINITION (5.33) *Let X be a curve of compact type. A limit linear series $\mathcal{D}$ of degree d and dimension r on X assigns to every irreducible component Y of X a linear series $|V_Y|$ of degree d and dimension r called the* aspect *of $\mathcal{D}$ on Y, such that for every pair of components Y, $Z \subset X$ meeting at a point p the aspects V_Y and V_Z satisfy*

$$(5.34) \qquad a_i(V_Y, p) + a_{r-i}(V_Z, p) \geq d.$$

We will say the limit linear series $\mathcal{D}$ is refined *if equality holds in (5.34) for each i; we'll call it* crude *otherwise.*

The notion of ramification is central to the theory of limit linear series. Given a smooth point p on a curve X of compact type, and a limit linear series $\mathcal{D}$ on X, we define the *ramification sequence* of $\mathcal{D}$ at p to be just the ramification sequence at p of the aspect of $\mathcal{D}$ on the component of X containing p.

The results of the previous section amount to saying that in a one-parameter family $\{X_t\}$ of curves of compact type with smooth total space, a linear series $\mathcal{D}_t$ on the general fiber specializes to a limit linear series $\mathcal{D}$ on the special fiber. We observe as well that a smooth point of the special fiber X_0 will be a ramification point for $\mathcal{D}$ if and only if it's the limit of ramification points of $\mathcal{D}_t$; further, the ramification index of $\mathcal{D}$ at p will be the sum of the ramification indices of $\mathcal{D}_t$ at points of X_t tending to p.

This ties in nicely with the Plücker formula (5.21): the linear series $\mathcal{D}_t$ will have a total of $(r+1)d + r(r+1)(g-1)$ ramification points (counting multiplicity), while if X_0 has components Y_i of genera g_i, we expect $(r+1)d + r(r+1)(g_i - 1)$ ramification points of the aspect of $\mathcal{D}$ on Y_i. Of course, some of the ramification of the V_{Y_i} will occur at the nodes of X_0 lying on Y_i: by (5.34), the sum of the ramification indices at a node of X_0 of the aspects of $\mathcal{D}$ on the two components containing the node will be at least $(r+1)(d-r)$; if X_0 has k components, and hence $k-1$ nodes, this will account for a total of $(k-1)(r+1)(d-r)$ ramification points of the aspects of $\mathcal{D}$. The total ramification of the aspects of $\mathcal{D}$ at smooth points of X_0 is thus at most

$$\sum_i \big((r+1)d + r(r+1)(g_i-1) \big) - (k-1)(r+1)(d-r)$$

$$= (r+1)d + r(r+1)(g-1);$$

comparing this with the number of ramification points of $\mathcal{D}_t$ we conclude that *the limit linear series $\mathcal{D}$ is refined if and only if no node of X_0 is a limit of ramification points of $\mathcal{D}_t$*. Note that as a consequence we can always achieve this state after blowing up our family at nodes of the special fiber, and then making further base changes and blowups to resolve the resulting singularities.

The calculation above suggests the following definition:

DEFINITION (5.35) *The* adjusted Brill-Noether number ρ *of a linear series $\mathcal{D}$ on a smooth curve X with respect to a given collection of points $p_j \in X$ is defined to be the ordinary Brill-Noether number of $\mathcal{D}$ minus the sum of the ramification indices of $\mathcal{D}$ at the points p_j.*

The idea behind this definition is to generalize the variety G_d^r parameterizing linear series of degree d and dimension r. Just as G_d^r is determinantal and either empty or of dimension at least ρ, so given any collection of points $p_1, \ldots, p_k \in C$ and, for each p_j, a ramification sequence $b^j = (b_0^j, \ldots, b_r^j)$ summing to β_j, we have a variety $G_d^r(p_1, \ldots, p_k; b^1, \ldots, b^k)$ parameterizing linear series with ramification sequence at least b_j at p_j, and this variety will again be determinantal and either empty or of dimension at least $\rho - \sum_j \beta_j$. When equality holds, we say the limit linear series is *dimensionally proper*.

We define in the same way the adjusted Brill-Noether number of a limit linear series $\mathcal{D}$ on a curve X of compact type with respect to a collection of smooth points. In these terms, the calculation above says that:

LEMMA (5.36) *The adjusted Brill-Noether number of a limit linear series $\mathcal{D}$ on a curve X of compact type with respect to a collection of smooth points $p_i \in X$ is equal to the sum, over the components Y of X, of the adjusted Brill-Noether numbers of the aspects of $\mathcal{D}$ on Y with respect to the union of the subset of the p_i on Y and the nodes of X lying on Y.*

This is useful because in the case of curves of genus 0 or 1, it's not hard to estimate adjusted Brill-Noether numbers. Two frequently useful estimates are

1. Since the sum of the ramification indices of a g_d^r on $\mathbb{P}^1$ at all points is $(r+1)(d-r)$, which is exactly equal to the Brill-Noether number of the g_d^r, the adjusted Brill-Noether number of any linear series on $\mathbb{P}^1$ with respect to any collection of points is nonnegative.

2. Similarly, the adjusted Brill-Noether number of a linear series on an elliptic curve with respect to any one point is nonnegative.

Combining these facts with the statement above, we deduce that if X is a curve of compact type formed by attaching g elliptic tails to a tree of rational curves, and $p_1, \ldots, p_k$ are points lying on rational components of X, then the adjusted Brill-Noether number of any limit linear series on X with respect to the points p_i is nonnegative. By specialization to such curves, we arrive at the following generalization of the Brill-Noether theorem:

THEOREM (5.37) Let C be a general curve of genus g, and $p_1, \ldots, p_k$ general points of C. Then the adjusted Brill-Noether number of any linear series on C with respect to the points p_j is nonnegative, i.e., for any linear series $\mathcal{D}$ on C the sum of the ramification indices of $\mathcal{D}$ at the p_j is at most $\rho = g - (r+1)(g-d+r)$. Moreover, for any collection of k ramification sequences, the variety $G_d^r(p_1, \ldots, p_k; b^1, \ldots, b^k)$ will have dimension exactly $\rho - \sum \beta_j^i$ when it's nonempty; that is, all limit linear series with nonnegative adjusted Brill-Noether number are dimensionally proper.

EXERCISE (5.38) Complete the argument given above by proving the dimension assertion in the last sentence.

EXERCISE (5.39) 1) Show that, if a general curve of genus g has a g_d^r with ramification sequence $(b_0(p), \ldots, b_r(p))$ at a general point p, then $\rho(g, r, d) \geq \beta(p)$. Give an example that shows that this condition is not sufficient for the existence of such a g_d^r.

2) Show that the inequality $\rho(g, r, d) \geq \beta(p)$ is equivalent to

$$\sum_{i=0}^{r} (b_i(p) + (g - d + r)) \leq g.$$

We may extend the results on limit linear series to a slightly larger class of stable curves, which we'll call *treelike* curves. Briefly, we call a node of a stable curve C an *interior* node if its two branches belong to the same irreducible component of C, and define a stable curve C to be *treelike* if the normalization of C at its interior nodes is of compact type. A nomenclatural warning is perhaps in order here. The dual graphs of most treelike curves are *not* trees; rather, their dual graphs become trees after removing all "loops" (edges joining a vertex to itself).

Treelike curves share with curves of compact type (i.e., those whose dual graphs *are* trees) the basic property that, if a family of line bundles of degree d on a family of smooth curves specializing to a treelike curve C_0 has a line bundle as limit, then it also has a limit with arbitrarily assigned degrees (adding up to d) on the components of C_0. (The difference is the need for the "if" clause: a family of line bundles on a family of smooth curves specializing to a treelike curve need not have a line bundle as limit.) We may then define a limit linear series on a treelike curve just as we do for curves of compact type, and likewise the ramification sequence of a limit linear series at a smooth point. As the following exercise shows, the basic property of curves of compact type — that a limit of smooth curves possessing linear series with negative Brill-Noether number must possess a limit linear series with negative ρ — holds as well for treelike curves, after a fashion.

EXERCISE (5.40) Let $\pi : \mathcal{X} \to B$ be a family of curves with smooth total space and treelike special fiber X_0. Show that if the general fiber of π possess a g_d^r with $\rho < 0$, then the partial normalization of X_0 at some subset of its interior nodes possesses a limit g_d^r with $\rho < 0$ as well. *Hint*: A family of line bundles on such a family of curves will have as limit a torsion-free sheaf on X_0. Consider the subset of interior nodes of X_0 at which this sheaf fails to be locally free.

Smoothing limit linear series

In the preceding and subsequent sections, we've used our analysis of the behavior of a linear series as the curve carrying it degenerates to obtain restrictions on the existence of individual series and on the dimensions of families of them. We would now like to turn these ideas around and use our analysis to show the existence of certain linear series on smooth curves. In other words, we ask: when does a limit linear series on a curve X_0 actually occur as the limit of linear series on a family X_t of smooth curves specializing to X_0.

The answer to this question in its full generality isn't known. On the one hand, there are examples of limit linear series that cannot be smoothed. On the other, we have techniques for proving the smoothability of such series under fairly mild hypotheses, which we'll now describe.

The key construction is that of a scheme parameterizing limit linear series. Just as for any smooth curve X there exists a scheme $G_d^r(X)$ parameterizing linear series of degree d and dimension r on X (cf. [7]), so there exists for any curve of compact type a scheme parameterizing limit linear series of degree d and dimension r on X, which we'll continue to denote by $G_d^r(X)$. Moreover, for any family $X \to B$ of curves of compact type, there exists a scheme $G_d^r(X/B) \to B$ parameterizing limit linear series of degree d and dimension r on the fibers of the family. Most importantly for our applications, $G_d^r(X/B) \to B$ has the expected local description: it's determinantal, and every component has dimension at least $\dim(B) + \rho$.

This dimension estimate is what allows us to assert the smoothability of certain limit linear series. For example, suppose that we have a curve X_0 of compact type, and that $G_d^r(X_0)$ has dimension exactly ρ. If $\pi : X \to B$ is any smoothing of X_0 — that is, a one-parameter family of curves with smooth general fiber and special fiber $\pi^{-1}(b_0) = X_0$ — then the dimension estimate tells us that $G_d^r(X/B)$ has dimension at least $\rho + 1$ everywhere, and in particular that no component of $G_d^r(X/B)$ can lie over b_0. In other words, *in every smoothing $\pi : X \to B$ of X_0, every limit linear series of degree d and dimension r on X_0 is a limit of linear series on the general fiber.*

More generally, we see that if $\pi : X \to B$ is any family of singular curves of compact type such that $\dim(G_d^r(X/B)) = \dim(B) + \rho$, then any limit linear series of degree d and dimension r on a fiber of π can be smoothed. The argument is the same: we simply enlarge (locally) the family to a family $X' \to B'$ of one larger dimension, such that $B \subset B'$ is the discriminant divisor of $X' \to B'$, and argue that for dimension reasons the subvariety $G_d^r(X/B) \subset G_d^r(X'/B')$ must lie in the closure of its complement.

We can also do the same thing for k-pointed curves, associating to a family $\pi : X \to B$ with k disjoint sections $\sigma_j : B \to X$ (with image in the smooth locus of π) and k ramification sequences $b^j = (b_0^j, \dots, b_r^j)$ a scheme $G_d^r(X/B; \sigma_1, \dots, \sigma_k; b^1, \dots, b^k)$ parameterizing limit linear series on fibers X_t of X with ramification sequence b^j at $\sigma_j(t)$. Arguing as above, it follows that, for any such family, the dimension of $G_d^r(X/B; \sigma_1, \dots, \sigma_k; b^i, \dots, b^k)$ is at least $\dim(B) + \rho - \sum_{i,j}(\beta_i^j)$; and if equality holds, then any limit linear series on a fiber X_t of X with ramification sequence b^j at $\sigma_j(t)$ is a limit of a linear series on a smooth curve X having ramification sequences b^j at points p_j specializing to $\sigma_j(t)$ as X specializes to X_t.

We will summarize the state of our knowledge as the following theorem, known as the Regeneration Theorem. We need one further bit of terminology. Let X' be a semistable curve stably equivalent to a stable curve X. We say a refined limit linear series on X' is *stably equivalent* to a refined limit linear series on X if the two series have the same aspects on corresponding components of X' and X. (Note that this determines completely the aspects on the components of X' contracted in X: the ramification sequences at the two nodes of X' lying on such a component are precisely complementary, and so force the aspect there to be monomial.)

THEOREM (5.41) (REGENERATION THEOREM) *Fix a family* $\pi : X \to B$ *and* $\sigma_j : B \to X$ *of k-pointed nodal curves and any k ramification sequences* $b^j = (b_0^j, \ldots, b_r^j)$. *Let* $G = G_d^r(X/B; \sigma_1, \ldots, \sigma_k; b^1, \ldots, b^k)$, *and let* $[V] \in G$ *be a point corresponding to a limit linear series* $V = \{V_Y\}$ *on a fiber* $X_0 = \pi^{-1}(0)$. *If the dimension of* G *at* $[V]$ *is exactly* $\dim(B) + \rho - \sum_{i,j}(\beta_i^j)$, *then there exists a family of smooth k-pointed curves* $(X_t; p_1(t), \ldots, p_k(t))$ *specializing to a k-pointed curve stably equivalent to* $(X_0, \sigma_1(0), \ldots, \sigma_k(0))$, *and a family* V_t *of linear series on the curves* X_t *specializing to a limit linear series stably equivalent to* V.

As we've indicated, the proof follows simply from the existence of parameter spaces $G_d^r(X/B; \sigma_i, \ldots, \sigma_k; b_i, \ldots, b_k)$ for limit linear series and the basic dimension estimate on them; for their construction, see Eisenbud and Harris ([39], [42]).

We will give some applications of the Regeneration Theorem in the following subsection. In addition, in Exercises (5.63)–(5.65), we'll deduce as a consequence the following converse to the generalized Brill-Noether theorem [Theorem (5.37)]. To state it, we introduce some notation: following Fulton [53, §14.7] for any sequence $\lambda = (\lambda_0, \ldots, \lambda_r)$ with $\lambda_0 \geq \lambda_1 \geq \ldots \geq \lambda_r$, we'll denote by $\{\lambda\}$ or $\{\lambda_0, \ldots, \lambda_r\}$ the class, in the cohomology or Chow ring of the Grassmannian, of the corresponding Schubert cycle: this is the class denoted $\sigma_{\lambda_0, \ldots, \lambda_r}$ in Griffiths and Harris [64]. In these terms, we have:

THEOREM (5.42) *Let* C *be a general smooth curve of genus* g, *let* $p_1, \ldots, p_k$ *be general points of* C *and let* $b^j = (b_0^j, \ldots, b_r^j)$ *for* $j = 1, \ldots, k$ *be any k ramification sequences. There exists a* g_d^r *on* C *having ramification at least* b^j *at* p_j *if and only if the product*

$$\prod_{j=1}^{k} \{b_r^j, \ldots, b_0^j\} \cdot \{1, 1, \ldots, 1, 0\}^g \neq 0$$

in the cohomology ring $H^*(G(r+1, d+1), \mathbb{Z})$ *of the Grassmannian* $G(r+1, d+1)$.

When $k = 1$, there is a simple condition on the ramification sequence $b = b^1$ equivalent to this: using the Littlewood-Richardson formula for products of Schubert cycles in the Grassmannian (see for example [53]), it's possible to see that the product $\{b_r, \ldots, b_0\} \cdot \{1, 1, \ldots, 1, 0\}^g \neq 0$ if and only if the sum $\sum(b_i + g - d + r)_+ \leq g$, where we use the notation $(x)_+$ to denote $\max\{x, 0\}$. We thus have the following:

COROLLARY (5.43) *Let C be a general curve of genus g, p be a general point of C and $b = (b_0, \ldots, b_r)$ be any ramification sequence. There exists a g_d^r on C having ramification at least b at p if and only if the sum*

$$\sum_{i=0}^{r} (b_i + g - d + r)_+ \leq g.$$

Limits of canonical series and Weierstrass points

As a first application of the theory in the preceding subsection, we'll describe the limiting positions of the Weierstrass points of a smooth curve degenerating to a reducible curve $X_0 = Y \cup Z$ where Y and Z are smooth curves of genera i and $g - i$ respectively with a point $p \in Y$ identified to a point $q \in Z$: for more on this topic, see [40]. In general, these limiting positions have nothing to do with — indeed, are disjoint from — the Weierstrass points of Y and Z themselves.

The key to the analysis here is to identify the Weierstrass points of a smooth curve X_t as the ramification points of the canonical linear series. This identifies their limits as the inflectionary points of the aspects of the limit (V_Y, V_Z) of the canonical linear series $V_t = |K_{X_t}|$ on X_t. Our task is thus to describe this limiting series. The key inequalities are:

$$a_{g-i}(V_Y, p) + a_{i-1}(V_Z, q) \geq 2g - 2 \quad \text{and}$$

$$a_{g-i-1}(V_Y, p) + a_i(V_Z, q) \geq 2g - 2.$$

The linear series $V_Y(-a_{g-i}(V_Y, p) \cdot p)$ has dimension $i - 1$ and hence, by Clifford's theorem, degree at least $2i - 2$ with equality holding only if it equals $|K_Y|$: therefore, we have $a_{g-i}(V_Y, p) \leq 2(g - i)$. Similarly, $a_{i-1}(V_Z, q) \leq 2(i - 1)$. This is only possible if equality holds throughout: in particular, $V_Y(-a_{g-i}(V_Y, p) \cdot p)$ is the complete canonical series of Y. Using the second inequality, we find that $a_{g-i-1}(V_Y, p) = 2(g - i - 1)$, $a_i(V_Z, q) = 2i$, and $V_Y(-a_i(V_Z, q) \cdot q)$ is the complete canonical series on Z.

Now, assume that p is not a Weierstrass point of Y. Then the linear series $V_Y(-a_{g-i}(V_Y, p) \cdot p) = |K_Y|$ will be unramified at p and in particular

$$a_{g-1}(V_Y, p) \leq a_{g-i}(V_Y, p) + i - 1 = 2g - i - 1.$$

This in turn forces

$$a_0(V_Z, q) \geq i - 1;$$

i.e., V_Z has a base point of order $i - 1$ at q, and after removing this base point, V_Z is a linear series of dimension $g - 1$ and degree $2g - 1 - i$. In other words, V_Z *is the complete series* $|K_Z((i + 1) \cdot q)|$, *plus the fixed divisor* $(i - 1) \cdot q$. Applying the same argument to the aspect V_Y, we see that in case p and q aren't Weierstrass points of Y and Z, the limit on X_0 of the canonical series $|K_{X_t}|$ can only be the limit linear series (V_Y, V_Z) with aspects

$$V_Y = |K_Y((g - i + 1)p)| + (g - i - 1)p$$

and

$$V_Z = |K_Z((i + 1)q)| + (i - 1)q.$$

Note in particular that in this case the limit of the series $|K_{X_t}|$ depends only on X_0, and not on the family of curves tending to it; in the general setting — that is, if p or q *is* a Weierstrass point — we'll see that the limit does depend on the family.

The argument so far shows that if p is not a Weierstrass point of Y, then a point of Z will be a limit of Weierstrass points of smooth curves X_t of genus g tending to X_0 only if it's a ramification point for the linear series $|K_Z((i + 1) \cdot q)|$. To obtain a converse to this result we need to suitably smooth this limit linear series; we do this by applying the Regeneration Theorem (5.41). To set it up, we take our base $B = Z \setminus \{q\}$ and consider the one-pointed family $X = X_0 \times B \to B$ with section $\sigma : B \to X$ given by the diagonal in $B \times B \subset X_0 \times B$ and ramification sequence $b = (0, \dots, 0, 1)$. Now, let $r \in Z \setminus \{q\} = B$ be any ramification point of the series $|K_Z((i + 1)q)|$ other than q. The argument above shows that the limit linear series (V_Y, V_Z), with V_Y and V_Z as above, will be an isolated point of the scheme $G^{g-1}_{2g-2}(X/B; \sigma; b)$. Since the expected dimension of $G^{g-1}_{2g-2}(X/B; \sigma; b)$ is indeed zero, we conclude from Theorem (5.41) that (V_Y, V_Z) is indeed a limit of g^{g-1}_{2g-2}'s on smooth curves tending to X_0, ramified at points $p_t \in X_t$ tending to r. In sum, then, we see that *every ramification point of* $|K_X((i+1) \cdot q)|$ *is indeed a limit of ramification points for canonical series on smooth curves — that is, of Weierstrass points.*

We turn now to the case of a curve $X_0 = Y \cup Z/p \sim q$ where the point of attachment *is* a Weierstrass point of a component. For simplicity, suppose that p is a simple Weierstrass point of Y — that is, the ramification sequence of the canonical series $|K_Y|$ at p is $(0, \dots, 0, 1)$ — and that q is not a Weierstrass point of Z. The inequality

$$a_{g-1}(V_Y, p) \le a_{g-i}(V_Y, p) + i - 1 = 2g - i - 1$$

on the vanishing sequence of the aspect V_Y of a limit g^{g-1}_{2g-2} on X_0 that we used in our previous analysis need no longer hold; instead, we have only

$$a_{g-1}(V_Y, p) \le 2g - i \quad \text{and} \quad a_{g-2}(V_Y, p) \le 2g - i - 2.$$

These in turn give

$$a_0(V_Z, q) \ge i - 2 \quad \text{and} \quad a_1(V_Z, q) \ge i.$$

In other words, the aspect V_Z of the limit linear series associated to the canonical series on the members X_t of a family of smooth curves tending to X_0 is not uniquely determined! Rather, it can a priori be any one of the one-parameter family of g_{2g-2}^{g-1}'s on Z satisfying

$$(5.44) \qquad |K_Z(iq)| + iq \subset V_Z \subset |K_Z((i+2)q)| + (i-2)q.$$

In fact, we claim that *all of these do occur* as (aspects of) limits of $|K_{X_t}|$ for suitable families of curves X_t tending to X_0. To see this, consider the family $\pi : \mathcal{X} \to B$ with base $B = Y$ obtained by identifying the fixed point $q \in Z$ with a variable point $p \in Y$. By our previous analysis, the fiber of $G_{2g-2}^{g-1}(\mathcal{X}/B)$ over points $p' \in Y = B$ that aren't Weierstrass points will consist of a single point, while the fiber over p will be one-dimensional. Thus, over a neighborhood of $p \in Y = B$, we have

$$\dim G_{2g-2}^{g-1}(\mathcal{X}/B) = 1 = \dim B + \rho,$$

and applying the Regeneration Theorem we deduce that for any linear series V_Z on Z satisfying the inclusions (5.44) above, the pair (V_Y, V_Z) (with $V_Y = |K_Y((g-i+1)p)| + (g-i-1)p$ as before) is the limit of the canonical series $|K_{X_t}|$ for some family of curves X_t tending to X_0.

By the same token, we see that for such a curve X_0, every point $s \in Z \subset X$ is a limit of Weierstrass points on (some) nearby smooth curves X_t. The point is that a general point $s \in Z$ will be a ramification point of a finite number $V_1, \ldots, V_m$ of the linear series V_Z satisfying the inclusions (5.44) above. Thus, we can take $\pi : \mathcal{X} \to B = Y$ as before, mark the family with the section σ corresponding to the point s in each fiber, and again set $b = (0, \ldots, 0, 1)$. Now in a neighborhood of the point $p \in Y = B$, the scheme $\dim G_{2g-2}^{g-1}(\mathcal{X}/B; \sigma; b)$ is zero-dimensional; and once more applying Theorem (5.41) we conclude that there exists a family of pointed curves (X_t, s_t) and g_{2g-2}^{g-1}'s on X_t ramified at s_t tending to (X_0, s) and (V_Y, V_i) — in other words, we conclude that s is a limit of Weierstrass points of smooth curves. In sum, then, we've proved the:

THEOREM (5.45) *Let $X_0 = Y \cup Z/p \sim q$ be a curve consisting of two smooth components meeting at a point as above.*

1) *If p is not a Weierstrass point of Y, then a point $s \in Z \setminus \{q\}$ will be a limit of Weierstrass points of smooth curves X_t of genus g tending to X_0 if and only if it's a ramification point for the linear series $|K_Z((i+1)q)|$.*

2) *If p is a Weierstrass point of Y, then every point of Z will be such a limit.*

Note finally that, according to the Plücker formulas, the result above does indeed account for all the limits of Weierstrass points: the linear series $|K_Z((i+1)q)|$ will have $(g-i)(g^2-1)$ ramification points other than q, and the series $|K_Y((g-i+1)\cdot p)|$ will have $i(g^2-1)$.

The limits of Weierstrass points on curves not of compact type are studied in [44].

As we indicated earlier, we can go further with this analysis to prove the existence of curves with Weierstrass points of higher weight. This is sketched in the following sequence of exercises. Before we begin, though, we need to introduce some terminology and notation relevant to Weierstrass points.

First of all, we have what we may call the *Weierstrass stratification* of the universal curve $\mathcal{M}_{g,1}$: for any ramification sequence $b = (b_0, \ldots, b_{g-1})$, we'll denote by $\mathcal{C}_b \subset \mathcal{M}_{g,1}$ the locally closed subset of $\mathcal{M}_{g,1}$ consisting of pairs (C, p) such that the canonical series $|K_C|$ of C has ramification sequence exactly equal to b at C; that is,

$$\mathcal{C}_b = \{[(C, p)] : b_i(|K_C|, p) = b_i\}.$$

This is just the locus of points with Weierstrass gap sequence $\{b_i + i + 1\}_{i=0,\ldots,g-1}$. We will say that b satisfies the *semigroup condition* if the complement $H(b) = \mathbb{N} \setminus \{b_i + i + 1 \mid i = 0, \ldots, g-1\}$ is indeed a semigroup; clearly, $\mathcal{C}_b$ will be empty otherwise.

One more bit of terminology. The expected codimension of $\mathcal{C}_b$ in $\mathcal{M}_{g,1}$ is simply the sum $\beta = \sum_{i=0}^{g-1} b_i$ (this is also an upper bound for the codimension of $\mathcal{C}_b$ whenever it's nonempty). We will say, then, that a Weierstrass point $(C, p) \in \mathcal{C}_b$ is *proper* if in fact $\dim_{[(C,p)]}(\mathcal{C}_b) = \beta$ — in English, the codimension of the Weierstrass stratum containing $[(C, p))]$ is equal to the weight. Note that any point of weight 0 or 1 is trivially proper. In these terms, our first result is a direct generalization of an argument above:

EXERCISE (5.46) Let $X_0 = Y \cup Z / p \sim q$ be as above the curve obtained by identifying $p \in Y$ with $q \in Z$, and assume that (Y, p) and (Z, q) are both proper, of weights w_1 and w_2. Show that the variety $G^{g-1}_{2g-2}(X_0)$ of limit g^{g-1}_{2g-2}'s on X_0 has dimension $w_1 + w_2$, and that any such limit on X_0 is indeed the limit of the canonical series $|K_{X_t}|$ on a family of smooth curves X_t tending to X_0.

Our first application will be to prove the existence of (both kinds of) Weierstrass points of weight 2.

EXERCISE (5.47) Let $X_0 = Y \cup Z / p \sim q$ be as above the curve obtained by identifying $p \in Y$ with $q \in Z$; but now assume $g \geq 4$, Y is a smooth curve of genus $g-1$, and Z an elliptic curve. Assume that p is a simple Weierstrass point of Y.

1) Let $r \in Z$ be a point such that $r - q$ is torsion of order exactly $g + 1$ in the group law on Z. Show that there is a limit g_{2g-2}^{g-1} on X_0 with ramification sequence $(0, \ldots, 0, 2)$ at r, and use the Regeneration Theorem to deduce that (X_0, r) is a limit of proper Weierstrass points with gap sequence $(1, 2, \ldots, g - 1, g + 2)$.

2) Now let $r \in Z$ be a point such that $r - q$ is torsion of order exactly $g - 1$ in the group law on Z. Show that there is a limit g_{2g-2}^{g-1} on X_0 with ramification sequence $(0, \ldots, 0, 1, 1)$ at r, and use the Regeneration Theorem to deduce that (X_0, r) is a limit of proper Weierstrass points with gap sequence $(1, 2, \ldots, g - 2, g, g + 1)$.

Note that in the last exercise, if r is a general point of Z then r will be a limit of simple Weierstrass points. We may apply the same techniques more generally:

EXERCISE (5.48) Let $X_0 = Y \cup Z/p \sim q$ be as above, and now assume Y is a smooth curve of genus $g - 1$ and Z is an elliptic curve. Assume only that (Y, p) is a proper Weierstrass point, with ramification sequence $(b_0, \ldots, b_{g-2})$. For any $j = 0, \ldots, g - 2$, let $r \in Z$ be a point such that $r - q$ is torsion of order exactly $b_j + j + 2$ in the group law on Z. Assuming that $b_j + j + 2$ doesn't divide $b_k + k + 2$ for any $k > j$, show that there is a limit g_{2g-2}^{g-1} on X_0 with ramification sequence $(0, b_0, \ldots, b_{j-1}, b_j + 1, b_{j+1}, \ldots, b_{g-2})$ and that (X_0, r) is a limit of proper Weierstrass points with this ramification sequence. Similarly, if $r \in Z$ is a general point, show that (X_0, r) is a limit of proper Weierstrass points with ramification sequence $(0, b_0, \ldots, b_{g-2})$.

We may deduce from this the following theorem and corollary.

THEOREM (5.49) If $\mathcal{C}_b \subset \mathcal{M}_{g-1,1}$ contains a proper point, then so does $\mathcal{C}_{b'} \subset \mathcal{M}_{g,1}$ if either

1) $b_0' = 0$ and $b_i' = b_{i-1}$ for $i = 1, \ldots, g - 1$; or

2) for some $j = 1, \ldots, g - 2$, we have $b_0' = 0$, $b_j' = b_{j-1} + 1$ and $b_i' = b_{i-1}$ for $i = 1, \ldots, g - 1, i \neq j$, and b satisfies the semigroup condition.

COROLLARY (5.50) If $H \subset \mathbb{N}$ is any semigroup of index $\#(\mathbb{N} \setminus H) = g$ and weight

$$ w = \left(\sum_{i \in \mathbb{N} \setminus H} i \right) - \frac{g(g + 1)}{2} \leq \frac{g}{2} $$

then there exists a Weierstrass point (C, p) with semigroup H.

EXERCISE (5.51) Use Exercise (5.48) to prove Theorem (5.49) and Corollary (5.50).

E Limit linear series on flag curves

As we've seen, the Brill-Noether theorem follows directly from inequalities on the total ramification indices of the aspects of a limit linear series on an arbitrary curve formed from a tree of rational curve and g elliptic tails. To get more subtle information about the behavior of linear series on a general curve, such as that expressed by the Gieseker-Petri theorem, we need to analyze in more detail limit linear series on a more special type of curve, namely the flag curves described in the first section of the chapter, and this is what we shall do in this final section. For the remainder of the chapter, then, X_0 will be the curve pictured in Figure (5.10), and $\mathcal{D}$ will be a limit linear series on X_0 of degree d and dimension r.

Inequalities on vanishing sequences

To analyze $\mathcal{D}$, we label the components of X_0 as in Figure (5.10), with $Y_1,\ldots,Y_N$ forming the main vertical chain and $p_i = Y_i \cap Y_{i-1}, for i = 2,\ldots,N$. In addition, for $m \in \{1,\cdots,g\}$, let us denote by E_m one of the elliptic tails, by $Y_{c(m)}$ the component of Y joined to E_m, and by q_m and s_m the respective points of attachment on $Y_{c(m)}$ and E_m.

We begin our analysis by looking at a fixed elliptic tail $E = E_m$, so we'll drop the subscript m's for the moment. Let $Z_0 = Y, Z_1,\ldots, Z_k$ be the chain of rational curves leading to E, with $q_i = Z_{i-1} \cap Z_i$. To start with, we observe that we must have

$$a_{r-1}(V_E, s) \le d - 2,$$

since if $a_{r-1}(V_E, s) = d - 1$, the pencil generated by the two sections of V_E vanishing to highest order at s would have $d - 1$ base points, and hence would contain a g_1^1. By Theorem (5.28), then

$$a_1(V_{Z_k}, s) \ge 2$$

and now by 1) of Lemma (5.30),

$$a_1(V_Y, q_1) \ge a_1(V_{Z_1}, q_2) \ge a_1(V_{Z_2}, q_3)$$
$$\vdots \qquad \vdots$$
$$\ge a_1(V_{Z_{k-1}}, q_k) \ge a_1(V_{Z_k}, s) \ge 2.$$

We deduce from this the:

PROPOSITION (5.52) *If $l = c_m$ for some m, the linear series V_{Y_l} can have at most one section vanishing only at p_l and p_{l+1}.*

PROOF. Two such sections σ, τ would generate a pencil totally ramified at p_l and p_{l+1}, *and consequently unramified elsewhere.* In particular, suitable linear combinations of σ and τ would vanish to orders 0 and 1 at q_m, violating $a_1(V_Y, q_1) \geq 2$. ∎

Note that, for like reasons, the equality $a_r(V_E, s) = d$ can hold only if $L_E \otimes \mathcal{O}_E = \mathcal{O}_E(ds)$ — equivalently, $L_{Y_l} \otimes \mathcal{O}_E = \mathcal{O}_E$. Mimicking the argument above, we see that, *unless $L_E \otimes \mathcal{O}_E = \mathcal{O}_E(ds)$, the point q_1 is a base point of V_{Y_L},* which implies that there can be *no* section of V_{Y_l} vanishing solely at p_l and p_{l+1}. Combining both parts of Lemma (5.30) with this observation, we see that in general we have the inequality

(5.53) $$a_i(V_{Y_{l+1}}, p_{l+1}) \geq a_i(V_{Y_l}, p_l)$$

and *if $l = c_m$ for some m, then equality can hold in (5.53) for at most one value of i.*

But now for any l and i we have trivially

$$i \leq a_i(V_{Y_l}, p_l) \leq d - (r - i)$$

and hence, in general,

(5.54) $$a_i(V_{Y_{l_1}}, p_{l_1}) - a_i(V_{Y_{l_2}}, p_{l_2}) \leq (d - r).$$

Taking $l_1 = N$ and $l_2 = 1$ and untelescoping the difference gives

$$\sum_{l=1}^{N-1} \left(a_i(V_{Y_{l+1}}, p_{l+1}) - a_i(V_{Y_l}, p_l) \right) \leq (d - r).$$

Finally, summing over i yields

$$\sum_{i=0}^{r} \sum_{l=1}^{N-1} \left(a_i(V_{Y_{l+1}}, p_{l+1}) - a_i(V_{Y_l}, p_l) \right) \leq (r + 1)(d - r).$$

Interchanging the summation, and dropping those (nonnegative) terms for which l isn't one of the $c(m)$'s gives the lower bound

(5.55)

$$(r + 1)(d - r) \geq$$

$$\sum_{m=1}^{g} \sum_{i=0}^{r} \left(a_i(V_{Y_{c(m)+1}}, p_{c(m)+1}) - a_i(V_{Y_{c(m)}}, p_{c(m)}) \right).$$

But, for fixed m, we know that all but one of the $(r + 1)$ terms in the sum over i is strictly positive. Thus, we recover the Brill-Noether inequality

$$(r + 1)(d - r) \geq rg$$

and the weak form of the Brill-Noether theorem: X_0, and hence a general curve of genus g, cannot possess a (limit) linear series with negative Brill-Noether number.

EXERCISE (5.56) The argument following Proposition (5.52) shows that, unless $L_{E_m} \otimes \mathcal{O}_{E_m} = \mathcal{O}_{E_i}(d \cdot s)$, every term in the mth inner sum in (5.55) will be positive. Use this sharpening to rederive the strong form of Brill-Noether as well.

The case $\rho = 0$

One case in which the above analysis gives us a more or less complete description of the limit linear series $\mathcal{D}$ is when $\rho = 0$. In this case, all the inequalities used in the argument must be equalities. In particular, this means that the vanishing sequences $a_{l,i} = a_i(V_{Y_l}, p_l)$ of the aspects V_Y of $\mathcal{D}$ do not merely satisfy $a_{l+1,i} \geq a_{l,i}$ for all l and i. More precisely, we have the:

LEMMA (5.57) In case $\rho = 0$,

1) for $l \neq c_m$, we have $a_{l+1,i} = a_{l,i}$ for all i; and

2) for $l = c_m$, we have $a_{l+1,i} = a_{l,i} + 1$ for all but one i.

There are only a finite number of collections of sequences $a_{l,i}$ satisfying this system of equations; in fact, as we'll see in Exercise (5.66), there are exactly as many as there are g_d^r's on a general curve of genus g. For example, in the case of g_3^1's on a curve of genus 4, we have the two solutions as shown in Table (5.59). Similarly, for the g_4^1's on a curve of genus 6 there are 5 solutions, shown in Table (5.60).

One circumstance in which there is always a unique solution is the case $d = 2g - 2$, $r = g - 1$ corresponding to the canonical series. The case of $g = 5$ is typical. Table (5.61) shows the only solution. In general, we claim that the unique solution of the constraints expressed in Lemma (5.57) is given, for l between c_{m-1} and c_m, by

$$(5.58) \qquad a_{l,i} = \begin{cases} m + i - 2 & \text{if } i < m - 1 \\ m + i - 1 & \text{if } i \geq m - 1. \end{cases}$$

To see this, observe that for each $m = 1, \ldots, g$ there is exactly one $i = i(m)$ such that $a_{c_m,i} = a_{c_m+1,i}$, every i occurs for exactly one m, and these indices completely determine the solution. Equation (5.58) amounts to the assertion that $i(m) = m - 1$ for all m. To see this for $m = g$, recall that the instances of (5.54) used in the proof are sharp only if $a_i(Y_N, p_N) = d - r + i = g - 1 + i$ for every i and hence, in particular, $a_{c_g,i}$ must be an increasing sequence. But now dropping the $a_{l,i}$ for $i = g - 1$ and/or $l \geq c_g$ we're left with the same system of equations with g replaced by $g - 1$ and can conclude by an induction. We remark that the uniqueness of the canonical series on X_0 is a property of this curve not shared by all curves of compact type.

	c_1	c_2	c_3	c_4	N	l
0	0	0	1	1	2	
1	1	2	2	3	3	
0	0	0	0	1	2	
1	1	2	3	3	3	
i						

TABLE (5.59)

	c_1	c_2	c_3	c_4	c_5	c_6	N	l
0	0	0	0	0	1	2	3	
1	1	2	3	4	4	4	4	
0	0	0	0	1	1	2	3	
1	1	2	3	3	4	4	4	
0	0	0	0	1	2	2	3	
1	1	2	3	3	3	4	4	
0	0	0	1	1	1	2	3	
1	1	2	2	3	4	4	4	
0	0	0	1	1	2	2	3	
1	1	2	2	3	3	4	4	
i								

TABLE (5.60)

	c_1	c_2	c_3	c_4	c_5	N	l
0	0	0	1	2	3	4	
1	1	2	2	3	4	5	
2	2	3	4	4	5	6	
3	3	4	5	6	6	7	
4	4	5	6	7	8	8	
i							

TABLE (5.61)

There is one particular consequence of (5.58) that will be crucial in the following argument.

LEMMA (5.62) *If $\mathcal{D}$ is a limit of the canonical series, there is, for $m = 1, \ldots, g$ and $l = c_m$, a unique section in V_{Y_l} vanishing only at p_l and p_{l+1}, and it vanishes at p_l to order exactly $2m - 2$.*

We will use this description of the limits of canonical series on flag curves in the following subsection to prove the Gieseker-Petri theorem. In the meantime, we'll indicate in the following series of exercises how this description of limit linear series on flag curves may be used to prove the converse to the generalized Brill-Noether theorem [Theorem (5.42)]. The one ingredient that doesn't involve limit linear series is Exercise (5.63), This is essentially an exercise in Schubert calculus which we express using the notation for Schubert classes introduced with Theorem (5.42); you may wish to simply assume this statement and proceed to the remaining exercises.

EXERCISE (5.63) Let r and $d \geq r$ be positive integers, and fix any three ramification sequences $b^j = (b_0^j, \ldots, b_r^j)$ for $j = 1, 2, 3$, satisfying

$$\sum_{i,j} b_i^j = (r + 1)(d - r).$$

Show that there exists a g_d^r on $\mathbb{P}^1$ having ramification sequences b^1, b^2 and b^3 at the points 0, 1 and ∞ respectively if and only if the intersection number

$$\prod_{j=1}^{3} \{b_r^j, \ldots, b_0^j\} \neq 0$$

in the cohomology ring $H^*(G(r + 1, d + 1), \mathbb{Z})$ of the Grassmannian $G(r + 1, d + 1)$. Moreover, in this case there exists a finite number of such g_d^r's.

EXERCISE (5.64) Now let C be a flag curve of genus g, and $p_1, \ldots, p_k$ smooth points of C lying on distinct components of the backbone of C. Let $b^j = (b_0^j, \ldots, b_r^j)$, $j = 1, \ldots, k$, be any k ramification sequences satisfying $rg + \sum b_i^j = (r + 1)(d - r)$. Assuming the result of Exercise (5.63), show that there exists a limit g_d^r on C having ramification at least b^j at p_j if any only if the product

$$\prod_{j=1}^{k} \{b_r^j, \ldots, b_0^j\} \cdot \{1, 1, \ldots, 1, 0\}^g \neq 0$$

in the cohomology ring $H^*(G(r + 1, d + 1), \mathbb{Z})$ of the Grassmannian $G(r + 1, d + 1)$, and that in this case there are a finite number of such limit linear series.

EXERCISE (5.65) Use the Regeneration Theorem (5.41) to deduce the converse of the generalized Brill-Noether theorem [Theorem (5.42)] from Exercise (5.64): you may need to impose extra ramification points with ramification sequence $(0, \ldots, 0, 1)$ to make up the equality $rg + \sum b_i^j = (r + 1)(d - r)$.

Here are a couple other applications of the description of limit linear series on flag curves.

EXERCISE (5.66) Again let C be a flag curve of genus g and $p_1, \ldots, p_k$ smooth points of C lying on distinct components of the backbone of C, and suppose that g, r and d satisfy the equality $g = (r+1)(g-d+r)$. Show that there are exactly as many limit g_d^r's on C as the expected number of g_d^r's on a general curve of genus g, and (using the Regeneration Theorem) deduce that for a general curve X of genus g the scheme $G_d^r(X)$ is reduced.

EXERCISE (5.67) The following somewhat lengthy exercise illustrates the use of many of the ideas we've seen so far to prove the existence of a stable curve C of compact type and of genus 23 possessing a dimensionally proper limit g_{12}^1 but no crude limit g_{17}^2. The existence of such a curve will be used in Section 6.F.

We take for C the curve in Figure (5.68) where (F_i, p_i) are general

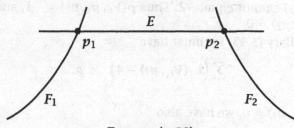

FIGURE (5.68)

pointed curves of genus 11, E is an elliptic curve, and p_1 and p_2 differ by a translation of order exactly 12; that is, $12(p_1 - p_2) \sim 0$, but $n(p_1 - p_2) \not\sim 0$ for $n < 12$.

1) Show that, if we set V_{F_i} to be the complete series $|12p_i|$ on F_i and take V_E to be the pencil spanned by $12p_1$ and $12p_2$ on E, then the V's are the aspects of a refined and dimensionally proper limit g_{12}^1 on C.

2) Now suppose that we're given a limit g_{17}^2 on C. Use the fact that the F_i are general and the additivity of the adjusted Brill-Noether number to deduce that $\rho(V_E, p_1, p_2) < 0$.

3) Use a dimension count to show that V_E contains sections σ_i vanishing on the divisors

$$A_i = a_i(V_E, p_1)p_1 + a_{2-i}(V_E, p_2)p_2,$$

so that

(5.69) $\deg(A_i) = a_i(V_E, p_1) + a_{2-i}(V_E, p_2) \le 17.$

Sum these to obtain the estimate

$$\beta_i(V_E, p_1) + \beta_{2-i}(V_E, p_2) \le 45.$$

Combine these with $\rho(1, 2, 17) = 43$ to show that

$$\rho(V_E, p_1, p_2) \ge -2,$$

and, moreover, that if $\rho(V_E, p_1, p_2) = -1$ [resp:$\rho(V_E, p_1, p_2) = -2$], then equality holds in (5.69) for at least two [resp: for all three] values of i.

4) Let σ_{i_1} and σ_{i_2} be two sections in V_E, with $i_1 < i_2$ with associated divisors A_{i_1} and A_{i_2}. Use the hypothesis on the points p_i to show that $0 \sim (\sigma_{i_1}) - (\sigma_{i_2}) = 12(p_2 - p_1)$ in $\mathrm{Pic}(E)$ and hence that

$$(\mathbf{5.70}) \qquad (\sigma_{i_1}) = B + 12p_2 \quad \text{and} \quad (\sigma_{i_2}) = B + 12p_1$$

for some effective divisor B of degree 5 supported on p_1 and p_2. Conclude that there cannot be three such sections and hence that $\rho(V_E, p_1, p_2)$ cannot equal -2. Thus $\rho(V_E, p_1, p_2) = -1$, and we must have $\rho(V_{F_i}, p_i) = 0$.

5) By Corollary (5.43) we must have

$$\sum_j \left(b_j(V_{F_i}, p_i) - 4 \right)_+ \le g.$$

Since $\rho(V_{F_i}, p_i) = 0$, we have also

$$\sum_j (b_j(V_{F_i}, p_i) - 4) = g,$$

from which we deduce $b_j(V_{F_i}, p_i) \ge 4$ for all i and j. Use the compatibility conditions (5.34) to show that, for each i and j, first $b_j(V_{F_i}, p_i) \le 11$, and then $b_j(V_E, p_i) \le 13$. Show that this contradicts at least one of the equations (5.70) and hence that the hypothetical limit g_{17}^2 cannot exist.

Proof of the Gieseker-Petri theorem

In this section, we'll use the methods of the previous sections to analyze products of linear systems on a family of curves. For the time being, we let $\pi : X \to B$ be a projective family and let t be a local parameter at a distinguished point $b_0 \in B$. We assume that the total space X is smooth and that the central fiber X_0 over b_0 is a nodal curve of compact type. We let L and M be line bundles on $X \setminus X_0$ of relative degrees d and e. As before, for each component Y of X_0, we denote by L_Y and M_Y the (unique) extensions of L and M to X whose restrictions to each component of X_0 other than Y have degree zero. Recall that a section $\sigma \in \Gamma(L_Y)$ vanishing on Y vanishes on all of X_0, and likewise for M_Y.

Our first task is to define, for each component Y of X_0 and point $p \in Y$, an "order of vanishing" at p along Y for elements μ of the tensor product $\Gamma(L_Y) \otimes \Gamma(M_Y)$. To do this, define a flag

$$\Gamma(L_Y) \otimes \Gamma(M_Y) = \Sigma_0 \supseteq \Sigma_1 \supseteq \Sigma_2 \supseteq \cdots$$

by letting Σ_k be the linear span of all tensor products $\sigma \otimes \tau$ such that

$$\mathrm{ord}_p(\sigma|_Y) + \mathrm{ord}_p(\tau|_Y) \geq k.$$

We adopt the convention that $\mathrm{ord}_p(\sigma|_Y) = \infty$ if $\sigma|_Y = 0$; thus, for all k, $\Sigma_k \supset \Sigma_\infty = t(\Gamma(L_Y) \otimes \Gamma(M_Y))$. Note also that $\Sigma_{d+e+1} = \Sigma_\infty$.

We now define the *order of vanishing* of an element $\mu \in \Gamma(L_Y) \otimes \Gamma(M_Y)$ to be the largest k such that $\mu \in \Sigma_k$; we write this as $\mathrm{ord}_p(\mu|_Y)$.

As in our analysis of a single linear series, our basic tool will be to relate the "order of vanishing" at p along Y of a given $\mu \in \Gamma(L_Y) \otimes \Gamma(M_Y)$ to its order of vanishing considered as an element of $\Gamma(L_Y) \otimes \Gamma(M_Y)$ for other components Z. Our basic lemma refers to the now familiar picture diagrammed in Figure (5.71) in which p' is any point of Y distinct from p. As before, we'll denote by E the sum of the irreducible

FIGURE (5.71)

components of X_0 lying in the connected component of $X_0 \setminus \{p\}$ containing Z, so that

$$L_Z = L_Y(-dE)$$

and

$$M_Z = M_Y(-eE).$$

We will use these relations to view L_Z, M_Z and $\Gamma(L_Z) \otimes \Gamma(M_Z)$ as subsheaves and a subspace of L_Y, M_Y and $\Gamma(L_Y) \otimes \Gamma(M_Y)$ respectively.

LEMMA (5.72) *If $\mu \in \Gamma(L_Y) \otimes \Gamma(M_Y)$, $\mu \notin \Sigma_\infty$, and α is the smallest integer such that*

$$\nu = t^\alpha \mu \in \Gamma(L_Z) \otimes \Gamma(M_Z),$$

then

$$\mathrm{ord}_{p'}(\mu|_Y) \leq \alpha \leq \mathrm{ord}_p(\nu|_Z).$$

PROOF. To begin with, choose sections $\sigma_0, \ldots, \sigma_r$ of L_Y and $\tau_0, \ldots, \tau_s$ of M_Y generating $\pi_* L_Y$ and $\pi_* M_Y$ respectively, as in Lemma (5.27). Suppose $\sigma_i' = t^{\alpha_i} \sigma_i$ and $\tau_i' = t^{\beta_i} \tau_i$ are the corresponding sections of L_Z and M_Z generating $\pi_* L_Z$ and $\pi_* M_Z$. We first write our given μ as

$$\mu = \sum f_{ij}(0)\, \sigma_i \otimes \tau_j$$

with $f_{ij}(t)$ a holomorphic function of t in B. (Here, and in the sequel, any implicit indexing is over the full set of pairs $\{(i,j) \mid 0 \le i \le r, 0 \le j \le s\}$.) Since this expression is unique and the elements of each of the σ and τ bases all vanish to distinct orders at p', we see that

(5.73) $\quad \operatorname{ord}_{p'}(\mu|_Y) = \min_{\{(i,j) \mid f_{ij}(0) \ne 0\}} \left(\operatorname{ord}_{p'}(\sigma_i|_Y) + \operatorname{ord}_{p'}(\tau_j|_Y) \right).$

The key point now is to express α. To do this, set

$$g_{ij}(t) = t^{\alpha - \alpha_i - \beta_j} f_{ij}(t)$$

and, note that, in these terms

$$\nu = t^{\alpha} \mu = \sum (t^{\alpha - \alpha_i - \beta_j} f_{ij}(0))(t^{\alpha_i} \sigma_i \otimes t^{\beta_j} \tau_j)$$

(5.74) $\quad\quad = \sum (t^{\alpha - \alpha_i - \beta_j} f_{ij}(0))(\sigma_i' \otimes \tau_j')$

$$= \sum g_{ij}(0)\, \sigma_i' \otimes \tau_j'.$$

Since, by hypothesis, $\nu \in \Gamma(L_Z) \otimes \Gamma(M_Z)$ but $\nu \notin t \cdot \Gamma(L_Z) \otimes \Gamma(M_Z)$, all the coefficient functions g_{ij} must be holomorphic and at least one must be nonzero at 0; thus

(5.75) $\quad\quad\quad\quad \alpha = \max(\alpha_i + \beta_j - \operatorname{ord}_0(f_{ij})).$

Now by Lemma (5.25),

$$\alpha_i \ge \operatorname{ord}_{p'}(\sigma_i|_Y) \quad \text{and}$$

$$\beta_j \ge \operatorname{ord}_{p'}(\tau_j|_Y)$$

and combining this with (5.73) and (5.75) yields

$$\operatorname{ord}_{p'}(\mu|_Y) \le \alpha.$$

On the other hand, if the pair (i,j) is chosen from among the subset for which $g_{ij}(0) \ne 0$ so as to minimize the sum

$$\operatorname{ord}_p(\sigma_i'|_Z) + \operatorname{ord}_p(\tau_j'|_Z),$$

then we have

$$\operatorname{ord}_p(\nu|_z) \geq \operatorname{ord}_p(\sigma_i'|_z) + \operatorname{ord}_p(\tau_j'|_z)$$

$$\geq \alpha_i + \beta_j \qquad \text{(by (5.25))}$$

$$= \alpha + \operatorname{ord}_0(f_{ij}) \qquad \text{(since } g_{ij}(0) \neq 0\text{)}$$

$$\geq \alpha,$$

which gives the other inequality. ∎

We now specialize. First, we go back to the situation where X_0 is the curve pictured in Figure (5.10) and $M \otimes L$ is the dualizing sheaf on the general fiber of π. We suppose that we have an element $\mu_1 \in \Gamma(L_{Y_1}) \otimes \Gamma(M_{Y_1})$ such that the image of μ_1 is zero under the map

$$\Gamma(L_{Y_1}) \otimes \Gamma(M_{Y_1}) \longrightarrow \Gamma(\omega_{Y_1}).$$

Hence, we may also suppose that $\mu_1 \in tbigl(\Gamma(L_{Y_1}) \otimes \Gamma(M_{Y_1})bigr)$.

We now plan to proceed as follows. First, as above, we inductively identify, for each l, $L_{Y_{l+1}}$ and $M_{Y_{l+1}}$ with subsheaves of L_{Y_l} and M_{Y_l} respectively and hence identify $\Gamma(L_{Y_{l+1}}) \otimes \Gamma(M_{Y_{l+1}})$ with a subspace of $\Gamma(L_{Y_l}) \otimes \Gamma(M_{Y_l})$. Then, for each l, we let γ_l be the smallest integer such that $t^{\gamma_l} \cdot \mu_1 \in \Gamma(L_{Y_l}) \otimes \Gamma(M_{Y_l})$, and we set

$$\mu_l = t^{\gamma_l} \mu_1.$$

Finally, we'll consider the orders of vanishing

$$\varepsilon_l = \operatorname{ord}_{p_l}(\mu_l|_{Y_l}).$$

These, by (5.72), form an nondecreasing sequence. In fact, we claim this can be sharpened as follows. If $Y_{c(m)}$ is the component of Y to which the mth elliptic curve is joined as in the preceding two subsections, then we claim:

(5.76) If $l > c_m$, then $\varepsilon_l \geq 2m$.

Since the sum of the relative degrees of L and M is $2g - 2$, so that we must have $\varepsilon_l \leq 2g - 2$ for all l, this will yield a contradiction by taking $l \geq c(m)$. Statement (5.76) will in turn follow from (5.72) once we establish the

LEMMA (5.77) *If $l = c_m$ for any $m = 1, \ldots, g$ and*

$$\varepsilon_l = \operatorname{ord}_{p_l}(\mu_l|_{Y_l}) \geq 2m - 2,$$

then

$$\gamma_{l+1} - \gamma_l \geq 2m.$$

PROOF. The proof amounts to a more careful examination of the situation of Lemma (5.72) whose notation we would like to reuse. We do this by fixing $l = c(m)$ and writing Y for Y_l, Z for Y_{l+1}, μ for μ_l, ν for μ_{l+1}, p' for p_l and p for p_{l+1}. We then have $\nu = t^\alpha \mu$ with $\alpha = \gamma_{l+1} - \gamma_l$. As in the proof, we also fix bases $\{\sigma_i\}$ of $\Gamma(L_Y)$ and $\{t_j\}$ of $\Gamma(M_Y)$ and write

$$\mu = \sum f_{ij}(0)\, \sigma_i \otimes \tau_j\,.$$

By (5.73), the hypothesis of the lemma then implies

$$2m - 2 \le \text{ord}_{p'}(\mu|_Y)$$
$$= \min_{\{(i,j)\,|\,f_{ij}(0)\ne 0\}} \left(\text{ord}_{p'}(\sigma_i|_Y) + \text{ord}_{p'}(\tau_j|_Y)\right)$$

and, by (5.75), its conclusion will follow if we show that

$$\alpha = \max\{\alpha_i + \beta_j - \text{ord}_0(f_{ij})\} \ge 2m\,.$$

We will prove the slightly more precise statement that *there exist i and j such that $f_{ij}(0) \ne 0$ and*

$$\alpha_i + \beta_j \ge 2m\,.$$

To find the pair (i, j), we use, for the first and only time, the assumption that μ is in the kernel of the product map

$$\Gamma(L_Y) \otimes \Gamma(M_Y) \longrightarrow \Gamma(L_Y \otimes M_Y)\,.$$

This implies that

$$\sum_{\{(i,j)\,|\,f_{ij}(0)\ne 0\}} (f_{ij}(0)\, \sigma_i \otimes \tau_j)|_Y \equiv 0\,.$$

Therefore, *the minimum value, among the pairs (i, j) such that $f_{ij}(0) \ne 0$, of* $\text{ord}_{p'}(\sigma_i|_Y) + \text{ord}_{p'}(\tau_j|_Y)$ *must be taken more than once.* Consequently, there exist at least two such pairs for which

$$2m - 2 \le \varepsilon_l = \text{ord}_{p'}\left(\mu|_Y\right) = \text{ord}_{p'}\left(\sigma_i|_Y\right) + \text{ord}_{p'}\left(\tau_j|_Y\right)\,.$$

Let (i_1, j_1) and (i_2, j_2) be two such pairs and note, further, that, since the σ_i and τ_j vanish to distinct orders at p', we must have both $i_1 \ne i_2$ and $j_1 \ne j_2$.

We claim now that $\alpha_i + \beta_j \ge 2m$ for either $(i, j) = (i_1, j_1)$ or $(i, j) = (i_2, j_2)$. To see this, recall that by Theorem (5.28),

$$\alpha_i \ge \text{ord}_{p'}\left(\sigma_i|_Y\right)$$

and that by 2) of (5.30) and (5.52), equality can hold for *at most one i*. Therefore, we'll have $\alpha_i + \beta_j \geq \varepsilon_l + 2 \geq 2m$ for either $(i,j) = (i_1, j_1)$ or (i_2, j_2) *unless* we have both

$$\varepsilon_l = 2m - 2$$

and, after interchanging 1 and 2 if necessary,

$$\alpha_{i_1} = \mathrm{ord}_{p'}\left(\sigma_{i_1}|_Y\right),$$

$$\alpha_{i_2} = \mathrm{ord}_{p'}\left(\sigma_{i_2}|_Y\right) + 1,$$

$$\beta_{j_1} = \mathrm{ord}_{p'}\left(\tau_{j_1}|_Y\right) + 1,$$

$$\beta_{j_2} = \mathrm{ord}_{p'}\left(\tau_{j_2}|_Y\right).$$

In this situation, the sections $\sigma_{i_1}|_Y$ and $\tau_{j_2}|_Y$ each vanish only at p' and p, and hence so does their product viewed as a section of $\omega_Y|_Y$. But by Lemma (5.62), this says that

$$\mathrm{ord}_{p'}\left(\sigma_{i_1}\tau_{j_2}|_Y\right) = \mathrm{ord}_{p'}\left(\sigma_{i_1}|_Y\right) + \mathrm{ord}_{p'}\left(\tau_{j_2}|_Y\right) = 2m - 2$$

and we arrive at

$$2m - 2 = \varepsilon_l = \mathrm{ord}_{p'}\left(\mu|_Y\right)$$
$$= \mathrm{ord}_{p'}\left(\sigma_{i_1}|_Y\right) + \mathrm{ord}_{p'}\left(\tau_{j_1}|_Y\right)$$
$$> \mathrm{ord}_{p'}\left(\sigma_{i_1}|_Y\right) + \mathrm{ord}_{p'}\left(\tau_{j_2}|_Y\right)$$
$$= 2m - 2,$$

a contradiction. ∎

Our conclusion is that μ cannot exist; i.e., the tensor product map is injective. We have thus established the:

THEOREM (5.78) (GIESEKER-PETRI THEOREM) *If C is a general curve and L any line bundle on C, the map*

$$\mu_0 : H^0(C, L) \otimes H^0(C, K \otimes L^{-1}) \to H^0(C, K)$$

is injective.

Chapter 6

Geometry of moduli spaces: selected results

Our aim in this chapter is to illustrate how the techniques we've developed so far may be used to prove theorems about the geometry of the various moduli and parameter spaces we've introduced. We have not aimed at completeness, even for the problems we discuss; rather, we want to briefly highlight a fairly broad range of examples.

A Irreducibility of the moduli space of curves

As a warm-up, we'll give here a proof of the classical fact that $\mathcal{M}_g$ is irreducible. This serves two purposes. First, it gives another example of the usefulness of the compactification of the Hurwitz scheme by admissible covers. Second, it'll also serve as something of a trial run for more complex arguments in later sections. For example, the series of reductions made here is in many ways analogous to the one used in the solution of the Severi problem in Section E of this chapter.

The basic plan is to work by induction, assuming it known that $\mathcal{M}_k$ is irreducible for $k < g$ and using an analysis of the behavior of $\overline{\mathcal{M}}_g$ near the boundary Δ to deduce irreducibility for $\mathcal{M}_g$. This of course requires some knowledge of the stable compactification $\overline{\mathcal{M}}_g$ of $\mathcal{M}_g$; in particular, we need to know that:

1. $\overline{\mathcal{M}}_g$ is everywhere locally irreducible;

2. The boundary Δ_g of $\overline{\mathcal{M}}_g$ is a normal crossing divisor.

3. The irreducible components of Δ_g are the varieties Δ_i, $i = 0, \ldots, \lfloor \frac{g}{2} \rfloor$.

4. For any distinct i and j, the components Δ_i and Δ_j intersect.

We verified the first point in (3.32) and the second and third on page 54. (The argument there used the irreducibility of $\overline{\mathcal{M}}_i$ but only for $i < g$.) The last point may be seen directly: Δ_i and Δ_j meet in the locus of curves of the form seen in Figure (6.1) if i and j are both

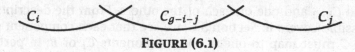

C_i $\qquad$ C_{g-i-j} $\qquad$ C_j

FIGURE (6.1)

positive, and in the locus of curves of the form seen in Figure (6.2) if

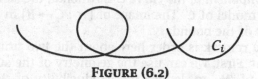

C_i

FIGURE (6.2)

$j = 0$. Given this, we may make a series of reductions of the original problem, as follows. First, since $\overline{\mathcal{M}}_g$ is locally irreducible everywhere, we have the:

First Reduction: It's sufficient to show that any two components Σ and Σ' of $\overline{\mathcal{M}}_g$ meet.

Second, since any two Δ_i's meet, we have the:

Second Reduction: It's sufficient to show that any component Σ of $\overline{\mathcal{M}}_g$ contains the boundary component Δ_i for some i.

Now, by the local irreducibility of $\overline{\mathcal{M}}_g$, any component of $\overline{\mathcal{M}}_g$ containing a point $[C] \in \Delta_i$ contains an open subset of Δ_i and hence, since the Δ_i's are themselves irreducible, all of Δ_i. We deduce the:

Third Reduction: It's sufficient to show that any component Σ of $\overline{\mathcal{M}}_g$ contains a point $[C] \in \Delta$.

With this said, how do we go about exhibiting a singular stable curve in a given component Σ of $\overline{\mathcal{M}}_g$? We could apply Diaz' theorem (2.34), which will be proved in the next section, but in fact there is a much simpler argument based on admissible covers. To set this up, choose an integer d large enough that every smooth curve C of genus g admits a map of degree d to $\mathbb{P}^1$ with simple branching — i.e., such that there exists a component $\mathcal{H}$ of the Hurwitz scheme $\mathcal{H}_{d,g}$ dominating Σ. Any component of the Hurwitz scheme maps surjectively onto the space P_b of b-pointed stable curves of genus 0, so that $\mathcal{H}$ will contain

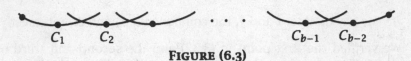

$$C_1 \qquad C_2 \qquad\qquad\qquad C_{b-1} \quad C_{b-2}$$

FIGURE (6.3)

a point $[\pi : C \to B]$ where B is the stable pointed curve consisting of a chain of rational curves $C_1, \ldots, C_{b-2}$ with two marked points on C_1 and C_{b-2} and one on each of the others. From the description of admissible covers in Section 3.G, we see that each component of the curve C must map to one of the components C_i of B; in particular it'll be a branched cover of $\mathbb{P}^1$ branched over only three points, with simple branching over at least one. By Riemann-Hurwitz, this implies that *every component of the curve C is rational*; the same is then true of the stable model of C. The image of $[\pi : C \to B]$ in $\mathcal{H}$ is thus the desired point of the boundary.

A historical remark is order here about the two principal ideas in this argument. First, we can use the geometry of the stable compactification $\overline{\mathcal{M}}_g$ of $\mathcal{M}_g$ to deduce the irreducibility of $\mathcal{M}_g$ if we show that any component must meet the boundary: this observation was made by Deligne and Mumford in their original study [29] of stable curves and was used by them to deduce the main result of that paper, the irreducibility of $\mathcal{M}_g$ in positive characteristic, from irreducibility in characteristic 0. Second, we use the properness of the Hurwitz scheme parameterizing admissible covers to find boundary curves in any component: this idea is due to Fulton in his appendix to [82].

B Diaz' theorem

Our next example is Diaz' proof that a complete subvariety of $\mathcal{M}_g$ has dimension at most $g - 2$.

The idea: stratifying the moduli space

The basic approach is one first proposed by Enrico Arbarello in [3]: to introduce a stratification of the moduli space $\mathcal{M}_g$ such that the (open) strata contain no complete curves, and such that closed strata of codimension $g-2$ likewise don't contain complete curves. The point is that if we can exhibit such a stratification, any complete subvariety Σ of dimension d will have to intersect the codimension 1 strata in complete subvarieties of dimension $d - 1$, hence the codimension 2 strata in complete subvarieties of dimension $d - 2$, and so on; ultimately we conclude that Σ will meet the codimension $g - 2$ strata in complete varieties of dimension $d - g + 2$, and hence that $d \leq g - 2$.

Arbarello, with this as one goal among others, introduced the sequence of subvarieties

$$W_2 \subset W_3 \subset \cdots \subset W_{g-1} \subset W_g = \mathcal{M}_g,$$

where W_α is defined to be the subvariety of smooth curves C of genus g expressible as a cover of $\mathbb{P}^1$ of degree at most α with one point of total ramification. Equivalently, W_α is the locus of curves C possessing a Weierstrass point p whose first nongap is at most α, that is, for which $h^0(C, \mathcal{O}(\alpha p)) \geq 2$. (In particular, the smallest subvariety W_2 is just the locus of hyperelliptic curves, which is affine.) Arbarello showed that these subvarieties were closed in $\mathcal{M}_g$ and that all the inclusions above are proper, or, in other words, that W_α is of codimension $g - \alpha$ in $\mathcal{M}_g$. In a deeper vein, he also showed that each W_α is irreducible. It remains an open question, though, whether or not the open strata $\mathcal{U}_\alpha = W_\alpha - W_{\alpha-1}$ of this stratification can contain complete curves.

To see how this question may be approached (and why the answer is unclear), consider a family $X \to B$ of curves with general fiber $X_b \in W_\alpha$. What we want to show is that if B is complete, then (possibly after a base change) some fibers must either be singular or lie in $W_{\alpha-1}$. As a first step, we can assume, after a base change, that the family has a cross-section — that is, a point $p_b = \sigma(b) \in X_b$ on each fiber such that $h^0(X_b, \mathcal{O}(\alpha \cdot p_b)) \geq 2$. After further base changes, we can pick out on each fiber X_b a linear series of dimension 1 in $|\alpha \cdot p_b|$, or in other words a map to $\mathbb{P}^1$, so that we can take the total space X to be an α-sheeted branched cover of a $\mathbb{P}^1$ bundle $Y \to B$, totally ramified over a cross-section $\Sigma = \sigma(B) \subset Y$. After one final base change, we can assume that the branch divisor of this map consists of the cross-section Σ (with multiplicity $\alpha - 1$, of course) plus a total of $2g - 3 + \alpha$ other cross-sections Σ_i.

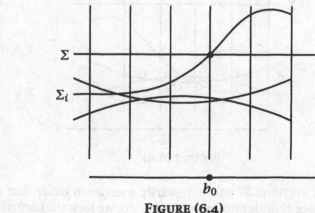

FIGURE (6.4)

The key now is to ask: what happens when one (or more) of the secondary branch divisors Σ_i crosses the primary branch divisor Σ over a point $b_0 \in B$ as shown in Figure (6.4)? The first point to note is that, since the cover $X_b \to Y_b \cong \mathbb{P}^1$ is totally ramified over the point $\sigma(b)$, the stable limit of the curve X_b as $b \to b_0$ cannot continue to be an α-sheeted branched cover of $\mathbb{P}^1$: given that there is no way to add any more ramification over the point $\sigma(b_0)$, we cannot produce enough total ramification in the limit to yield such a cover. Using the theory of admissible covers we can determine what type of degeneration must occur. This is the subject of the following:

EXERCISE (6.5) Show that in the above circumstances, either

1) the stable limit as $b \to b_0$ of the curves X_b is singular, or

2) the stable limit X of the curves X_b is smooth, so that the limit of the admissible covers $X_b \to Y_b \cong \mathbb{P}^1$ is an admissible cover $X_0 \to Y_0$ with X_0 stably equivalent to X, and the map has degree strictly less than α on the component of X_0 isomorphic to X.

Given this exercise, why is the result not proved? The reason is simply that *we have no assurance that any of the secondary branch divisors Σ_i will meet Σ*. What might happen is for Σ to be a section of the $\mathbb{P}^1$ bundle $\mathcal{Y} \to B$ having negative self-intersection $-n$, and for all the Σ_i to be sections of self-intersection n disjoint from Σ, giving instead the picture in Figure (6.6). (More precisely, $\mathcal{Y} \to B$ will be the projectivization $\mathbb{P}(E)$ of a split vector bundle $E \cong L \oplus L^{-1}$, with $\deg(L) > 0$; Σ will be the cross-section corresponding to the unique summand L of positive degree; and the Σ_i will be cross-sections corresponding to certain of the infinitely many summands isomorphic to L^{-1}.) Of course, in this

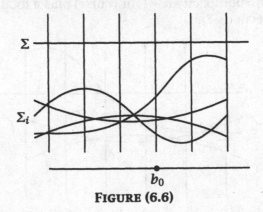

FIGURE (6.6)

situation, the sections Σ_i will necessarily meet each other. But while we expect to see singular fibers when they do, we lack the control over which admissible covers will arise at such points to be sure. The upshot is that *it remains an open question whether or not the open strata of the Arbarello flag admit complete curves.*

How can we revive the argument? Diaz' idea was this: while one section of a $\mathbb{P}^1$ bundle may have negative self-intersection and be disjoint from a collection of others, *two such sections cannot exist*: if three sections of a $\mathbb{P}^1$ bundle are disjoint, the bundle must be trivial and the sections constant. So if we keep track of the branching of a cover over not one but two points of $\mathbb{P}^1$, we're sure to see some degeneracy in a complete, one-parameter family.

Now, we cannot look just at curves that are branched covers of $\mathbb{P}^1$ totally ramified at two points, since such curves will never fill up the moduli space $\mathcal{M}_g$ (whatever the degree of the map, the variety of such curves will, by Riemann-Hurwitz, have dimension $2g - 1$). What we can do, however, is to take more generally a branched cover of $\mathbb{P}^1$, fix its total ramification index at two points (or, equivalently and more conveniently, the cardinality of two fibers), and track the behavior of this total. This motivates us to define the (closed) subvariety $\mathcal{D}_k \subset \mathcal{M}_g$ to be the locus of curves C that admit maps $\pi : C \to \mathbb{P}^1$ such that $\deg(\pi) \le g$ and

$$\#(\pi^{-1}\{0, \infty\}) \le k.$$

(We take $\mathcal{D}_1$ to be the empty set.) The upper bound g for $\deg(\pi)$ is taken purely for technical convenience. We need some bound on the degree of π or the locus $\mathcal{D}_k$ would have countably many components, but we could replace g by any other integer $d_0 \ge g$ without affecting the argument that follows.

EXERCISE (6.7) Show that $\mathcal{D}_k$ is a closed subvariety of $\mathcal{M}_g$, of pure codimension $g - k$ for $k \ge 2$.

Note that $\mathcal{D}_2$ is simply the locus of curves expressible as branched covers of $\mathbb{P}^1$ (of degree at most g) with two points of total ramification. At the other extreme, $\mathcal{D}_g$ is all of $\mathcal{M}_g$: for example, every curve may be expressed as a branched cover of $\mathbb{P}^1$ of degree $d \le g$ totally ramified over 0 — that is, every curve has a Weierstrass point — and then it suffices to choose ∞ to be another branch point of this cover. An alternative proof is given in:

EXERCISE (6.8) 1) If g is even, show that $\mathcal{D}_g = \mathcal{M}_g$ by arguing that for any curve C of genus g we can find a branched cover $C \to \mathbb{P}^1$ of degree $(g + 2)/2$ branched over 0 and ∞.

2) In general, for which triples (n, α, β) of integers with $n \ge \alpha, \beta \ge 0$ and $2n - \alpha - \beta + 2 \ge g$, can we find a map $\pi : C \to \mathbb{P}^1$ of degree n with ramification indices α and β at two points p and q of C?

The proof

The basic theorem to be proved, from which Diaz' theorem will follow immediately, is the:

THEOREM (6.9) *The open strata $\mathcal{D}_k - \mathcal{D}_{k-1}$ of the Diaz stratification do not contain complete curves.*

PROOF. Let $B \subset \mathcal{D}_k$ be a complete curve; we'll show that B must meet $\mathcal{D}_{k-1}$. The basic setup for doing this has already been described: after making a base change, we may assume we have a family of curves $\mathcal{X}' \to B$ and for some open subset $U \subset B$ a family of branched covers $\pi_U : \mathcal{X}_U \to \mathcal{Y}_U$ of degree k, where $\mathcal{Y}_U$ is a $\mathbb{P}^1$-bundle over U. Suppose that for a general point $b \in U$ the restriction $\pi_b : X_b \to \mathbb{P}^1$ has branch points $p_1, \ldots, p_b$ with the monodromy around p_i given by the conjugacy class τ_i in the symmetric group $\mathfrak{S}_k$. We may take p_1 and p_2 to be the points 0 and ∞. Let $\mathcal{H}$ be the Hurwitz scheme of branched covers of $\mathbb{P}^1$ of degree k branched over b points with branching $\tau_1, \ldots, \tau_b$, and let $\overline{\mathcal{H}}$ be the compactification of $\mathcal{H}$ by pseudo-admissible covers (see the end of Section 3.G for the definition of these covers). We may then complete $\mathcal{Y}_U$ to a birationally ruled surface $\mathcal{Y} \to B$ with disjoint sections Σ_i, $\mathcal{X}_U$ to a surface $\mathcal{X} \to B$ birationally equivalent to X, and π_U to a family $\pi : \mathcal{X} \to \mathcal{Y}$ of pseudo-admissible covers branched over the sections Σ_i.

The point is now simply that since the family $\pi : \mathcal{X} \to \mathcal{Y}$ is nontrivial, we must have degeneracies involving Σ_1 or Σ_2 and one of the other sections Σ_i, that is, a fiber $X_0 \to Y_0$ of the map π such that the points $p_1(0) = \Sigma_1 \cap Y_0$ and $p_2(0) = \Sigma_2 \cap Y_0$ lie on different irreducible components of Y_0 — say Y' and Y''. Now, since we've assumed that the original family $B \subset \mathcal{D}_k$ is a complete curve in $\mathcal{M}_g$, all the fibers of $\mathcal{X} \to B$ are stably equivalent to smooth curves; in particular, the fiber X_0 of $\mathcal{X}$ over 0 will consist of a smooth curve X with trees of rational curves attached.

We want to show that the point $[X] \in \mathcal{D}_{k-1}$. To do this we look at the map π restricted to X. Let Y be the image of X in Y_0; Y_0 can also be pictured as the curve $Y \cong \mathbb{P}^1$ with one or more trees of rational curves attached (each at one point). If Y is distinct from both Y' and Y'', then we let p' and p'' be the points in Y where the trees containing Y' and Y'' meet Y; if $Y' = Y$, we take p' to be the point $p_1(0)$, and similarly in case $Y'' = Y$.

We claim that *the covering map $X \to Y$ has total ramification index of at most $2g - 3 + k$ outside $\{p', p''\}$*. Indeed, the ramification of X over Y occurs only over those points $p_j(0)$ lying on Y, and over each $p_j(0) \in Y$ the ramification index of $X \to Y$ is at most the ramification index of τ_j. By construction, the sum of the ramification indices of the classes $\tau_1, \ldots, \tau_b$ is $2g + 2d - 2$. Since the ramification indices of τ_1 and

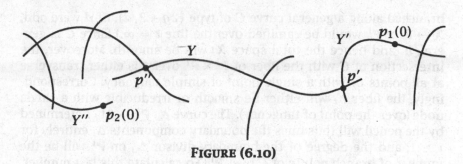

FIGURE (6.10)

τ_2 add up to $2d - k$, the sum of the ramification indices of $\tau_3, \ldots, \tau_b$ is $2g - 2 + k$. Finally, since the locus $Y - \{p', p''\}$ must contain a *proper* subset of the points $\{p_3, \ldots, p_b\}$, it follows that the total ramification index of $X \rightarrow Y$ over $Y - \{p', p''\}$ is strictly less than $2g - 2 + k$. Now, whatever the degree of the cover $X \rightarrow Y$, it follows from the claim that the cardinality of the inverse image of $\{p', p''\} \subset Y$ is at most $k - 1$ so that $X \in \mathcal{D}_{k-1}$ as desired. ∎

C Moduli of hyperelliptic curves

Next, we show how the techniques we've assembled may be used to study suitably geometric subvarieties of the moduli space of curves with the example of the moduli space $H_g \subset \mathcal{M}_g$ of smooth hyperelliptic curves, and its closure $\overline{H}_g$ in the moduli space of stable curves. The space H_g is particularly tractable, since it can be viewed as a subvariety of the moduli space of all curves of genus g, as a finite quotient of the Hurwitz scheme of double covers, or, what is, in this case, the same, as the moduli space of stable $(2g + 2)$-pointed rational curves.

We will be mostly concerned with the divisor theory of $\overline{H}_g$: we would like to describe the Picard group of $\overline{H}_g$, understand the restriction map from $\mathrm{Pic}(\overline{\mathcal{M}}_g)$ to $\mathrm{Pic}(\overline{H}_g)$, and get some idea which line bundles are positive on $\overline{H}_g$.

Fiddling around

To get a feel for the problem, we start by simply writing down a "general" one-parameter family of hyperelliptic curves: that is, we consider the family of curves given by the equation

$$y^2 = a_{2g+2}(t) \cdot x^{2g+2} + \cdots + a_1(t) \cdot x + a_0(t)$$

where the a_i are general polynomials in t of some even degree d. The total space X of this family is simply the double cover of $\mathbb{P}^1 \times \mathbb{P}^1$

branched along a general curve C of type $(2g+2, d)$. (If d were odd, $X \to \mathbb{P}^1 \times \mathbb{P}^1$ would be ramified over the line $t = \infty$.) Since C is general, it (and hence the total space X) will be smooth. Moreover, the intersection of C with the fiber of $\mathbb{P}^1 \times \mathbb{P}^1$ over t is either transverse at all points or with a single point of simple tangency. Correspondingly, the fiber X_t will either be smooth or irreducible with a single node (over the point of tangency). The curve $\alpha : \mathbb{P}^1 \to \overline{\mathcal{M}}_g$ determined by the pencil will thus miss the boundary components Δ_i entirely for $i \geq 1$; and the degree of the boundary divisor Δ_0 on $\mathbb{P}^1$ will be the number of branch points of C over $\mathbb{P}^1$. To calculate this last number, use Riemann-Hurwitz: C has genus $(d-1)(2g+1)$, and so a map expressing C as a simply branched cover of degree $2g+2$ of $\mathbb{P}^1$ will have $2d(2g+1)$ branch points; thus

$$\deg_{\mathbb{P}^1}(\delta_0) = 2d(2g+1).$$

To calculate the degree of the line bundle λ on $\mathbb{P}^1$ there are several approaches. The simplest is to write down a frame for the Hodge bundle: for general (finite) t, a basis for the space of holomorphic one-forms on X_t is given by

$$\omega_1 = \frac{dx}{y}, \quad \omega_2 = \frac{x\,dx}{y}, \quad \cdots, \quad \omega_g = \frac{x^{g-1}dx}{y}.$$

These give sections of the Hodge bundle E that are everywhere independent except at the point $t = \infty$. At this point, each section ω_α has a zero of order $\left(\frac{d}{2}\right)$ and after multiplying ω_α by $t^{d/2}$ these sections do in fact form a frame for E in a neighborhood of $t = \infty$; thus

$$E \cong \left(\mathcal{O}_{\mathbb{P}^1}\left(\frac{d}{2}\right)\right)^{\oplus g}$$

and in particular

$$\deg_{\mathbb{P}^1}(\lambda) = g \cdot \frac{d}{2}.$$

For future reference, we note the relation

(6.11) $$\deg_{\mathbb{P}^1}(\delta_0) = \left(8 + \frac{4}{g}\right)\deg_{\mathbb{P}^1}(\lambda).$$

The next exercise gives another method for calculating the degree of λ.

EXERCISE (6.12) 1) Apply Riemann-Hurwitz to the map $X \to \mathbb{P}^1 \times \mathbb{P}^1$ to find the canonical class K_X of X.

2) Use this to find the class of the dualizing sheaf $\omega_{X/\mathbb{P}^1} = K_X/K_{\mathbb{P}^1}$ of the family, and hence the degree of the class κ_1 on $\mathbb{P}^1$.

3) Finally, apply formula (3.110) — $12 \cdot \lambda = \kappa_1 + \delta$ — to recover $\deg_{\mathbb{P}^1}(\lambda)$.

You may have noticed that we used quotes when describing the pencil above as general. Of course, these families don't give general curves in the hyperelliptic locus: taking the branch curve C to be general gives us curves that avoid all the Δ_i for $i > 0$. But the pencils above are not even general amongst families with such branching. The reason is that not every family of hyperelliptic curves over a base B will have total space X a double cover of $B \times \mathbb{P}^1$; X will in general be a double cover of a $\mathbb{P}^1$-bundle over B but only if this bundle is trivial is X itself a double cover of $B \times \mathbb{P}^1$. This exercise shows that, nonetheless, the relation (6.11) continues to hold.

EXERCISE (6.13) Let $S \rightarrow B$ be a ruled surface and let D be a suitably ample divisor class on S of *even* degree $d = 2g + 2$ over B. Let C be a general curve in the linear system $|D|$, and let $X \rightarrow S$ be the double cover of S branched along C. Mimic the calculations above to determine the degrees of λ and δ_0 for the family of hyperelliptic curves $X \rightarrow B$. In particular, show that we'll always have

$$\deg_B(\delta_0) = \left(8 + \frac{4}{g}\right) \cdot \deg_B(\lambda_0).$$

The calculation for an (almost) arbitrary family

The next logical question to ask is: how do we go from the "general" one-parameter families described (those in the last exercise are, at least, general amongst families that meet only the boundary component of Δ_0) to more arbitrary families. One natural way to attempt this would be to consider double covers of ruled surfaces $S \rightarrow B$ branched over curves that have more general branching behavior over B. Unfortunately, this quickly gets very complicated: the fibers over B of the double covers we get aren't in general stable, and the process of stable reduction makes the formulas messy. A more tractable approach is to use the description of the closure $\overline{H}_g$ of H_g given by the Hurwitz scheme of admissible covers of degree 2 of stable $(2g + 2)$-pointed rational curves — that is, to consider one-parameter families of such admissible covers, over one-parameter families of stable pointed rational curves. We will do this, deriving in this subsection a relation among the degrees of various divisor classes associated to such a family, and then in the next one, laying out what this means in terms of the rational Picard group of $\overline{H}_g$.

Let's set things up. Given a one-parameter family of curves whose general member is smooth and hyperelliptic, we may after base change assume we have a family of admissible covers of degree 2 of stable $2g + 2$-pointed rational curves: that is, a surface $S \rightarrow B$ fibered over a curve B with rational nodal fibers and $2g + 2$ everywhere disjoint

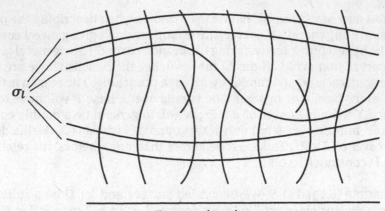

FIGURE (6.14)

sections $\sigma_l : B \rightarrow S$, and a double cover $X \rightarrow S$ that restricts on each fiber S_b to an admissible cover of S_b branched at the points $\sigma_l(b)$. In fact, since we're primarily interested in the divisor theory of the hyperelliptic locus in moduli, we can simplify by working only with families that don't meet any of the strata of the Hurwitz scheme of codimension greater than 1. This amounts to making the additional assumption that each fiber S_b of $S \rightarrow B$ has either one or two components. The schematic picture of a typical family $S \rightarrow B$ is therefore as shown in Figure (6.14).

Next, we ask what the double cover $X_b \rightarrow S_b$ looks like over one of the singular curves S_b. A first observation is that the answer depends on whether the numbers of points $\sigma_l(b)$ on each of the two components of S_b are even or odd. The cover cannot be branched over the node of S_b in the former case and must be branched over this node in the latter.

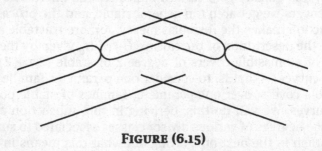

FIGURE (6.15)

In the even case, the admissible cover $X_b \rightarrow S_b$ looks, near the node of S_b, like that shown in Figure (6.15). There are two possibilities for the curve X_b. If each component S_b contains at least four of the branch points $\sigma_l(b)$, then X_b is a stable curve consisting of two hyperelliptic curves attached to each other at two points conjugate on each. If one of the components S_b contains only two of the $\sigma_l(b)$, then the corresponding component of X_b will be a semistable rational curve and the main component will be a hyperelliptic curve of genus $g - 1$. The stable model of X_b is thus obtained by taking this hyperelliptic curve and identifying two conjugate points. In both cases, the stable model of the curve X_b will lie in the boundary component Δ_0. Note that the curve $B \rightarrow \overline{\mathcal{M}}_g$ associated to the family $X \rightarrow B$ will have intersection number 2 with Δ_0 at such points $b \in B$ if the total space S (and hence likewise X) is smooth over b. More generally, it will have intersection number $2k$ with Δ_0 if the local equation of S at the node of S_b is $(xy - t^k)$.

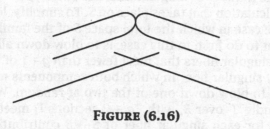

FIGURE (6.16)

In the odd case, there must be at least three branch points on each component since both are stable. This means that X_b is always stable and consists of two hyperelliptic curves joined by identifying a Weierstrass point on each. In this case, the curve X_b will lie in the boundary component of Δ_i if and only if the components of S_b contain $2i + 1$ and $2g + 1 - 2i$ of the points $\sigma_l(b)$. Moreover, if the local equation of S at the node p of S_b is $(xy - t^{2k})$ — the power of t *must* be even since $X \rightarrow S$ will be ramified at the isolated point p — then X will have local equation $(xy - t^k)$ at the point lying over p, and b will be a point of intersection multiplicity k of B with Δ_i.

To count the number of such fibers, we'll let d_j be the number of singular fibers S_b of $S \rightarrow B$ in which one component contains j of the points $\sigma_l(b)$ and the other $2g + 2 - j$, each such fiber counted with multiplicity k if the local equation of S at the node of S_b is $xy - t^k$. We can then write

$$(6.17) \qquad \deg_B(\delta_0) = 2 \cdot \sum_j d_{2j};$$

while for each $i > 0$ we have

$$(6.18) \qquad \deg_B(\delta_i) = \frac{d_{2i+1}}{2}.$$

Note that we can define the integers d_j for general families $S \rightarrow B$ of pointed rational curves: let d_j be the number of *nodes* p of fibers S_b such that the connected components of the complement $S_b - \{p\}$ contain j and $2g + 2 - j$ of the points $\sigma_l(b)$.

EXERCISE (6.19) Show that with this convention, if $X \to S \to B$ is the family of hyperelliptic curves given as admissible covers of the curves S_b branched in the points $\sigma_l(b)$, then the relations above on the degrees $\deg_B(\delta_i)$ still hold.

The next task is to find the self-intersection of the relative dualizing sheaf of X over B in terms of the numbers d_j. By Riemann-Hurwitz, the relative dualizing sheaf of X over B is expressed in terms of the relative dualizing sheaf of S over B and the classes of the branch divisors $\Gamma_l = \sigma_l(B) \subset S$. To use this we must thus relate the self-intersections of the sections, the class of the relative dualizing sheaf of the surface $S \to B$, and the number of singular fibers of each type.

This is a calculation that takes place on S. To simplify, let's first consider only the case in which the total space S of the family is smooth. What we want to do first, in this case, is to blow down all the components of the singular fibers that meet fewer than $g + 1$ of the sections; and for every singular fiber in which both components meet $g + 1$ of the sections, to blow down one of the two at random. We then have a nice $\mathbb{P}^1$ bundle T over B, with $2g + 2$ sections Γ_l meeting pairwise transversely; for each singular fiber of $S \to B$ contributing to d_j we have a point where j of the sections meet. In particular,

$$\sum_{l<m} \Gamma_l \Gamma_m = \sum_j \frac{j(j-1)}{2} \cdot d_j.$$

On the other hand, there is a basic relation between the pairwise intersections of the Γ_l and their self-intersections. For any two sections Γ_l and Γ_m, the difference $\Gamma_l - \Gamma_m$ is numerically equivalent to a sum of fibers, and so has self-intersection 0; thus

$$\Gamma_l^2 + \Gamma_m^2 = 2\Gamma_l \cdot \Gamma_m.$$

Combining this with the last equality, we have

$$(2g + 1) \cdot \sum_l \Gamma_l^2 = \sum_j j(j-1) \cdot d_j.$$

What happens when we pass from the $\mathbb{P}^1$ bundle back to our original surface S? Each time we blow up a point where j of the sections Γ_l meet, the sum of the self-intersections of the Γ_l decreases by j, so that on S the sum of the self-intersections of the proper transforms of the sections $\tilde{\Gamma}_j$ satisfies

(6.20)
$$(2g + 1) \cdot \sum_l \tilde{\Gamma}_l^2 = (2g + 1)(\sum_l \Gamma_l^2 - \sum_j j \cdot d_j)$$

$$= -\sum_j j(2g + 2 - j) \cdot d_j.$$

For use later in this chapter, it's convenient to observe that this inequality generalizes in two ways.

EXERCISE (6.21) Verify that this last relation holds even if the family $S \to B$ has singular points as follows. Suppose that, at a point $p \in S_b$, S has local equation $xy - t^k$. Show that:

1) after blowing down the component of S_b containing $j \leq g + 1$ of the points $\sigma_l(b)$, the resulting surface will be smooth at the image of p;

2) the corresponding sections Γ_l will meet pairwise with intersection number k, contributing $k \cdot j(j - 1)$ to the sum of the intersections $\Gamma_l \cdot \Gamma_m$; and,

3) to recover the original surface S we must first blow up k times to separate the sections Γ_l passing through the image of p, and then blow down the first $(k - 1)$ exceptional divisors lowering from the sum of the self-intersections Γ_l^2 by $k \cdot j^2$.

EXERCISE (6.22) Suppose $S \to B$ is a family of rational curves with smooth total space and such that each singular fiber contains exactly two components and that we're given n pairwise disjoint sections Γ_l. Define, as above, d_j to be the number of singular fibers with i of the sections passing through one component and $(n - i)$ passing through the other. Show that the argument above now yields

$$(n - 1) \cdot \sum_l \tilde{\Gamma}_l^2 = - \sum_j j(n - j) \cdot d_j.$$

Next, consider the class $\omega_{S/B}$ of the relative dualizing sheaf of $S \to B$. Before we blew up, the relative dualizing sheaf had self-intersection 0. This follows, for example, from the fact that any $\mathbb{P}^1$-bundle over B is obtained from $B \times \mathbb{P}^1$ by a like number of blowups and blowdowns; or, alternately, by applying $(K_T)^2 = 4 - 4g(B)$. Since there is one blowup for each singular fiber there are a total of $\sum_j (d_j)$, giving

(6.23) $$(\omega_{S/B})^2 = - \sum_j d_j.$$

Finally, we can use this to calculate the self-intersection of the class $\omega_{X/B}$ of the relative dualizing sheaf of a family $X \to B$ of hyperelliptic curves realized as a double cover $\varphi : X \to S$ of S branched along the sections $\tilde{\Gamma}_\alpha$. We have, first of all, by Riemann-Hurwitz

$$\omega_{X/B} = \varphi^* \omega_{S/B} + R_X$$

$$= \varphi^* \left(\omega_{S/B} + \frac{R_S}{2} \right)$$

where R_X is the class of the ramification divisor of $X \to S$ on X and $R_S = \sum_i \tilde{\Gamma}_i$ is the corresponding the branch divisor on S. We thus have

$$(\omega_{X/B})^2 = 2 \cdot \left(\omega_{S/B} + \sum_i \frac{\tilde{\Gamma}_i}{2} \right)^2$$

$$= 2 \, (\omega_{S/B})^2 + 2 \sum_i \left(\omega_{S/B} \cdot \tilde{\Gamma}_i \right) + \frac{1}{2} \left(\sum_i \tilde{\Gamma}_i^2 \right)$$

since $\tilde{\Gamma}_l \cdot \tilde{\Gamma}_m = 0$ for $l \neq m$. On the other hand, the intersection number of $\omega_{S/B}$ with a section of the map $S \to B$ is just minus the self-intersection of that section, so that after clearing denominators, the last two terms yield

$$(\omega_{X/B})^2 = 2 \, (\omega_{S/B})^2 - \frac{3}{2} \left(\sum_i \tilde{\Gamma}_i^2 \right).$$

Using (6.20) and (6.23), this gives the relation

$$2(2g+1) \cdot (\omega_{X/B})^2 = 4(2g+1) \cdot (\omega_{S/B})^2 - 3(2g+1) \left(\sum_i \tilde{\Gamma}_i^2 \right)$$

$$= \sum_j \left(-4(2g+1) - 3j(j - 2g - 2) \right) \cdot d_j$$

$$= \sum_j \left(6jg - 3j^2 + 6j - 8g - 4 \right) \cdot d_j.$$

On the other hand, by (6.17) and (6.18),

$$\deg_B(\delta) = 2 \left(\sum_{j \text{ even}} d_j \right) + \frac{1}{2} \left(\sum_{j \text{ odd}} d_j \right).$$

Combining both of these with the basic formula $12\lambda = \kappa_1 + \delta$ (cf. (3.110)), we see that we can write

$$\deg_B(\lambda) = \sum_j c_j \cdot d_j$$

where, for even $j = 2\alpha$, we have

$$c_j = \left(\frac{1}{12} \right) \left(\frac{1}{(4g+2)} (6jg - 3j^2 + 6j - 8g - 4) + 2 \right)$$

$$= \left(\frac{1}{12(4g+2)} \right) \left(12\alpha g - 12\alpha^2 + 12\alpha \right)$$

$$= \frac{\alpha(g + 1 - \alpha)}{(4g+2)},$$

and, for odd $j = 2\alpha + 1$, we have

$$c_j = \left(\frac{1}{12}\right)\left(\frac{1}{(4g+2)}(6jg - 3j^2 + 6j - 8g - 4) + \frac{1}{2}\right)$$

$$= \frac{\alpha(g - \alpha)}{(4g + 2)}.$$

To summarize, then, we have

$$\deg_B(\lambda) = \sum_{j=2\alpha} \frac{\alpha(g + 1 - \alpha)}{4g + 2} d_j + \sum_{j=2\alpha+1} \frac{\alpha(g - \alpha)}{4g + 2} d_j.$$

Since the numbers $d_{2\alpha}$ aren't determined by the degrees of the divisor classes δ_i on B, we cannot express this as a relation among the restrictions to the hyperelliptic locus $\overline{H}_g \subset \overline{\mathcal{M}}_g$ of the generators $\lambda, \delta_0, \dots, \delta_{\lfloor g/2 \rfloor}$ of $\mathrm{Pic}(\overline{\mathcal{M}}_g)$. The exception is our original "general" case, however, where all $d_i = 0$ for $i \geq 3$. Then we have

$$d_2 = \deg_B(\delta_0) = \deg_B(\delta)$$

and so we recover the relation

$$\deg_B(\delta) = \left(8 + \frac{4}{g}\right) \cdot \deg_B(\lambda).$$

In general, we obtain only an inequality. If we first note that

$$c_{2\alpha} \geq c_2 \quad \text{and} \quad 2 \cdot c_{2\alpha+1} \geq \frac{c_2}{2},$$

we obtain:

COROLLARY (6.24) *Let $X \to B$ be any one-parameter family of curves whose general member is smooth and hyperelliptic. Then*

$$\deg_B(\delta) \leq \left(8 + \frac{4}{g}\right) \cdot \deg_B(\lambda).$$

We will have occasion to use this inequality in determining the ample cone of $\overline{\mathcal{M}}_g$ in the following section.

The Picard group of the hyperelliptic locus

What conclusions can we draw from the relations derived in the preceding discussion? We will try to shed some light on them by expressing them in terms of the rational Picard group $\mathrm{Pic}(\overline{H}_g) \otimes \mathbb{Q}$ of the locus $\overline{H}_g \subset \overline{\mathcal{M}}_g$ of hyperelliptic curves.

To begin with, we can describe the locus $\overline{H}_g \subset \overline{\mathcal{M}}_g$ as the quotient of the Hurwitz scheme $\mathcal{H}_{2,g}$ of admissible covers of degree 2 by the action of the symmetric group $\mathfrak{S}_{2g+2}$ on $2g + 2$ letters. Thus, to find the rational Picard group of $\overline{H}_g$ we'll start by describing the Picard group of $\mathcal{H}_{2,g}$. This is reasonably straightforward, since the Hurwitz scheme $\mathcal{H}_{2,g}$ of admissible covers of degree 2 is the same thing as the moduli space $\overline{\mathcal{R}} = \overline{\mathcal{M}}_{0,2g+2}$ of stable $2g + 2$-pointed rational curves.

We have an open subset $\mathcal{R}$ of $\overline{\mathcal{R}}$, consisting of smooth curves $C \cong \mathbb{P}^1$ with $2g + 2$ distinct marked points. This is isomorphic to an open subset in $(\mathbb{P}^1)^{2g-1}$: specifically, if we choose for each C the unique isomorphism $\varphi : C \to \mathbb{P}^1$ carrying p_1, p_2 and p_3 to $0, 1$ and ∞ respectively, we can identify $\mathcal{R}$ with the complement in $(\mathbb{P}^1)^{2g-1}$ of the divisors determined by the conditions $p_i \in \{0, 1, \infty\}$ or $(p_i = p_j)$. This clearly has trivial Picard group, so that the Picard group of $\overline{\mathcal{R}}$ is generated by the classes of the boundary divisors. For every partition of $I \subset \{1, 2, \ldots, 2g + 2\}$ into two sets I and J such that both I and J have cardinality at least 2, we have a divisor $\mathcal{D}_I \subset \overline{\mathcal{R}}$ whose general point corresponds to a curve $C = C_1 \cup_p C_2$ consisting of two smooth rational curves C_i meeting at a point p, with the marked points p_i in C_1 for $i \in I$ and in C_2 for $i \in J$. Of course, swapping I and J gives rise to the same divisor.

The rational Picard group of the hyperelliptic locus $\overline{H}_g$ can be described in these terms. To begin with, $\overline{H}_g$ is the quotient of $\overline{\mathcal{R}}$ by the symmetric group on $2g + 2$ letters, with the open subset H_g of smooth curves corresponding to the open subset $\mathcal{R}$ (see Exercise (6.25) below); thus

$$\text{Pic}(H_g) \otimes \mathbb{Q} = (0),$$

and $\text{Pic}(\overline{H}_g) \otimes \mathbb{Q}$ is generated by those linear combinations of the divisors $\mathcal{D}_I$ invariant under the symmetric group. These are just the sums $\mathcal{E}_i$ over all subsets $I \subset \{1, 2, \ldots, 2g + 2\}$ of given cardinality i of the divisors $\mathcal{D}_I$. The upshot is that we get one class $\mathcal{E}_i$ for each $i = 2, \ldots, g + 1$, and $\mathcal{E}_i$ is simply the closure of the locus of admissible covers of curves of the form $C = C_1 \cup_p C_2$ with i marked points p on C_1 and $2g + 2 - i$ marked points on C_2.

Next, observe that these divisor classes are indeed independent. One way to see this is to exhibit a collection of curves in $\overline{\mathcal{M}}_{0,2g+2}$ whose images in $\overline{H}_g$ have independent intersection numbers with them. For example, $2g + 2$ general sections of $\mathbb{P}^1 \times \mathbb{P}^1 \to \mathbb{P}^1$ will meet pairwise but not triply; thus, the corresponding curve $\mathbb{P}^1 \to \overline{H}_g$ will meet the boundary component $\mathcal{E}_2$ but will not meet $\mathcal{E}_i$ for $i > 2$. On the other hand, we could simply fix a point $p \in \mathbb{P}^1 \times \mathbb{P}^1$ and take i general sections passing through this point followed by $2g + 2 - i$ general sections; the corresponding curve will meet $\mathcal{E}_2, \mathcal{E}_i$ and no divisor $\mathcal{E}_j$ with $j \neq 2, i$. This shows that the divisor classes $\mathcal{E}_i$ are independent.

To relate this to the standard divisor classes on the moduli space $\overline{\mathcal{M}}_g$, we can express this as saying that the Picard group of the locus $\overline{H}_g \subset \overline{\mathcal{M}}_g$ of hyperelliptic curves is generated by boundary components $\Xi_0, \ldots, \Xi_{\lfloor (g-1)/2 \rfloor}$ and $\Theta_1, \ldots, \Theta_{\lfloor g/2 \rfloor}$, where the general point of $\Xi_\alpha = \mathcal{E}_{2\alpha+2}$ corresponds to a double cover of $\mathbb{P}^1 \cup \mathbb{P}^1$ branched over $2\alpha + 2$ points in one component and $2g - 2\alpha$ in the other, and a general point of $\Theta_\alpha = \mathcal{E}_{2\alpha+1}$ corresponds to a double cover of $\mathbb{P}^1 \cup \mathbb{P}^1$ branched over $2\alpha + 1$ points in one component and $2g + 1 - 2\alpha$ in the other.

Now, let $X \to B$ be any family of hyperelliptic curves. After making a base change, we can associate to $X \to B$ a family of admissible covers with base B'; and the numbers d_i introduced in the preceding part are just the intersection numbers of the corresponding curve $B' \subset \overline{\mathcal{R}}$ with the divisor classes $\mathcal{E}_i$. It follows by our previous analysis that the pullback map

$$\iota^* : \mathrm{Pic}\left(\overline{\mathcal{M}}_g\right) \to \mathrm{Pic}\left(\overline{H}_g\right)$$

sends Δ_0 to $2 \cdot \sum \Xi_\alpha$ and Δ_α to $\Theta_\alpha/2$ for each α. Finally, our previous relation says that the pullback of the divisor class λ to $\mathrm{Pic}(\overline{H}_g)$ is given as the linear combination

$$\iota^*(\lambda) = \sum_{i=0}^{\lfloor (g-1)/2 \rfloor} \frac{\alpha(g + 1 - \alpha)}{4g + 2} \Xi_\alpha + \sum_{i=1}^{\lfloor g/2 \rfloor} \frac{\alpha(g - \alpha)}{4g + 2} \Theta_\alpha.$$

EXERCISE (6.25) Prove that if $(B; p_1, \ldots, p_{2g+2})$ is any stable $(2g + 2)$-pointed rational curve, there exists a *unique* admissible cover $C \to B$ branched at the p_i. In this way, we have a bijection between $\overline{H}_g$ and the quotient of the moduli space $\overline{\mathcal{M}}_{0,2g+2}$ by the symmetric group on $2g + 2$ letters. Is this, in fact, an isomorphism?

D Ample divisors on $\overline{\mathcal{M}}_g$

Our goal in this section is to describe the cone of ample divisors on the moduli space of stable curves. The main tool here is a theorem that translates the stability (in the G.I.T. sense) of the Hilbert point of a general fiber of a family of curves in $\mathbb{P}^r$ into an inequality relating certain intersection numbers of the first Chern class of the line bundle that embeds the family. We begin by setting up the statement of this inequality. Its proof then takes up most of the section. The third subsection translates this inequality into one relating standard divisor classes and the fourth then combines this with the results of the preceding section to determine the ample cone.

An inequality for generically Hilbert stable families

We start with a very naive and general question. Suppose we're given a proper flat family $\pi : X \to B$ of varieties, and a family of line bundles on the fibers $X_b = \pi^{-1}(b)$ of the family — that is, a line bundle L on X, considered modulo pullbacks of line bundles on B. If L is sufficiently ample, its direct image $\pi_* L$ will be a vector bundle E of some rank $r + 1$: we'll assume this from now on. What we would like is to estimate the twisting of this vector bundle, as encoded in its first Chern class.

Of course, this question isn't well-posed as it stands, because tensoring L with the pullback $\pi^* M$ of a line bundle M on B will have the effect of tensoring E by M (and so in particular adding $(r + 1)c_1(M)$ to $c_1(E)$). There is a way to fix this, though: we simply consider the difference between the pullback $\pi^* c_1(E)$ of the first Chern class of E to X, and $r + 1$ times the first Chern class of L itself. Thus, we consider the divisor class

$$D = (r + 1)c_1(L) - \pi^* c_1(E);$$

by what we've said, this is invariant under tensoring L with pullbacks of line bundles from B.

The class D has the drawback of being a divisor class on the total space X of our family, and not on the base B as desired, but we can fix this too. If the family $X \to B$ has fiber dimension k, we can define a divisor class F on B by raising the class $D \in A^* X$ to the $(k + 1)^{\text{st}}$ power and taking the Gysin image

$$F = \pi_*(D^{k+1}).$$

Thus, for example, if B is one-dimensional, then all terms involving $c_1(E)$ to a power greater than 1 will vanish and the degree of F will equal

$$(r + 1)^{k+1} c_1(L)^{k+1} - (k + 1)(r + 1)^k c_1(E)c_1(L)^k.$$

With all this said, what can we say about the class F? The answer in general seems to be: nothing. However, with one relatively mild hypothesis we have a straightforward inequality:

THEOREM (6.26) (CORNALBA-HARRIS [27]) *Assume that B is one dimensional, and that for a general point $b \in B$ the line bundle $L_b = L|_{X_b}$ is very ample and embeds X_b as a Hilbert stable variety in $\mathbb{P}^r$. Then $\deg(F) \geq 0$, i.e.,*

$$(r + 1) \cdot c_1(L)^{k+1} \geq (k + 1) \cdot c_1(L)^k \cdot c_1(E).$$

Note that, since B is one-dimensional, it suffices to exhibit one value of b for which $h^0(X_b, L_B) = r + 1$, L_b is very ample and $\varphi_{L_b}(X_b)$ is

Hilbert stable. In general, if we assume that these conditions are met for every $b \in B$ (or for all but a finite number), we may deduce that F has nonnegative intersection number with every curve in B and hence that F lies in the closure of the cone of ample divisors on B.

Before proving this theorem, let's see what its consequences are for curves. We apply it in the simplest way possible: we assume we have a one-parameter family $\pi : \mathcal{X} \to B$ of stable curves, with the general fiber X_b smooth and nonhyperelliptic; and we take the line bundle L to be simply the relative dualizing sheaf $L = \omega_{\mathcal{X}/B}$. The degree of L on the fibers of π is $2g - 2$ and the degree of L on $\mathcal{X}$ is the degree of the line bundle κ on B, so we have

$$g\kappa \geq 2(2g - 2)\lambda.$$

On the other hand, we know that $\kappa = 12\lambda - \delta$; plugging this in and collecting terms, we have the:

COROLLARY (6.27) *If $\pi : \mathcal{X} \to B$ is any one-parameter family of stable curves, not all hyperelliptic or singular, then the degree λ of the Hodge bundle and the number δ of singular fibers satisfy the inequality*

$$\deg_B(\delta) \leq \left(8 + \frac{4}{g}\right) \cdot \deg_B(\lambda).$$

This, combined with a separate analysis of families of hyperelliptic and/or singular curves, will allow us to say when a linear combination of the divisor classes λ and δ is ample on $\overline{\mathcal{M}}_g$.

Proof of the theorem

The first step is to observe that for any cover $B' \to B$, the divisor class F' associated to the pullback of L to the pullback family $\mathcal{X}' = \mathcal{X} \times_B B'$ is just the pullback of F to B'. It's thus sufficient to prove the inequality after such a base change; in particular, we may assume, if we like, that the first Chern class $c_1(E)$ is divisible by $r + 1$. Next, since the divisor class F was specifically chosen to be invariant under tensoring L with pullbacks of line bundles on B, we may choose a line bundle M on B with first Chern class $c_1(E)/(r + 1)$ and replace L by $L \otimes \pi^{-1} M^\vee$. Thus we may assume to begin with that $c_1(E) = 0$ and what we have to show under this hypothesis is that $c_1(L)^{k+1} \geq 0$.

Now consider the natural map

$$\varphi : \mathrm{Sym}^m(E) \to \pi_*(L^m).$$

For sufficiently large values of m, $\mathrm{Sym}^m(E)$ and $\pi_*(L^m)$ will be vector bundles of ranks $O_r(m) = \binom{r+m}{m}$ and $P(m)$ respectively where

$P = P_{X_b}$ is the Hilbert polynomial of the fiber X_b of π and the map φ will be generically surjective. We thus have an induced map

$$\psi : W = \Lambda^{P(m)}\left(\mathrm{Sym}^m(E)\right) \longrightarrow \Lambda^{P(m)}\left(\pi_*(L^m)\right)$$

which is likewise generically surjective: since the right-hand side is a line bundle this simply means the map isn't identically zero.

Let us now choose a point $b \in B$ such that on the fiber X_b the line bundle L_b is very ample and embeds X_b as a Hilbert stable variety $\overline{X}_b \subset \mathbb{P}^r = \mathbb{P}(E_b^\vee)$, and consider these maps just over that point. The kernel of φ_b is just the mth graded piece of the ideal of $\overline{X}_b$; so the kernel of ψ_b, viewed as a point in the projective space $\mathbb{P}(W_b)$ is just the Hilbert point $[\overline{X}_b]$ of $\overline{X}_b$ in

$$G = \mathbb{G}(P(m), \mathrm{Sym}^m(E_b)) \subset \mathbb{P}(W_b).$$

Now, by the hypothesis that $\overline{X}_b$ is stable, there exists a polynomial f_b of some degree n on the vector space $V := W_B^\vee$, with the properties that

1. f_b is invariant under the action of the group $\mathrm{SL}(E_b)$ on $\mathrm{Sym}^n(V)$;

2. $f_b([\overline{X}_b]) \neq 0$.

The first of these properties states that: *there is a global holomorphic section f of the bundle $\mathrm{Sym}^n(W)$ whose value at b is f_b*. To see this, observe that, because the vector bundle E has zero first Chern class, we can choose a collection of trivializations $\varphi_\alpha : E_{U_\alpha} \overset{\cong}{\longrightarrow} \mathcal{O}_{U_\alpha}$ whose transition functions $g_{\alpha\beta}$ take values in $\mathrm{SL}(n, \mathbb{C})$ rather than $\mathrm{GL}(n, \mathbb{C})$. Such trivializations induce trivializations on all the multilinear algebra relatives of E; in particular, we get trivializations $\tilde{\varphi}_\alpha$ of $\mathrm{Sym}^n(W)$ whose transition functions $\tilde{g}_{\alpha\beta}$ preserve f. Thus, if $b \in U_\alpha$ we can simply take f to be given in each coordinate patch by the constant polynomial $f_\alpha = \tilde{\varphi}_\alpha(f_b)$ and the compatibilities $f_\beta = \tilde{g}_{\alpha\beta}f_\alpha$ on the overlaps are automatic.

The second property above says simply that *the image of the section f under the map*

$$\mathrm{Sym}^n(\psi) : \mathrm{Sym}^n(W) \longrightarrow \mathrm{Sym}^n\left(\Lambda^{P(m)}\left(\pi_*(L^m)\right)\right)$$

is nonzero at the point b. In particular, $\mathrm{Sym}^n\left(\Lambda^{P(m)}\left(\pi_*(L^m)\right)\right)$ has a nonzero global holomorphic section and hence

$$c_1\left(\mathrm{Sym}^n\left(\Lambda^{P(m)}\left(\pi_*(L^m)\right)\right)\right) \geq 0.$$

This is all we really need to know. To start with, this implies that

$$c_1\left(\Lambda^{P(m)}\left(\pi_*(L^m)\right)\right) = c_1\left(\pi_*(L^m)\right) \geq 0.$$

What is this last quantity? We can try to estimate it by applying the Grothendieck-Riemann-Roch formula to the line bundle L^m on $\mathcal{X}$. Of course, this formula describes, not the Chern class of the direct image, but the alternating sum

$$c_1(\pi_!(L^m)) = \sum_i (-1)^i \cdot c_1(R^i \pi_*(L^m)).$$

In the present circumstances, though, the higher cohomology of L^m vanishes on every fiber of $\mathcal{X} \to B$, so that the higher direct images of L^m are zero. Grothendieck-Riemann-Roch then tells us that

$$c_1(\pi_* L^m) = [\pi_*(\operatorname{td}(\mathcal{X}/B) \cdot \operatorname{ch}(L^m))]_1$$

$$= \pi_*([\operatorname{td}(\mathcal{X}/B) \cdot \operatorname{ch}(L^m)]_{k+1})$$

$$= \pi_* \left(\frac{c_1(L^m)^{k+1}}{(k+1)!} + \frac{c_1(L^m)^k}{k!} \cdot \operatorname{td}_1(\mathcal{X}/B) + \cdots \right)$$

$$= \pi_* \left(m^{k+1} \frac{c_1(L)^{k+1}}{(k+1)!} + m^k \frac{c_1(L^m)^k}{k!} \cdot \operatorname{td}_1(\mathcal{X}/B) + \cdots \right).$$

This last expression is a polynomial in m so, if it's nonnegative for all sufficiently large m, then the leading coefficient must be nonnegative. Thus, as desired, we see that

$$c_1(L)^{k+1} = \deg(f) \geq 0.$$

EXERCISE (6.28) What inequality on the degrees of λ and δ do you get for an arbitrary family $\mathcal{X} \to B$ of stable curves by applying this theorem to powers of the relative dualizing sheaf?

EXERCISE (6.29) Now let $\pi : \mathcal{X} \to B$ be a family of stable curves, not all singular or hyperelliptic, and let $\sigma : B \to \mathcal{X}$ be a section with image $\Sigma = \sigma(B)$. Assume that $K_{X_b}^m(-n\sigma(b))$ is very ample and embeds X_b as a stable curve for some b so that Theorem (6.26) applies to the line bundle $\omega^m(-n\Sigma)$. What inequality on the degrees of λ and δ and the intersection number $\omega \cdot \Sigma$ does the theorem yield? In particular, is it possible to improve the ratio $(8 + \frac{4}{g})$ for families of curves not all hyperelliptic?

The discussion that follows will shed light on this last question and you may want to return to it after reading the remainder of this section.

An inequality for families of pointed curves

We will consider here inequalities among the degrees of the three divisor classes λ, δ and ω on the moduli space $\overline{\mathcal{C}}_g = \overline{\mathcal{M}}_{g,1}$ of pointed curves. Recall that λ here stands for the pullback of the class of the Hodge bundle on $\overline{\mathcal{M}}_g$, $\omega = c_1(\omega_{\overline{\mathcal{C}}_g/\overline{\mathcal{M}}_g})$ is the class of the relative dualizing sheaf of $\overline{\mathcal{C}}_g$ over $\overline{\mathcal{M}}_g$, and the total boundary class δ is the sum of the boundary components $\Delta_0, \dots, \Delta_{g-1} \subset \overline{\mathcal{C}}_g$, or equivalently the pullback of the total boundary δ of $\overline{\mathcal{M}}_g$. (The methods we'll use would allow us to obtain more precise estimates in terms of λ, ω and the individual divisor classes δ_α, but we won't go into these here).

As we indicated at the beginning of this section, the argument here is elementary. Start with a curve $B \to \overline{\mathcal{C}}_g$. After a base change, which of course just multiplies the degrees of λ, ω and δ by a common integer, we can assume that B arises from an actual family of stable pointed curves, that is, a family $X \to B$ of nodal curves and a section $\sigma : B \to X$ such that for each $b \in B$ the pair $(X_b, \sigma(b))$ is a stable pointed curve. Replacing X by its minimal desingularization, we can assume instead that X is smooth and that the fibers $(X_b, \sigma(b))$ are semistable pointed curves — that is, nodal curves X_b with a marked smooth point such that every component of the normalization $\tilde{X}_b$ of genus 0 contains at least two points lying over the marked point or the nodes of X_b.

Consider now the three divisor classes on X given by

- the class f of a fiber of $X \to B$;
- the class y of the section $\Gamma = \sigma(B) \subset X$; and,
- the class $\eta = c_1(\omega_{X/B})$.

Note that the self-intersection y^2 is simply minus the intersection number $y \cdot \eta$, since the relative dualizing sheaf of X/B restricts to the normal bundle of Γ in X (in general, the intersection number of the relative dualizing sheaf of a family $X \to B$ with a curve $C \subset X$ will be the self-intersection of C plus the number of branch points of $C \to B$, properly counted). Next, the intersection number $y \cdot \eta$ is just the degree of the line bundle ω on $\overline{\mathcal{C}}_g$ pulled back to B. Finally, the self-intersection η^2 is just the degree of the line bundle κ on $\overline{\mathcal{M}}_g$ pulled back to B. In particular, the reduction above in which the total space of the stable model is replaced by its minimal desingularization doesn't affect the self-intersection of the relative dualizing sheaf.

We can thus write out the intersection matrix of the three classes f, y and η as shown in Table (6.30). Now, the subspace of the Neron-Severi group $NS(X)$ spanned by these three classes contains divisors of positive self-intersection: as an example, take y plus a large multiple of f. It follows from the Hodge index theorem that the intersection form on this subspace has one positive and two nonpositive eigenvalues; in particular, *its determinant must be nonnegative.*

	f	γ	η
f	0	1	$2g-2$
γ	1	γ^2	$-\gamma^2$
η	$2g-2$	$-\gamma^2$	η^2

TABLE (6.30)

Writing this out, we have

$$-2g(2g-2)\gamma^2 - \eta^2 \geq 0,$$

which translates into the basic inequality

(6.31) $4g(g-1) \cdot \deg_B(\omega) \geq 12 \cdot \deg_B(\lambda) - \deg_B(\delta).$

Note that if $X \to B$ has general fiber smooth (and of genus $g > 1$), then the term on the right is positive. In other words, *the self-intersection of a section of a family of stable curves, not all singular, is nonpositive; if the family is nonconstant, it's negative.* Another corollary follows from the observation that if $X \to B$ is a nonconstant family of stable curves, not all singular, and $C \subset X$ is any curve of degree m over B, then the degree of the relative dualizing sheaf $\omega_{X/B}$ restricted to C is greater than or equal to $m \cdot (12 \deg_B(\lambda) - \deg_B(\delta))$. In particular, we can invoke:

SESHADRI'S CRITERION (6.32) *Let X be a projective variety and let L be a line bundle on X. If for some $\varepsilon > 0$, L satisfies*

$$\deg\left(L|_C\right) > \varepsilon \cdot \mathrm{mult}_p(C)$$

for all curves $C \subset X$ and points $p \in C$, then L is ample.

Since the multiplicity of a singular point $p \in C$ on a curve $C \subset X$ is at most the degree of C over B (and since we know the relative dualizing sheaf has positive degree on the components of the fibers of $X \to B$), we may deduce the:

THEOREM (6.33) *If $X \to B$ is any nonconstant family of stable curves, not all of whose fibers are singular, then the relative dualizing sheaf $\omega_{X/B}$ of X over B is ample.*

PROBLEM (6.34) Is the inequality

$$4g(g-1) \cdot \deg_B(\omega) \geq 12 \cdot \deg_B(\lambda) - \deg_B(\delta)$$

optimal? Does any sharper inequality hold generally, or do there instead exist one-parameter families of pointed curves for which the ratio of the two sides is arbitrarily close to 1?

PROBLEM (6.35) Can we find a more exact collection of inequalities on the degrees of the divisor classes λ, ω and δ_i on a one-parameter family of pointed curves? To be more precise, can we describe the cone in $\mathbb{R}^{g-2}$ of linear combinations of λ, ω and the δ_i that have nonnegative degree on every family?

PROBLEM (6.36) Do the results above have analogues for multiply-pointed curves? That is, can we find inequalities on the degrees of the restrictions to one-parameter families of the various divisor classes on the moduli space $\overline{\mathcal{M}}_{g,n}$? This is already a fairly substantial problem in case $g = 0$ (cf. [105]). A first step, however, would be to look for inequalities in which the set of divisor classes δ_i corresponding to degenerations of the underlying curve, as well as the set of boundary components corresponding to points coming together are grouped much as the δ_i are in this section.

EXERCISE (6.37) We can now make the question asked at the end of Exercise (6.29) slightly more precise. For example, we ask: can we do better than (6.31) for a family $\pi : \mathcal{X} \rightarrow B$ of stable pointed curves with section $\sigma : B \rightarrow \mathcal{X}$ if we assume that the general curve X_b is embedded as a stable curve by the line bundle $\omega_{X_b}(-\sigma(b))$?

Ample divisors on $\overline{\mathcal{M}}_g$

Combining the results of the preceding section and this one, we see that the inequality

$$\left(8 + \frac{4}{g}\right) \deg_B(\lambda) \geq \deg_B(\delta)$$

holds for any family $\mathcal{X} \rightarrow B$ of stable curves whose general member is smooth. What about families $\mathcal{X} \rightarrow B$ whose general member is singular? We can use the inequalities of the last subsection to estimate the degrees of the line bundles λ and δ on these, and ultimately to show the ampleness of certain linear combinations of these two line bundles.

To set this up, let $\mathcal{X} \rightarrow B$ be a family of stable curves whose general fiber has d nodes. By way of terminology, we'll call those nodes of a fiber X_b that are specializations of the nodes on a general fiber the *general nodes* of X_b, and call those nodes of X_b that aren't limits of nodes on nearby fibers the *special nodes* of X_b. Thus, every fiber will have exactly d general nodes and a finite number will have some special nodes as well.

Let $\mathcal{Y} \to \mathcal{X}$ be the normalization of the total space of $\mathcal{X}$: that is, $\mathcal{Y} \to B$ is the family whose fiber Y_b over any $b \in B$ is the partial normalization of X_b at its general nodes. After making a base change, we can assume that there are $2d$ sections $\sigma_1, \dots, \sigma_{2d} : B \to \mathcal{Y}$ whose images Γ_l meet a fiber Y_b in the points lying over the general nodes of the corresponding fiber X_b. Note that the general fiber Y_b of $\mathcal{Y} \to B$ will be reducible if the general fiber of $\mathcal{X} \to B$ is. If so, then after a further base change we may assume that $\mathcal{Y}$ itself is the disjoint union of a collection of families $\mathcal{Y}_i \to B$ with connected fibers. The exercise below shows that any fiber of one of the $\mathcal{Y}_i$, together with those marked points $\sigma_i(b)$ lying on it, is a *stable* pointed curve. Finally, we replace each $\mathcal{Y}_i$ by its minimal desingularization (so that now each fiber of $\mathcal{Y}_i$ is a semistable pointed curve).

EXERCISE (6.38) Let $(C, p_1, \dots, p_n)$ be a stable n-pointed curve. Let $\pi_S : \tilde{C}_S \to C$ be the partial normalization of C at a set S of nodes. Make each connected component D of $\tilde{C}_S$ into a pointed curve by marking the points on D that map under π to either a marked point of C or a node lying in S. Show that each such component D is then *stable* as a pointed curve.

We're now ready to describe the degrees of the divisor classes λ and δ on B associated to the family $\mathcal{X} \to B$ in terms of the corresponding classes λ_i and δ_i associated to the families $\mathcal{Y}_i \to B$ and the self-intersections $(\Gamma_l)^2$ of the images of the sections $\sigma_l : B \to \mathcal{Y}_i$. We have

$$\deg(\lambda) = \sum_i \deg(\lambda_i) \quad \text{and} \quad \deg(\delta) = \sum_i \deg(\delta_i) + \sum_l (\Gamma_l)^2.$$

Given this, what is the largest possible ratio of $\deg(\delta)$ to $\deg(\lambda)$? The first thing to notice is that components $\mathcal{Y}_i \to B$ whose general fiber has large genus g_i do not help maximize this ratio: for such a component we'll have $\deg(\delta_i) \leq (8 + 4/g_i) \cdot \deg(\lambda_i)$, and the sections Γ_l lying on $\mathcal{Y}_i$ will have negative self-intersection, bringing the total degree of δ down further. Components $\mathcal{Y}_i$ with fibers of genus 1 do better: we have

$$\deg(\delta_i) = 12 \cdot \deg(\lambda_i);$$

and while we do have to have at least one section Γ_α lying on $\mathcal{Y}_i$, its self-intersection will be simply $-\deg(\lambda_i)$. We can thus make up a family of any genus g with

$$\deg(\delta) = 11 \cdot \deg(\lambda) :$$

just take a constant family $C \times B \to B$ of smooth curves of genus $g - 1$, with constant section $\Gamma = \{p\} \times B$, and attach any family of semistable curves of genus 1 — for example, take a pencil $\{E_\lambda\}$ of plane cubics, choose a base point q of the pencil, and attach each E_λ to C by identifying p with q as shown, schematically, in Figure (6.39).

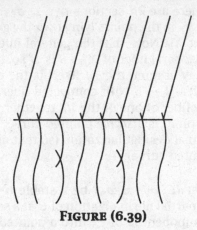

FIGURE (6.39)

Can we do better than 11? Clearly, we can do this only by including components $\mathcal{Y}_i$ whose general fiber is rational; so we have to investigate the contributions of these. But we've already done this in effect in the preceding section: each family $\mathcal{Y}_i \to B$ is a family of nodal rational curves with smooth total space (and at most two components in each fiber). If n_i of the disjoint sections Γ_l lie on $\mathcal{Y}_i$, then applying Exercise (6.22), we have

$$(n_i - 1) \cdot \sum_l \Gamma_l^2 = - \sum_j j(n_i - j) \cdot d_j$$

where d_j is the number of singular fibers with j of the n_i sections passing through one component and $n_i - j$ passing through the other. Given that in our present circumstance each $n_i \geq 3$, we see that *the sum of the self-intersections of the sections Γ_l is less than or equal to minus the number of singular fibers.* A component $\mathcal{Y}_i \to B$ with rational fiber thus contributes nothing to λ and a negative quantity to δ, so that in fact the ratio of 11 obtained above is the best (or worst, depending on your point of view) we can do. We deduce the:

THEOREM (6.40) (CORNALBA-HARRIS [27]) *For any positive integers a and b, the divisor class $a\lambda - b\delta$ is ample on $\overline{\mathcal{M}}_g$ if and only if $a > 11b$.*

The following exercise is a warning against the temptation to conclude that λ itself is ample on $\overline{\mathcal{M}}_g$: that, in other words, we can let $b = 0$ above.

EXERCISE (6.41) Let $\mathcal{Y} \to B$ be the family of stable curves obtained by identifying a fixed point on a fixed smooth curve C_1 of genus g_1 with a variable point on a fixed curve C_2 of genus $g_2 = g - g_1$. Show that $\deg_B(\lambda) = 0$ and hence that the linear system given by any multiple of λ contracts the image of B in $\overline{\mathcal{M}}_g$.

Theorem (6.40) suggests numerous variations that, to our knowledge, have never been worked out. First, we can, as usual, ask what happens if we consider the boundary components individually:

PROBLEM (6.42) What linear combinations of the classes λ and $\delta_0, \ldots, \delta_{\lfloor g/2 \rfloor}$ are ample on $\overline{\mathcal{M}}_g$?

Next, we can ask for extensions to moduli spaces of n-pointed curves. A first question is:

PROBLEM (6.43) What linear combinations of λ, δ and ω are ample on the moduli space $\overline{\mathcal{C}}_g$ of one-pointed curves?

Note finally that, among all generically smooth one-parameter families of stable curves, the ones that we've seen achieve the maximum ratio of $\deg(\delta)$ to $\deg(\lambda)$ consist entirely of hyperelliptic curves. We could ask: does a stronger inequality hold for families not contained in the hyperelliptic locus? How about trigonal curves, and so on? All of this motivates the:

PROBLEM (6.44) Define, as usual, a stable curve to be hyperelliptic, trigonal, etc. if it's the limit of smooth curves in the corresponding locus in $\mathcal{M}_g$. What linear combinations of λ and δ are ample when restricted to the locus of hyperelliptic, or of trigonal curves in $\overline{\mathcal{M}}_g$?

The best results to date on this question are due to Stankova-Frenkel [143].

E   Irreducibility of the Severi varieties

In this section we'll sketch a proof of the result stated in Chapter 1 as the third part of Theorem (1.49): that the family of irreducible plane curves of given degree and (geometric) genus is irreducible. This is a topic that draws upon many of the ideas we've developed in the preceding chapters; in fact, it represents one of the best examples of how we can combine the insights obtained from both the parameter space and moduli-theoretic viewpoints to analyze a family of curves. The reason why the abstract and projective viewpoints are both involved may not be clear from a first glance at the problem, but it should emerge in the course of the following reductions.

Initial reductions

Let's first ask the naive question, why should we expect such a theorem to be true? After all, we've seen that the Hilbert schemes parameterizing irreducible, nondegenerate curves of fixed degree and genus in higher-dimensional projective spaces are only very rarely irreducible. What about the geometry of plane curves should make them different?

The first answer was given by Severi [140]. (Indeed, although there are now a number of proofs given of the theorem, they all go back to this answer in the end.) Severi's idea was to look at *degenerations* of the curves parameterized by a given component of the Severi variety. Specifically, Severi claimed that if we let $\mathbb{P}^N$ be the space of all plane curves of degree d, let $m = \binom{d-1}{2}$ be the genus of a smooth curve of degree d, and $V_{d,g} \subset \mathbb{P}^N$ be the variety of irreducible nodal plane curves of degree d and geometric genus g (or, equivalently, with exactly $\delta = m - g$ nodes), then *the closure of any component Σ of $V_{d,g}$ must contain the variety $V_{d,0}$ of rational nodal plane curves.* (Observe that a rational nodal plane curve of degree d is the same thing as an irreducible nodal curve of degree d with exactly m nodes).

To see why this implies the irreducibility of $V_{d,g}$, observe first that the variety $V_{d,0}$ is irreducible, since all rational nodal curves of degree d are simply projections of the rational normal curve $X \subset \mathbb{P}^d$ from various subspaces $\Lambda \cong \mathbb{P}^{d-3} \subset \mathbb{P}^d$ to a plane $\Gamma \cong \mathbb{P}^2$. This gives a dominant rational map from the product of the Grassmannian $\mathbb{G}(d-3, d)$ with the variety PGL_3 of isomorphisms of Γ with $\mathbb{P}^2$ to the Severi variety $V_{d,0}$, showing that $V_{d,0}$ is irreducible.

In fact, more is true. Given a family of nodal curves $C_t \in V_{d,0}$, the $m = \binom{d-1}{2}$ nodes $p_1, \ldots, p_m$ vary continuously, tracing out arcs $p_1(t), \ldots, p_m(t)$. We claim that not only can we find such a family $\{C_t\}$ of nodal curves joining any two given ones C_0 and C_1 in $V_{d,0}$, but we can find one such that *the arcs $p_1(t), \ldots, p_m(t)$ induce an arbitrary bijection between the nodes $p_i(0)$ of C_0 and the nodes $p_i(1)$ of C_1.* In other words, the monodromy in the family $V_{d,0}$ acts on the nodes of C_0 as the full symmetric group on m letters. To see this, note that a nodal curve $C \subset \mathbb{P}^2$ may be represented as above as the projection of a rational normal curve $X \subset \mathbb{P}^d$ from a plane $\Lambda \cong \mathbb{P}^{d-3} \subset \mathbb{P}^d$; the nodes of C correspond to the points of intersection of Λ with the *chordal variety* of X. Our monodromy assertion then follows from the:

UNIFORM POSITION PRINCIPLE (6.45) *If Z is any nondegenerate irreducible variety of dimension k and degree d in $\mathbb{P}^n$, and $\Lambda \cong \mathbb{P}^{n-k} \subset \mathbb{P}^n$ is a subspace meeting Z transversely at points $p_1, \ldots, p_d$, then, as we vary $\Lambda \in \mathbb{G}(n-k, n)$ in the open subset U of planes transverse to Z, the monodromy acts on the points p_i as the symmetric group on d letters.*

For a proof (and other applications) of this statement, see [7] or [78]. In our present circumstances, it gives us a picture of what the closure $\overline{V}_{d,g}$ of the variety $V_{d,g}$ must look like in a neighborhood of $V_{d,0}$. To begin with, let $[C_0] \in V_{d,0}$ correspond to a curve $C_0 \subset \mathbb{P}^2$ with nodes $p_1, \ldots, p_m$. We've seen in our discussion of deformation theory that a neighborhood of $[C_0]$ in $\mathbb{P}^N$ will map to the product of the deformation spaces of the singularities (C_0, p_i) (which are each smooth and one-dimensional). This means that we have local coordinates $z_1, \ldots, z_N$ on $\mathbb{P}^N$ in a neighborhood of $[C_0]$ in terms of which $V_{d,0}$ is the codimension m coordinate subspace with equations $z_1 = \cdots = z_m = 0$, and the closure $\overline{V}_{d,g}$ is the union of the codimension δ-coordinate planes $z_{i_1} = \cdots = z_{i_\delta} = 0$ determined by all δ-element subsets $\{i_1, \ldots, i_\delta\}$ of $\{1, \ldots, m\}$ — in other words, for any subset I of $m - \delta$ of the nodes of C_0, there is a (smooth) branch Σ_I of $\overline{V}_{d,g}$ near $[C_0]$ such that a deformation of C_0 in Σ_I smooths all the nodes in I and none of the δ others. The schematic picture is that like that in Figure (6.46) (where we have taken $m = 2$ and $\delta = 1$).

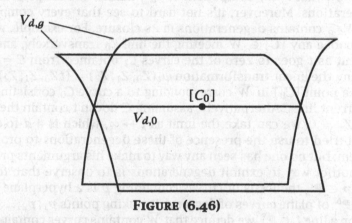

FIGURE (6.46)

Thus, $\overline{V}_{d,g}$ isn't locally irreducible at $[C_0]$; rather, it has $\binom{m}{\delta}$ branches. But, as $[C_0]$ varies in $V_{d,0}$, the monodromy acts transitively on these branches. Since any irreducible component of $V_{d,g}$ containing $V_{d,0}$ in its closure must contain one of the branches of $V_{d,g}$ near $V_{d,0}$, it must therefore contain them all. Thus, in order to demonstrate the irreducibility of $V_{d,g}$ it's enough to show that any irreducible component of $V_{d,g}$ must contain $V_{d,0}$ in its closure. As an obvious extension, we have the:

First Reduction: It's sufficient to show that any irreducible component of $V_{d,g}$ contains in its closure an irreducible component of $V_{d,g'}$ for some $g' < g$.

We can get away with even less because the picture of $V_{d,g}$ and $V_{d,g'}$ in a neighborhood of a point $[C_0] \in V_{d,g'}$ corresponding to a curve C_0 with $m - g'$ nodes looks very much like the one above — that is, $V_{d,g}$ is locally a union of sheets on each of which some $(g - g')$-element subset of the nodes of C_0 smooth. Thus:

Second Reduction: It's sufficient to show that any irreducible component of $V_{d,g}$ contains in its closure a point $[C_0]$ corresponding to a nodal curve C_0 of genus $g' < g$.

This reduction seems to be what convinced many that $V_{d,g}$ is irreducible: it certainly seems plausible enough that any component of $V_{d,g}$ must include in its closure, for example, curves with $\delta + 1$ nodes. However, it turns out to be something of a red herring for us: our proof will use a statement that reduces the irreducibility of $V_{d,g}$ to the existence of a more general degeneration in its closure.

We will eventually produce such a degeneration, but our path to it will be circuitous. Why can't we proceed directly? All we need to do is to show that every component of $V_{d,g}$ admits the simplest kind of degenerations. Moreover, it's not hard to see that every component $\mathcal{W}$ of $V_{d,g}$ contains degenerations in its closure. For example, we can just choose any $[C] \in \mathcal{W}$ meeting the line Z_0 transversely, and take the limit as t goes to zero of the curves C_t obtained from $C = C_1$ by applying the linear transformation $\psi_t(Z_0, Z_1, Z_2) = (tZ_0, Z_1, Z_2)$. This yields a point $[C_0]$ in $\overline{\mathcal{W}}$ corresponding to a curve C_0 consisting of d concurrent lines. Alternatively, assuming C doesn't contain the point $Z_0 = Z_1 = 0$, we can take the limit as $t \to \infty$, which is a d-fold line. (Severi tried to use the presence of these degenerations to prove the theorem; but no one has seen any way to make his arguments precise.) Yet another way to exhibit degenerations is to observe that, for any point $p \in \mathbb{P}^2$, the locus of curves containing p is a hyperplane in the space $\mathbb{P}^N$ of plane curves of degree d. By taking points $p_1, p_2, \ldots, p_{d+1}$ lying on a line $L \subset \mathbb{P}^2$, we deduce that $\overline{\mathcal{W}}$ contains curves containing all the p_i and hence containing L. The problem here is that the presence of these possibly wild degenerations doesn't guarantee that $\mathcal{W}$ admits the milder degenerations to nodal curves required by the reductions.

Another approach to showing that every component of $V_{d,g}$ admits degenerations is to apply Diaz' theorem (2.34) that $\mathcal{M}_g$ doesn't contain any complete subvarieties of dimension $g - 1$. Consider the rational map

$$\varphi : V_{d,g} \dashrightarrow \mathcal{M}_g .$$

Since any component of $W^r_d(C)$ whose general member is birationally very ample has dimension at most $d - 3r$ (cf. Exercise IV-E-2 of [7]), the fibers of φ cannot have dimension greater than $d + 2$. Since the dimension of $V_{d,g}$ is $3d + g - 1$, Diaz concluded that the closure in

$\overline{\mathcal{M}}_g$ of the image of φ on any component $\mathcal{W}$ of $V_{d,g}$ necessarily meets the boundary $\Delta \subset \overline{\mathcal{M}}_g$. An arc in $\mathcal{W}$ whose image in $\overline{\mathcal{M}}_g$ tended to Δ would certainly induce a family of curves degenerating in moduli. The existence of even such families had not previously been known. Here again, however, there seems to be no way to control the singularities of the limiting model *in* $\mathbb{P}^2$ and hence to conclude the irreducibility of $V_{d,g}$. A priori, every such arc in $\mathcal{W}$ might tend, for example, to a nonreduced curve.

These considerations make it clear that what we really need here isn't simply to degenerate, but to do so retaining some control over the limiting curve, i.e., to exhibit in the closure of any $\mathcal{W}$ reduced curves having lower geometric genus but still reasonably mild singularities. The question is, how do we exert such control? For example, we've seen that in $\overline{\mathcal{W}}$ there are curves that contain a line L, but how do we rule out the possibility that every such curve is just a d-tuple of concurrent lines?

To surmount these difficulties, the key extra idea is *to keep track of dimensions*, or degrees of freedom, in our families. Using this, we'll make two further reductions. In the end, these reductions won't be used in the proof; however, they help us come to grips with the difficulties discussed above and lead to the kind of degenerations we'll study in the next subsection.

How does keeping track of dimensions help us rule out, for example, the possibility that every curve in $\overline{\mathcal{W}}$ containing L is a d-tuple of concurrent lines? Simply, we know that the dimension of $\mathcal{W}$ is $3d + g - 1$, and so the locus in $\overline{\mathcal{W}}$ of curves containing L is, by the argument above, of dimension at least $\dim(\mathcal{W}) - (d + 1) = 2d + g - 2$. But the family of d-tuples of concurrent lines including L has dimension just d.

As another example, recall that, by Lemma (3.45), any locus $\mathcal{W}' \subset \mathbb{P}^N$ consisting of curves of degree d and genus $g' < g$ and having dimension $3d + g - 2$ must be open in a component of $U_{d,g-1}$. We conclude that:

Third Reduction: It's sufficient to show that if $\mathcal{W}$ is any component of $V_{d,g}$, then $\overline{\mathcal{W}} - \mathcal{W}$ has codimension 1: i.e., that $\overline{\mathcal{W}}$ contains a locus of codimension 1 consisting of curves other than reduced curves of geometric genus g.

Now, clearly the first two constructions of degenerations given above will not produce such loci. Diaz' construction, on the other hand, seems a much better bet: after all, all we have to show is that the inverse image, under the map φ, of the boundary $\Delta \subset \overline{\mathcal{M}}_g$ has codimension 1 in $\overline{\mathcal{W}}$. Since Δ has codimension 1 in $\overline{\mathcal{M}}_g$ (and is even the support of a Cartier divisor), this, at first, seems promising. The difficulty now is that φ isn't a regular map, only a rational one. When we talk loosely about the inverse image of Δ, we really mean $\varphi_1(\varphi_2^{-1}(\Delta))$, where φ_1 and φ_2 are the projection maps from the graph $\mathcal{T}$ of φ to $\overline{W}$ and $\overline{\mathcal{M}}_g$ respectively. And, while $\varphi_2^{-1}(\Delta)$ will necessarily have codimension 1 in $\mathcal{T}$, the fibers of φ_1 on $\varphi_2^{-1}(\Delta)$ may be positive-dimensional.

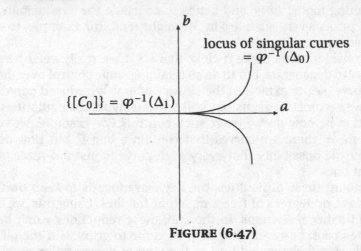

locus of singular curves
$= \varphi^{-1}(\Delta_0)$

$\{[C_0]\} = \varphi^{-1}(\Delta_1)$

FIGURE (6.47)

To see that this really happens, consider a family of plane curves over a two-dimensional base $\mathcal{W}$, generically smooth in a neighborhood of a point $p \in \mathbb{P}^2$ and specializing to a curve C_0 with a cusp at p — e.g., the family given, in terms of affine coordinates x, y on $\mathbb{P}^2$ near p and a, b on $\mathcal{W}$ near $[C_0]$, by the equation $y^2 = x^3 + ax + b$ as sketched in Figure (6.47). Assuming the curves C_λ are well-behaved away from p, the map $\varphi : \mathcal{W} \to \overline{\mathcal{M}}_g$ will be defined everywhere except at the origin $a = b = 0$ in $\mathcal{W}$. There, however, it's undefined, and we'll have to blow up three times — once at the origin and then at the intersection of the first two exceptional divisors with the proper transform on the a-axis — to resolve it. The map will then blow down the first two exceptional divisors, so the graph $\mathcal{T}$ of φ will be as shown in Figure (6.48). This is the basic example of a situation where a divisor in $\overline{\mathcal{M}}_g$ — in this case, Δ_1 — may have inverse image of codimension greater than 1 in $\overline{W}$.

A second example, not involving singularities, can be obtained by considering the rational map $\varphi : \mathcal{W} = \mathbb{P}^{14} \dashrightarrow \overline{\mathcal{M}}_3$ from the space of plane quartics to their moduli. In $\overline{\mathcal{M}}_3$, the locus H of hyperelliptic curves is a divisor; but anytime we have a family of smooth curves of genus 3 approaching a hyperelliptic one, the canonical models will tend to a double conic, and the locus $\mathcal{S}$ in $\mathbb{P}^{14}$ of double conics is of course five-dimensional. What's going on here is again simple to describe, at least over a general point of H: the map φ is blowing up the locus $\mathcal{S}$, replacing points $[2C] \in \mathcal{S}$ with pairs $(2C, D)$ where D is a normal direction to H in $\mathbb{P}^{14}$, represented by a divisor $D \in |\mathcal{O}_{2C}(4)|$, and

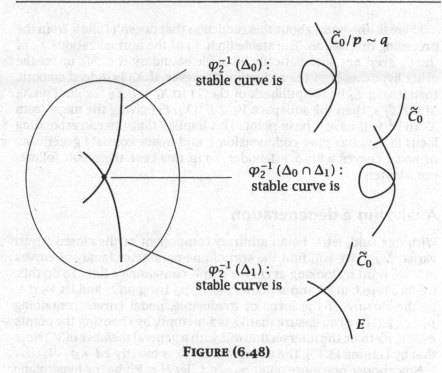

$\varphi_2^{-1}(\Delta_0)$:
stable curve is

$\tilde{C}_0/p \sim q$

$\tilde{C}_0$

$\varphi_2^{-1}(\Delta_0 \cap \Delta_1)$:
stable curve is

$\varphi_2^{-1}(\Delta_1)$:
stable curve is

$\tilde{C}_0$

E

FIGURE (6.48)

mapping the pair $(2C, D)$ to the curve of genus 3 given as the double cover of $C \cong \mathbb{P}^1$ branched over the eight points of D. (Incidentally, the inverse image $\varphi^{-1}(H)$ doesn't have codimension 9 in $\mathbb{P}^{14}$, but codimension 3: the locus in $\overline{M}_3$ of curves consisting of two elliptic curves meeting at two points lies in H, and its inverse image in $\mathbb{P}^{14}$ contains the locus of quartics with a tacnode.)

In any event, it's clear that to find a locus of codimension 1 in $\overline{W}$ consisting of curves of lower geometric genus, we need to find a point $[C] \in \varphi^{-1}(\Delta) \subset \overline{W}$ such that the fiber of $\varphi_2^{-1}(\Delta) \subset \mathcal{T}$ over $[C]$ is (ideally) finite or, failing this, at least contains an isolated point. Now, one circumstance in which a point $(C, D) \in \mathcal{T}$ will be an isolated point of $\varphi_2^{-1}(\Delta) \cap \varphi_1^{-1}(\{[C]\})$ is when the stable curve D is stably equivalent to a nodal partial normalization of the plane curve C — there are only finitely many nodal partial normalizations of a given curve. We have thus arrived at the final reduction:

Fourth Reduction: It's sufficient to exhibit in any component W of $V_{d,g}$ an arc $B \subset \overline{W}$ with parameter t such that $B - \{0\}$ lies in W but $\{0\}$ does not, and such that the stable limit of the normalizations of the curves C_t as $t \to 0$ is a singular curve (i.e., maps to a point of the boundary Δ of $\overline{M}_g$) that is stably equivalent to a partial normalization of the curve C_0.

There is one point about this reduction that doesn't follow from the preceding discussion. The stable limit X_0 of the normalizations X_λ of the C_λ need not automatically lie in the boundary $\Delta \subset \overline{\mathcal{M}}_g$ under the other hypotheses of the reduction. However, if X_0 is indeed smooth, then letting $\mathcal{L}_t$ be the pullback of $\mathcal{O}_{\mathbb{P}^2}(1)$ to X_t and $\mathcal{L}_0$ the limit on X_0 of the $\mathcal{L}_t$'s, then the subspace $V_0 \subset H^0(X_t, \mathcal{L}_t)$ giving the maps from X_t to $\mathbb{P}^2$ will have a base point. This implies that the corresponding locus in $V_{d,g}$ has pure codimension 1, and hence consists generically of nodal curves with $\delta + 1$ nodes, so in this case the result follows immediately.

Analyzing a degeneration

With this said, let V be an arbitrary component of the closed Severi variety $\overline{V}_{d,g}$. We will find the sort of one-parameter family of curves in V we want by looking at curves C in V containing a line. To do this, we fix a line L in $\mathbb{P}^2$ and d points $p_1, \ldots, p_d$ lying on L, and let $V_0 \subset V$ be the closure of the locus of irreducible, nodal curves containing $p_1, \ldots, p_d$. (We can ensure that V_0 is nonempty by choosing the points $p_1, \ldots, p_d$ to be the intersection of L with a general member of V.) Note that by Lemma (3.45), the dimension of V_0 is exactly $2d + g - 1$.

Now choose one more point $p_{d+1} \in L$, let $H \subset \mathbb{P}^N$ be the hyperplane of plane curves containing P_{d+1} and let $W = V_0 \cap H$. Of course, W will consist entirely of curves containing L. Note also that, being the intersection of V_0 with a hyperplane, W will have dimension exactly $2d + g - 2$.

Let $[C_0] \in W$ be a general point of any component of W. Our initial goal will be to describe the curve $C = C_0$, and our tool will be the method of semistable reduction.

To carry this out, let $\{[C_t]\}_{t \in \Delta}$ be an arc in $\overline{V}_0$ meeting W transversely at $[C_0]$, with C_t irreducible and nodal for $t \neq 0$. Let X_t be the normalization of C_t for $t \neq 0$; the curves X_t form the fibers of a family $\mathcal{X}^* \to \Delta^*$ over the punctured disc, and the normalization maps $\eta_t : X_t \to C_t$ string together to form a map $\eta : \mathcal{X}^* \to \mathbb{P}^2$. Applying nodal reduction [Proposition (3.49)], we can (after base change) complete this family to a family $\pi : \mathcal{X} \to \Delta$ of nodal curves, proper over Δ and satisfying the conditions that

1. the total space of $\mathcal{X}$ is smooth;

2. there is a regular map $\eta : \mathcal{X} \to \mathbb{P}^2$ restricting to the map $\eta_t : X_t \to C_t$ on each fiber X_t of π over Δ^* ; and

3. $\mathcal{X}$ is minimal with respect to these properties; i.e., there are no rational components of the central fiber X_0 meeting the rest of the central fiber only once and on which the map η is constant.

We have, of course, no a priori idea of what the central fiber X_0 of this family looks like, or how the map η_0 behaves. However, we'll see that the information we do have — that η_0 has degree d, that X_0 has arithmetic genus g, and that η_0 is the limit of maps η_λ whose images contain the points p_i — combined with the basic dimension estimates above allow us to describe X_0 completely.

We start by introducing some notation. By construction, the image C_0 of η_0 contains the line L. Let $Y_0 \subset X_0$ be the union of the components of X_0 mapping to L, let Y_1 be the union of the remaining components, and let $q_1, \ldots, q_k$ be the points of intersection of Y_0 and Y_1. Denote by α the degree of η_0 on the curve Y_0, so that we can write

$$C_0 = \alpha \cdot L + D$$

where D is the image of Y_1 under η_0. Clearly a crucial part of our analysis must be to control α: it turns out that α must equal 1. A con-

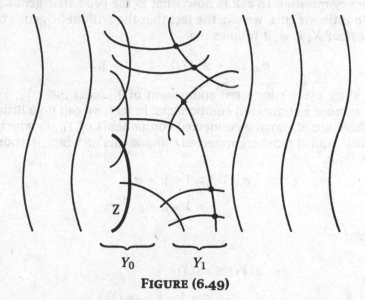

FIGURE (6.49)

sequence will be that X_0 looks schematically as shown in Figure (6.49): Y_0 is a tree of curves (all, in fact, rational) whose root Z (shown thickened) maps isomorphically under η to L and whose leaves each meet Y_1 in one of the points q_i (shown as dots).

To express the fact that η_0 is the limit of maps whose images contain the points p_i so that we can apply this hypothesis, let Σ_i^* be the inverse image of the point p_i in X^*, and let Σ_i be its closure. Observe that since the total space X is smooth, the sections Σ_i of π must meet the central fiber X_0 in smooth points of X_0. We can now write

(6.50) $\eta^* L = \Sigma_1 + \cdots + \Sigma_d + M,$

where M is a divisor whose support is exactly the curve Y_0. In particular, the *only* points of Y_1 lying over a point of L other than one of the p_i are the points $q_1, \ldots, q_k$ of intersection of Y_1 with Y_0. In fact, we can refine this a little: of the k points q_i, suppose that β lie on connected components of Y_0 on which η isn't constant, and $\gamma = k - \beta$ on connected components of Y_0 on which η is constant. A connected component Y of Y_0 on which η is constant must meet one of the sections Σ_i and hence map to one of the points p_i: if Y were disjoint from all Σ_i, then the part M_Y of the divisor M supported on Y would have self-intersection $(M_Y \cdot M_Y) = (M_Y \cdot \eta^*L) = 0$, contradicting the fact that any divisor supported on a proper subset of a fiber of π must have negative self-intersection. Thus, *there can be at most β points of Y_1 lying over points of L other than the p_i.* (Note that, since every connected component of Y_1 must meet Y_0, η can't be constant on any connected component of Y_1.)

The key question to ask is now: what is the geometric genus g_1 of Y_1? To estimate this, we use the fact that the arithmetic genus of the whole fiber X_0 is g; it follows that

$$p_a(Y_0) + p_a(Y_1) + k - 1 = g.$$

Now, since every connected component of Y_0 must meet Y_1, Y_0 can have at most k connected components. In fact, we can do a little better: there are at most γ connected components of Y_0 on which η is constant, and at most α connected components on which it's nonconstant. Thus

$$p_a(Y_0) \geq 1 - \gamma - \alpha$$

$$= 1 - k + \beta - \alpha$$

and hence

$$g(Y_1) \leq p_a(Y_1)$$

(6.51) $$= g + 1 - k - p_a(Y_0)$$

$$\leq g + \alpha - \beta.$$

Now let's assemble what we know about $D = \eta(Y_1)$. It's the image of the nodal curve Y_1 of geometric genus at most $g + \alpha - \beta$ via the map η of degree $d - \alpha$. Moreover, η isn't constant on any connected component of Y_1, and it takes at most β points of Y_1 to points of L other than $p_1, \ldots, p_d$. By Corollary (3.46), such curves form a family of dimension at most

(6.52) $$2(d - \alpha) + (g + \alpha - \beta) + \beta - 1 = 2d + g - 1 - \alpha.$$

On the other hand, the curve D moves in the $(2d+g-2)$-dimensional family of plane curves W. We conclude that $2d+g-2 \leq 2d+g-1-\alpha$. Therefore, equality is only possible if $\alpha = 1$ and if, moreover, all the upper bounds used in obtaining the estimate (6.52), and in the application of Corollary (3.46), are in fact exact. A whole series of consequences follow:

1. Y_0 is connected: it consists of a tree with one irreducible component Z mapping isomorphically to L, plus chains of rational curves joining Z to the points $q_1, \ldots, q_k$ on Y_1. In particular, $\gamma = 0$ and $\beta = k$.

2. Y_1 is smooth of genus $g + 1 - k$.

3. The stable limit of the curves X_t is the union $\tilde{X}_0$ of Y_1 and $Z \cong \mathbb{P}^1$, joined at the k points $q_1, \ldots, q_k$ (or the curve we get from this by contracting Z if $k = 1$ or 2).

First, the sharpness of the estimate (6.51) for the genus of Y_1 implies that any connected component of Y_0 has arithmetic genus 0, so is a tree of rational curves. Moreover, any connected component of Y_0 on which η is constant can meet Y_1 in only one point, since equality in the dimension statement of Corollary (3.46) implies that the k points of $Y_0 \cap Y_1$ map via η to distinct points of L. Since each connected component of Y_0 is a tree, the minimality hypothesis in the construction of X implies that there are no connected components of Y_0 on which η is constant. Equivalently, we must have $\gamma = 0$ and $\beta = k$. Since $\alpha = 1$, there is exactly one component of Y_0 on which η is nonconstant. This gives the first consequence. The chain of inequalities in (6.51) must also be sharp: i.e., $g(Y_1) = p_a(Y_1) = g + 1 - k$, which gives the second consequence. The third is then immediate from the first two.

Our analysis also gives us a fairly precise picture of what is happening to the plane models. Since the k points $q_1, \ldots, q_k$ don't map to points p_i, the points of Y_1 that do map to points p_i are points of intersection of Y_1 with the sections Σ_i; in particular, since each Σ_i occurs with multiplicity 1 in $\eta^* L$, we see that whenever the image D of Y_1 meets L at a point p_i, it does so transversely. Thus D is a nodal curve of degree $d-1$ (thus having $\delta - d + k + 1$ nodes), meeting L transversely at a subset of l the points p_i and at k further points $r_i = \eta(q_i)$ which are smooth on D.

In sum, the picture of the degeneration is this: as C_t tends to $C_0 = L \cup D$, we see a node tending to each of the l points p_i lying on D. The curve D meets L in k other points r_i, and if m_i is the order of contact of D with L at r_i, then we'll see $m_i - 1$ nodes of C_t tending to r_i. Note that l may be zero — i.e., D may miss all of the points p_i — but k cannot be; D must contain at least one point of L other than the p_i. Also, observe that there is only one other restriction on the numbers k and l: l must be at least δ. As for the multiplicities m_i of intersection of D with L at the points r_i, they have to satisfy only the restriction that $\sum(m_i - 1) \le \delta - l$. The components of the variety W are thus classified by the sequences $(l, m_1, \ldots, m_k)$ satisfying these inequalities.

We repeat, for emphasis, that this analysis applies only to a family of curves in V_0 tending to a *general* point of a component of the hyperplane section W of V_0; obviously, in an arbitrary family tending to a point of W, things can happen that would curl your toenails.

We have now almost arrived at a family meeting the conditions of the fourth reduction. Before completing the argument, however, we'll pause to give an example making explicit the general constructions of this subsection.

An example

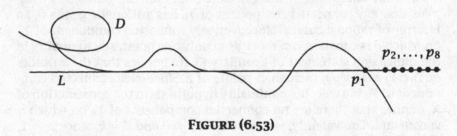

FIGURE (6.53)

To make the preceding general analysis a bit more concrete, consider a family C_t of plane curves of degree 8 and genus 16 meeting a line L in fixed points $p_1, \ldots, p_8$, and specializing to a curve C_0 containing L as above — that is, with C_0 a general point of a component of the locus of curves in the closure of $V_{8,16}$ containing L. The limiting curve will then consist of the line L plus an irreducible nodal septimic D meeting L at smooth points. One possibility, consistent with all our restrictions, is that D will pass through p_1 but none of the points $p_2, \ldots, p_8$, will meet L three times away from p_1, once transversely, once in a point of simple tangency, and once with contact of order 3, and will have single node not lying on L: a schematic diagram is shown in Figure (6.53)

In this case, the central fiber X_0 of the family $\pi : \mathcal{X} \rightarrow \Delta$ constructed above will look like Figure (6.54). There, Y_1 is the normalization of D and all the other curves shown are components of Y_0 and are rational with the unique component Z mapping onto L under η shown at the left. If we write $\eta^* L = \sum \Sigma_i + M$, then the numbers next to components of Y_0 are the multiplicities with which each appears in M; these are

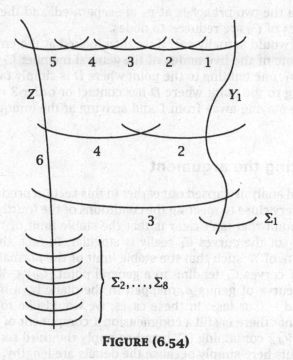

FIGURE (6.54)

dictated by the requirements that η be constant on all but one component, so that the degree of $\eta^*L = \sum \Gamma_i + M$ on all other components of Y_0 is zero, and that M meet Y_1 with multiplicities 1, 2 and 3.

EXERCISE (6.55) Verify the multiplicities in Figure (6.54)

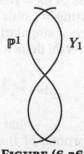

FIGURE (6.56)

The stable reduction of X_0, shown in Figure (6.56) thus is the union of the curve Y_1 with a copy of $\mathbb{P}^1$ — the image of Z — meeting it at three points (which, in the notation of the preceding subsection would be q_1, q_2 and q_3) or in other words the partial normalization of the curve

C_0 in which the two branches at p_1 are separated, and the remaining singularities of C_0 are reduced to nodes.

What we would actually see, if we then looked at the entire family, would be one of the five nodes of the general member C_t tending to the point p_1, one tending to the point where D is simply tangent to L, two tending to the point where D has contact or order 3 with L, and, finally, one staying away from L and arriving at the (unique) node of D itself.

Completing the argument

The general analysis carried out earlier in this section produces a family that is very close to meeting the conditions of the fourth reduction. The only point that isn't clear is that the stable limit of the normalizations X_t of the curves C_t really is singular. In fact, there will be components of W such that the stable limit of the normalizations of a family of curves C_t tending to a general point $[C_0] \in W$ is simply a smooth curve of genus g, mapped to the plane by a linear series of degree $d - 1$ or less. In these cases, we would like to be able to conclude that there is still a codimension 1 component of the boundary $\overline{V}_{d,g} \setminus V_{d,g}$ containing $[C_0]$, and so apply the third reduction. We won't do this here simply because the details are lengthy. We leave it as an exercise for you to carry out the modifications of the previous arguments necessary to make this approach work, if you're interested. Instead, we'll give an alternative argument in which we use a local analysis based on facts about deformations of tacnodes as a shortcut to deducing the existence of a degeneration to a nodal curve of lower geometric genus from the existence of the degenerations constructed above.

Recall that the singularities of $C_0 = L \cup D$ along the line L are two-branch double points, consisting of L plus a smooth arc having contact of order say m with L. Such a singularity may be given by a local equation $y^2 + x^{2m}$, and its versal deformation given by the family of curves

$$(6.57) \qquad y^2 + x^{2m} + a_{2m-2}x^{2m-2} + \cdots + a_1 x + a_0$$

parameterized by $a_0, \ldots, a_{2m-2}$ — that is, the versal deformation space A is an étale neighborhood of the origin in affine $(2m - 1)$-space with coordinates $a_0, \ldots, a_{2m-2}$.

Within this deformation space, there are two loci of particular interest to us. One is the *equigeneric locus* B, defined to be the locus of δ-constant deformations, or deformations that preserve the total contribution of the singularity to the geometric genus. In general, for a singularity (C, p) with contribution $\delta = \text{length}(\mathcal{O}_{\tilde{C},p}/\mathcal{O}_{C,p}) = m$, this will be the closure of the locus of curves with m nodes; in our present case, it's the locus of $(a_0, \ldots, a_{2m-2})$ such that the polynomial $f(x) = x^{2m} + a_{2m-2}x^{2m-2} + \cdots + a_1x + a_0$ has m double roots. Using the vanishing of the x^{2m-1} coefficient in (6.57), this locus is given parametrically by

$$f(x) = (x - \lambda_1)^2 (x - \lambda_2)^2 \cdots (x - \lambda_{m-1})^2 \left(x + \sum_{i=1}^{m-1} \lambda_i\right)^2;$$

in particular, we see that it's irreducible of dimension $m - 1$ in A.

EXERCISE (6.58) Show, more generally, that the codimension of the δ-constant locus in the versal deformation space of a plane curve singularity is always δ.

The second locus $B' \subset A$ of interest to us is the closure of the locus of curves with $m - 1$ nodes nearby. This is given parametrically as

$$f(x) = (x - \lambda_1)^2 (x - \lambda_2)^2 \cdots (x - \lambda_{m-1})^2 (x - \mu) \left(x + \mu + 2 \sum_{i=1}^{m-1} \lambda_i\right)$$

and so is irreducible of dimension m.

We now consider how this relates to deformations of C_0 as a plane curve. Specifically, we take U to be the component of an étale neighborhood of the point $[C_0]$ in $\overline{V}$ containing the arc $\{[C_t]\}$ constructed in the preceding subsections, and look at the induced map ψ from U to the product of the deformation spaces A_i of the k singular points $r_i \in C_0$. We first observe that the fiber of U over the point $[C_0]$ consists entirely of reducible curves $D = L' \cup D'$ where L' is a line and D' has degree $d - 1$. Moreover, the singularities of D' consist of exactly $\delta - d + k + 1$ nodes (i.e., D' is nodal of geometric genus $g - k + 1$), and D' meets L' at smooth points with multiplicities $m_1, \ldots, m_k$.

By our earlier dimension counts, D' has $3(d - 1) + (g - k + 1) - 1$ moduli. Adding 2 moduli for L' and subtracting the $(m_i - 1)$ conditions imposed by the ith multiplicity, the dimension of this fiber is thus

$$3(d - 1) + (g - k + 1) - 1 + 2 - \sum_{i=1}^{k} (m_i - 1) = 3d - 1 + g - \sum_{i=1}^{k} m_i.$$

On the other hand, the map ψ carries U into the product of the subspaces B_i' corresponding to deformations of C_0 to curves with $m_i - 1$ nodes near r_i. Since this product has dimension $\sum_i m_i$ and U itself has dimension $3d + g - 1$, we deduce that ψ *maps U surjectively onto it*. In particular, the image of ψ must also contain the product of the equigeneric subspaces B_i.

Therefore, *the deformations of C_0 in $U \subset \overline{V}$ include nodal curves with $\delta + k$ nodes.* This is exactly the statement we need to apply the last of the initial reductions above and complete the proof of the irreducibility of $V_{d,g}$.

F Kodaira dimension of $\mathcal{M}_g$

In this section, we'll show how the theory of limit linear series, in combination with our description of the Picard group of $\overline{\mathcal{M}}_g$, may be used to prove that:

THEOREM (6.59) (HARRIS-MUMFORD THEOREM, [82] AND [79]) *The moduli space of curves of genus g is of general type if $g \geq 24$, and has Kodaira dimension at least 1 if $g = 23$.*

Before undertaking this, we should say a few words about the background of the problem.

Writing down general curves

If someone put a gun to your head and demanded that you show him this "general curve of genus 2" that everyone was proving theorems about — in other words, that you write down the equation of a general curve of genus 2 — you would have no problem: you would whip out your pen, write

$$y^2 = x^6 + a_5 x^5 + \cdots + a_0$$

and say, "where the a_i are general complex numbers". Likewise, if the challenge were to write down a general curve of genus 3, you could write the equation of a plane quartic

$$\sum_{i+j\leq 4} a_{ij} x^i y^j = 0$$

and again take the coefficients a_{ij} to be general. For genus 4 and 5, there is a similar solution: in each case, the canonical model of a general curve is a complete intersection, and you can just write down a homogeneous quadric and a cubic in four variables (for genus 4) or three homogeneous quadrics in five variables (for genus 5) and once more let the coefficients vary freely.

In genus 6 you might have to stop and think. Here the canonical model $C \subset \mathbb{P}^5$ of a general curve of genus 6 isn't a complete intersection; it's the intersection of six quadric hypersurfaces in $\mathbb{P}^5$, and you can't just take six general quadric polynomials and expect their intersection to be anything but empty. But Brill-Noether theory provides an answer: a general curve of genus 6 is representable as a plane sextic curve with four double points, no three collinear. So everything is OK: we can take those points to be the coordinate points in $\mathbb{P}^2$ together with the point $[1, 1, 1]$, write down a basis for the subspace $V \subset H^0(\mathbb{P}^2, \mathcal{O}(6))$ of sextics vanishing to order 2 at those points, and simply take a general linear combination of these basis elements.

Genus 7 presents a new challenge: a general such curve can most simply be represented as a plane octic with eight double points, and those points can't be put in a fixed position. But it's really no problem: if we simply let the points $p_1, \ldots, p_8$ be general, we see that there is a 21-dimensional vector space of octic polynomials vanishing doubly on $p_1, \ldots, p_8$, and as these points vary the corresponding vector spaces form a vector bundle over an open subset of $(\mathbb{P}^2)^8$. Trivialize this vector bundle, and we once more have a family of curves, parametrized by an open subset of an affine space, that includes the general curve of genus 7. Of course, the gunman may get a little anxious at this point: the nodes of the octic plane model of a general curve of genus 7, he may point out, need not be eight general points in the plane. But you can handle this one: simply refer him to the paper of Arbarello and Cornalba [4], where this is verified. Moreover, analogous constructions work as well for genera $g = 8$, 9 and 10.

Unfortunately, it stops working with $g = 11$. Chang-Ran [22] and Sernesi [137] prove the existence of such families for genera 11, 12 and 13, but far more subtle methods are needed for these cases, and that's the end of the line, at least as far as our present knowledge is concerned. If your mugger wants more, you might as well tell him to go ahead and shoot.

What's going on here? Basically, to say that there exists a family of curves, parametrized by an open subset of an affine space, that includes the general curve of genus g, is exactly to say that the moduli space $\mathcal{M}_g$ is *unirational*, that is, there is a dominant rational map from a projective space $\mathbb{P}^N$ to $\mathcal{M}_g$. In particular, it implies that the Kodaira dimension of $\overline{\mathcal{M}}_g$ is negative; that is, there are no pluricanonical forms on $\overline{\mathcal{M}}_g$. Thus, one consequence of Theorem (6.59) is the fact that for $g \geq 23$, such a family cannot exist.

Other facts are known about the birational geometry of $\overline{\mathcal{M}}_g$ for small values of g. It's known to be rational for $g = 2$, 4 and 6 (see the articles of Dolgachev [35] and Shepherd-Barron [141]). In addition, Kollár and Schreyer [101] prove that $\overline{\mathcal{M}}_g$ is actually rational for all $g \leq 6$. Also, it's known that $\mathcal{M}_{15}$ has negative Kodaira dimension. The proofs employ a variety of methods but there doesn't seem to be much chance of using similar ideas to fill in the missing g. We conjecture, however, that *for $g \leq 22$, $\mathcal{M}_g$ has Kodaira dimension $-\infty$.*

More precisely, the authors conjecture that the third condition in the criterion at the start of the next subsection holds for such g. This would have a number of other consequences of interest (for further details, see [81]). But whatever our guesses about the intermediate cases, Theorem (6.59) remains the extent of our knowledge about the cases where $\overline{\mathcal{M}}_g$ isn't unirational, and the remainder of this section will be devoted, more or less, to its proof.

Basic ideas

It turns out that the computations required for the proof of Theorem (6.59) are significantly simpler if we assume that $g + 1$ is a composite integer. Since this case reveals all of the ideas needed in general, we'll deal only with it, using a strategy laid out by Eisenbud and Harris in [41]. The variants needed to deal with general g can also be found there.

We should establish one point of notation before we begin. The essential ingredient in our argument is a calculation in the Picard group of the moduli space $\overline{\mathcal{M}}_g$; and while the divisors we'll be considering will come to us as subschemes $D \subset \overline{\mathcal{M}}_g$ of the moduli space, we'll find it much more convenient to carry out the necessary calculations in the group $\mathrm{Pic}_{\mathrm{fun}}(\overline{\mathcal{M}}_g) \otimes \mathbb{Q}$ of rational divisor classes on the moduli stack, and to express the results in terms of the standard generators λ and δ_i of $\mathrm{Pic}_{\mathrm{fun}}(\overline{\mathcal{M}}_g) \otimes \mathbb{Q}$. Since it would be burdensome to introduce a separate symbol each time, we'll abuse notation and use the same letter D to denote an effective divisor $D \subset \overline{\mathcal{M}}_g$ and the class in $\mathrm{Pic}_{\mathrm{fun}}(\overline{\mathcal{M}}_g) \otimes \mathbb{Q}$ associated to it in Proposition (3.91). (Recall that by Proposition (3.92), this coincides with the the class $\pi^*([D]) \in \mathrm{Pic}_{\mathrm{fun}}(\overline{\mathcal{M}}_g) \otimes \mathbb{Q}$ associated to $[D] \in \mathrm{Pic}(\overline{\mathcal{M}}_g)$ by the isomorphism in Proposition (3.88) except in the cases of genus 2, of the divisor Δ_1 in general, and of the divisor $H_3 \subset \overline{\mathcal{M}}_3$ of hyperelliptic curves of genus 3.)

With this said, the starting point of our analysis is a criterion that relates the Kodaira dimension of $\overline{\mathcal{M}}_g$ to the existence of certain effective divisors $D \subset \overline{\mathcal{M}}_g$.

CRITERION (6.60) *For any effective divisor $D \subset \overline{\mathcal{M}}_g$, express the class of D in terms of standard classes as*

$$D = a\lambda - \sum_{i=0}^{\lfloor g/2 \rfloor} b_i \delta_i.$$

1) $\mathcal{M}_g$ is of general type if there exists an effective divisor D with

$$\frac{a}{b_i} < \frac{13}{2} \quad \text{for all } i, \text{ and} \quad \frac{a}{b_1} < \frac{13}{3}.$$

2) $\mathcal{M}_g$ has Kodaira dimension ≥ 1 *if there are two effective divisors with distinct support in $\mathcal{M}_g$ satisfying the weaker inequalities*

$$\frac{a}{b_i} \leq \frac{13}{2} \quad \text{for all } i, \text{ and,} \quad \frac{a}{b_1} \leq \frac{13}{3}.$$

3) $\mathcal{M}_g$ has Kodaira dimension $-\infty$ *if there is no effective divisor satisfying the inequality*

$$\frac{a}{b_0} \leq \frac{13}{2}.$$

The general type statement follows almost immediately from two facts established earlier. The first is the computation of the canonical class of $\overline{\mathcal{M}}_g$ in (3.113):

$$K_{\overline{\mathcal{M}}_g} = 13\lambda - 2[\Delta_0] - \frac{3}{2}[\Delta_1] - 2[\Delta_2] - \cdots$$

$$= 13\lambda - 2\delta_0 - 3\delta_1 - 2\delta_2 - \cdots.$$

The second is the calculation of the cone of ample divisors in Theorem (6.40), which shows that $a\lambda - b\delta$ is ample on $\overline{\mathcal{M}}_g$ whenever $a > 11b > 0$. Together these show that if there is an effective divisor D as in the criterion, then for suitably divisible m we can find an effective divisor E and an ample divisor H such that

$$K_{\overline{\mathcal{M}}_g}^{\otimes m} = H + E.$$

In particular, this shows that the Hilbert function

$$h^0(\overline{\mathcal{M}}_g, K_{\overline{\mathcal{M}}_g}^{\otimes m})$$

has order in m at least that of the ample divisor H: this is just another way to say that this order is maximal, or, equivalently, that $\overline{\mathcal{M}}_g$ is of general type.

EXERCISE (6.61) Prove the second and third assertions of the criterion.

There is one other point that needs to be addressed before we may conclude Criterion (6.60). As we remarked when it was first introduced, since $\overline{\mathcal{M}}_g$ is singular, it doesn't have a canonical bundle per se; the canonical bundle on $\overline{\mathcal{M}}_g$ is defined simply as the unique (rational) line bundle on $\overline{\mathcal{M}}_g$ extending the canonical bundle on its smooth locus. There is thus no guarantee that a global regular section of a power of $K_{\overline{\mathcal{M}}_g}$ will yield a pluricanonical form on a desingularization of $\overline{\mathcal{M}}_g$. In order to ensure that this is in fact the case, we need to study more closely the singularities of $\overline{\mathcal{M}}_g$. What must be checked is what was stated classically as the property that "the singularities of $\overline{\mathcal{M}}_g$

don't impose adjunction conditions", or, in the language of contemporary birational geometry, that "$\overline{\mathcal{M}}_g$ has only *terminal singularities*". Fortunately, the Reid-Tai criterion (cf., [134] and [145]) provides a very effective method of checking whether any finite quotient singularity — and recall that all singularitites of $\overline{\mathcal{M}}_g$ are of this type — is terminal. We will give no details here and simply refer to Mumford's argument in [82]. You should be aware, however, that this verification involves some lengthy and nontrivial combinatorial complications, since, for each g, we find a different menagerie of such singularities on $\overline{\mathcal{M}}_g$. Indeed, the last step in the argument requires a computer verification whose Basic program listing must surely be the only one ever to appear in *Inventiones*!

The idea of the proof of Theorem (6.59) is clear: show that for all $g \geq 24$ there are divisor classes on $\overline{\mathcal{M}}_g$ that satisfy the first part of the criterion. Those that are easiest to work with are usually dubbed *Brill-Noether* divisors. Informally, a Brill-Noether divisor is the locus of curves that carry a g_d^r for which the Brill-Noether number $\rho = -1$. More carefully, these are the union of the codimension 1 components of the closure of the locus of smooth curves possessing such a linear series. The one defect these divisors have is that they exist only for certain g. Since we're assuming that

$$\rho = g - (r+1)(g-d+r) = -1,$$

$g + 1$ must be composite. This is why our proof, which will use only these divisors, applies only in this case. For other g, their role can be taken by certain *Petri divisors* but the computations become much more complicated. Again, see [41]. We will loosely refer to loci of curves possessing exceptional linear series as loci of "special" curves.

We can rewrite the condition $\rho = -1$ in terms of r and the projective dimension $s = g - d + r - 1$ of the linear series residual to the given one in the canonical series as

$$g = (r+1)(s+1) - 1.$$

Under this assumption, d, r and s are also related by

$$d = r(s+2) - 1.$$

Of course, in view of these constraints, once g is fixed any of the quantities r, d and s determines the other two. However, it'll simplify statements of several propositions to index these divisors by both r and s. We will thus define $D_s^r \subset \overline{\mathcal{M}}_g$ to be union of the codimension 1 components of the closure of the locus of smooth curves possessing such a g_d^r.

The aim of the remainder of this section is to compute the class of D_s^r, up to a positive rational multiple, for all (r,s) with $r, s \geq 1$ or, more precisely, to prove:

THEOREM (6.62) (BRILL-NOETHER RAY THEOREM) *Whenever $s \geq 3$ and $g = (r + 1)(s - 1) - 1$, the class of D_s^r on $\overline{\mathcal{M}}_g$ is given, for some rational number $c > 0$, by*

$$D_s^r = c\left(\left(g + 3\right)\lambda - \left(\frac{g+1}{6}\right)\delta_0 - \sum_{i=1}^{\lfloor g/2 \rfloor}\left(i(g - i)\right)\delta_i\right).$$

Note the remarkable fact that the coefficients (apart from c) depend only on g, not on r or s.

A little arithmetic with the coefficients should quickly convince you that the only ratio a/b_i in this expression for D_s^r that is substantially larger than 1 for large g is

$$\frac{a}{b_0} = \frac{g + 3}{\left(\frac{g+1}{6}\right)} = 6 + \frac{12}{g + 1}.$$

This is less than $13/2$ for $g \geq 24$ so we're done by applying criterion (6.60) for such (composite) g.

When $g = 23$, $a/b_0 = 13/2$. We can therefore only conclude that the Kodaira dimension is positive and then only if we find two D_s^r's that have distinct support. Since the Hurwitz scheme is irreducible, the locus of smooth curves of genus 23 possessing a g_{13}^1 is of pure codimension 1; i.e., D_{13}^1 contains every such curve. Thus, we can show that D_{13}^1 isn't contained in D_9^2 by producing a smooth curve with a g_{13}^1 but no g_{17}^2. This in turn follows from the existence of a curve of compact type possessing a smoothable (i.e., dimensionally proper) limit g_{13}^1 but no limit g_{17}^2. Exercise (5.67) constructs such a curve.

For genera $g < 23$, according to Theorem (6.62), $a/b_0 < 13/2$ and so the first two parts of Theorem (6.62) give no information. In the opposite direction. However, all known examples suggest that Brill-Noether divisors minimize this ratio amongst all effective ones:

CONJECTURE (6.63) (SLOPE CONJECTURE, [81]) *If D is any effective divisor on $\overline{\mathcal{M}}_g$, the ratio*

$$\frac{a}{b_0} \geq 6 + \frac{12}{g + 1}.$$

Applying the third part this would imply that $\mathcal{M}_g$ has Kodaira dimension $-\infty$ (at least modulo analogous results for the other ratios). The conjecture is known to be true only for $g \leq 6$.

Theorem (6.62) is deduced by studying the pullbacks of D_s^r to smaller spaces. These pullbacks lie in certain special subloci and general results show that the coefficients of divisors whose pullbacks lie in these subloci satisfy various relations. Together these relations are enough to yield Theorem (6.62).

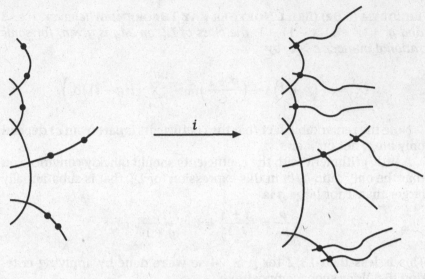

FIGURE (6.64)

The first space we use is the moduli space $\overline{\mathcal{M}}_{0,g}$ of stable g-pointed rational curves. We let $i : \overline{\mathcal{M}}_{0,g} \to \overline{\mathcal{M}}_g$ be the map obtained by attaching a copy of a fixed pointed elliptic curve at each of the g marked points as in Figure (6.64). The second space we use is $\overline{\mathcal{M}}_{2,1}$, the moduli space of stable one-pointed curves of genus 2 equipped with the map $j : \overline{\mathcal{M}}_{2,1} \to \overline{\mathcal{M}}_g$ obtained by attaching a fixed general smooth one-pointed curve of genus $g - 2$ at the marked point.

It seems to be rather common that loci of special curves in $\overline{\mathcal{M}}_g$ meet $j(\overline{\mathcal{M}}_{2,1})$ only along the closure $W \subset \overline{\mathcal{M}}_{2,1}$ of the locus in which the marked point is a Weierstrass point of the underlying curve. This is the case for both the D_s^r and the Petri divisors mentioned above. Similarly, the curves in $i(\overline{\mathcal{M}}_{0,g})$ seem to be rather general. This time D_s^r — but not the more general Petri divisors — misses $i(\overline{\mathcal{M}}_{0,g})$ entirely. These yield relations on coefficients by applying:

THEOREM (6.65) *Let $D \subset \overline{\mathcal{M}}_g$ be an effective divisor, with class*

$$D = a\lambda - \sum_{i=0}^{\lfloor g/2 \rfloor} b_i \delta_i$$

1) *If j^*D is supported on W, then*

$$a = 5b_1 - 2b_2 \quad \text{and} \quad b_0 = \frac{b_1}{2} - \frac{b_2}{6}.$$

Further, if we write $j^*D = qW$ for some (rational) number q, then $b_2 = 3q$.

2) *If $i^*D = 0$, then*

$$b_i = \frac{i(g-i)}{g-1}b_1 \quad \text{for} \quad i = 2,\dots,\lfloor\tfrac{g}{2}\rfloor.$$

EXERCISE (6.66) Show that if a divisor D satisfies the relations in both parts of this theorem, then it satisfies Theorem (6.62) for some c.
Hint: Use the second relation to write b_2 in terms of b_1. Then use the first to show that $a/b_0 = 6 + (12/(g+1))$. Then show that if $a = g+3$, then $b_0 = (g+1)/6$ and $b_1 = 1$. The remaining coefficients are then immediate from the second set of relations.

Thus, three tasks remain. First, show that the divisors D_s^r meet $j(\overline{\mathcal{M}}_{2,1})$ only along the closure of the image of W and miss $i(\overline{\mathcal{M}}_{0,g})$ entirely. Second, show that the constant of proportionality c in Theorem (6.62) is in fact positive. These will follow from considerations about limit linear series that follow fairly directly from the results of Chapter 5. We do this in the next subsection. Third, we must prove Theorem (6.65). This will take most of the effort and be carried out in the last two subsections.

Pulling back the divisors D_s^r

In this section, we'll apply results about limit linear series to reduce Theorem (6.62) to Theorem (6.65).

The key point is that the treelike curves in D_s^r are limits of smooth curves possessing certain linear series with negative ρ, so they all possess generalized crude limit series with negative ρ by Exercise (5.40).

On the other hand, no curve in $i(\overline{\mathcal{M}}_{0,g})$ possesses a series with negative ρ by the argument on page 275. Hence, no curve in D_s^r can lie in $i(\overline{\mathcal{M}}_{0,g})$. For the same reason, D_s^r cannot contain any treelike curve in $j(\overline{\mathcal{M}}_{2,1} - W)$. But the generic points of the boundary components of $\overline{\mathcal{M}}_{2,1}$ are seen in Figure (6.67) where all components have elliptic

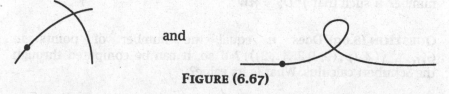

and

FIGURE (6.67)

normalizations, and these are treelike curves. Thus the locus of non-treelike curves is of codimension > 1 in $j(\overline{\mathcal{M}}_{2,1})$, and the intersection of $j(\overline{\mathcal{M}}_{2,1} - W)$ with a divisor, were it nonempty, could not consist only of non-treelike curves. Thus, we've shown all but the last assertion of:

PROPOSITION (6.68) *If $g = (r+1)(s+1)$ with $r, s \geq 1$, then the divisor $D_s^r \subset \overline{\mathcal{M}}_g$ doesn't meet either $i(\overline{\mathcal{M}}_{0,g})$ or $j(\overline{\mathcal{M}}_{2,1} \setminus W)$. Moreover, $j^*(D_s^r)$ is a positive multiple of the class of W.*

This provides the necessary reduction. The first statement shows that Theorem (6.62) must be true for some c by Exercise (6.66). By Theorem (6.65), the last statement shows that the coefficient cb_2 of δ_2 in Theorem (6.62) is positive. Since D_s^r is effective, this, in turn, implies that $c > 0$ as required.

To verify the last statement of the proposition, we'll show that if (Y, p) is a curve of genus 2 with Weierstrass point p and (Z, p) is a general pointed curve of genus $g - 2$ with $g = (r + 1)(s + 1) - 1$, then the curve $C = Y \cup_p Z$ possesses a smoothable limit $g_{r(s+2)-1}^r$ that extends to a codimension 1 family of nearby smooth curves. This will show that C, which is a general point of $j(W)$, lies in D_s^r as required.

We will construct the desired limit series aspect by aspect. Leaving the easy case $r = s = 1$ to you, we may assume $r \geq 2$ or $s \geq 2$, so that $rs + r - 3 \geq 0$. On Y we take the aspect V_Y to be $|(r+2)p| + (rs+r-3)p$. One computes easily that

$$b(V_Y, p) = (rs - r - 3, \ldots, rs - r - 3, rs - r - 2, rs - r - 1),$$

and hence that $\rho(L_Y, p) = -1$, and V_Y is dimensionally proper with respect to p.

A g_{rs-1}^r on Z with ramification sequence $(0, 1, 2, \ldots, 2)$ at p will have adjusted Brill-Noether number 0. By Theorem (5.37), there are finitely many (dimensionally proper) g_{rs-1}^r's on Z with this ramification sequence at p. We may take any of these to be V_Z.

We have $\rho(L) = -1$ by additivity. Since both the aspects constructed above are dimensionally proper, the discussion of smoothings on page 267 shows that this limit linear series smooths to a codimension 1 family. Thus $Y \cup_p Z = j(Y) \in D_s^r$, as required.

Remark. The constant c of Theorem (6.62) can be computed from the number n such that $j^*D_s^r = nW$.

QUESTION (6.69) Does n equal the number of points in $G_{r(s+2)-1}^r(Z, (p, (0, 1, 2, \ldots, 2)))$? If so, it can be computed through the Schubert calculus. What is its value?

Divisors on $\overline{\mathcal{M}}_g$ that miss $j(\overline{\mathcal{M}}_{2,1} \setminus W)$

Next, we want to verify the first part of Theorem (6.65). We claim this follows from the following computation of W in terms of the standard divisor classes $\overline{\mathcal{M}}_{2,1}$.

PROPOSITION (6.70) *W is irreducible, and its class in $\overline{\mathcal{M}}_{2,1}$ is given by*

$$W = 3\omega - \lambda - \delta_1.$$

Let's admit this for a moment and complete the argument. To do this, we need to know the pullbacks to $\overline{\mathcal{M}}_{2,1}$ of the standard classes on $\overline{\mathcal{M}}_g$. Clearly, a curve in $\Delta_0(\overline{\mathcal{M}}_g) \cap j(\overline{\mathcal{M}}_{2,1})$ must come from a curve in $\Delta_0(\overline{\mathcal{M}}_{2,1})$ so $j^*(\delta_0) = \delta_0$. Likewise, $j^*(\delta_1) = \delta_1$. Next, $j^*(\delta_i) = 0$ if $i \geq 3$ since then $\Delta_i(\overline{\mathcal{M}}_g)$ is disjoint from $j(\overline{\mathcal{M}}_{2,1})$. To determine $j^*(\delta_2)$, observe that, given any family of genus 2 curves $\pi : \mathcal{C} \to B$ with a marked point given by a section $\sigma : B \to \mathcal{C}$ of π, $j^*(\delta_2)|_B$ is the pullback from $\overline{\mathcal{M}}_g$ of the normal bundle $\mathcal{O}_{\delta_2}(\delta_2)$. This may in turn be identified with the normal bundle to the section σ. Therefore, by adjunction $j^*(\delta_2) = -\omega$. Finally, if C is a curve of compact type with components C_i, then $H^0(C, \omega_C)$ is naturally the direct sum $\oplus_i H^0(C_i, \omega_{C_i})$. In our situation, this means that $j^*(\lambda_{\overline{\mathcal{M}}_g}) = \lambda_{\overline{\mathcal{M}}_{2,1}}$. Since λ, δ_0 and δ_1 on $\overline{\mathcal{M}}_{2,1}$ are pullbacks of the analogous classes on $\overline{\mathcal{M}}_2$, where, by Exercise (3.143), they satisfy the relation

$$\lambda = \frac{1}{10}\delta_0 + \frac{1}{5}\delta_1,$$

this relation will continue to hold on $\overline{\mathcal{M}}_{2,1}$.

Using the pullbacks, we see that if $j^*D = qW$ then the coefficients of D satisfy the relation

$$a\lambda - b_0\delta_0 - b_1\delta_1 + b_2\omega = q(3\omega - \lambda - \delta_1)$$

in $\mathcal{A}(\overline{\mathcal{M}}_{2,1})$. This immediately gives $b_2 = 3q$ as claimed. Moreover, making this substitution for q and the one above for λ gives

$$\frac{a}{10} - b_0 = \frac{b_2}{30} \quad \text{and} \quad \frac{a}{5} - b_1 = -\frac{2}{5}b_2$$

from which the relations in the first part of Theorem (6.65) follow by solving for a and b_0.

We are thus left with the proposition. Recall that W is defined as the closure of $W \cap \overline{\mathcal{M}}_{2,1}$, the locus of Weierstrass points on smooth genus 2 curves. By the usual construction of curves of genus 2 as hyperelliptic covers, the monodromy of $W \to \overline{\mathcal{M}}_{2,1}$ is transitive, and thus W is irreducible.

As for the class of W, it's enough to prove the relation of (6.70) after restricting to families

$$\pi : \mathcal{C} \to B, \quad \sigma : B \to \mathcal{C}$$

of stable genus 2 curves pointed by a section σ for which B is a complete smooth curve. Further, we may harmlessly assume that B avoids

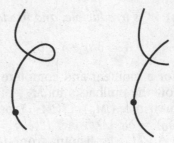

FIGURE (6.71)

any codimension 2 phenomena in $\overline{\mathcal{M}}_{2,1}$ that would be inconvenient. Thus, we may assume that all the finitely many singular fibers C_b of $\mathcal{C}/B$ are of the forms given in Figure (6.71). We let R be the finite set of points of B where the fiber is of the reducible type on the right. For such a fiber C_b, we let Z_b be the component of C_b containing $\sigma(b)$ and Y_b be the other component, and let p and q denote the points of Y and Z respectively that are identified.

The key to our argument is that we can identify the Weierstrass points on any fiber C_b in terms of ramification of the corresponding canonical or limit canonical series. A smooth point of a smooth curve is a Weierstrass point if it's a ramification point of the canonical series. On the other hand, Theorem (5.49) shows that, on a reducible curve $X_0 = Y \cup Z/p \sim q$ consisting of two smooth components meeting at a point that is *not* a Weierstrass point on Y, a smooth point $s \in Z \setminus \{q\}$ is the limit of Weierstrass points on nearby smooth curves if and only if s is a ramification point of the Z-aspect of the (unique) limit canonical series on X.

Thus, the condition that $\sigma(b)$ be a Weierstrass point on C_b can be reexpressed as a degeneracy condition on the matrix giving the Taylor expansion of the sections in the canonical series (or its limiting aspects). Our next goal is to fit these matrices together into map of bundles over B. To do this, we first let

$$\omega_{\lim} = \omega_{C/B}\left(-\sum_{b \in R} Z_b\right).$$

Thus, $\omega_{\lim}|_{C_b} = \omega_{C_b}$ if C_b is irreducible. If C_b is reducible, then $\omega_{\lim}|_{Z_b} = \mathcal{O}_{Z_b}(2q)$ (i.e., the restriction is the Z-aspect of the limit canonical series), and $\omega_{\lim}|_{Y_b} = \mathcal{O}_{Y_b}$. Then, let

$$\mathcal{E} = \pi_* \omega_{\lim}.$$

Since $h^0(C_b, \omega_{\lim}|_{C_b}) = 2$ for all b, $\mathcal{E}$ is a rank 2 vector bundle on B.

Next, let $\Sigma = \sigma(B)$ be the section, $\mathcal{I}$ its ideal sheaf, and

$$\mathcal{F} = \pi_*\left(\omega_{\lim} \otimes \mathcal{O}_C/\mathcal{I}^2\right).$$

It's easy to see that $\mathcal{F}$ is also a vector bundle of rank 2: in fact, we have an exact sequence

$$0 \longrightarrow \sigma^* \omega_{\lim} \otimes \omega \longrightarrow \mathcal{F} \longrightarrow \sigma^* \omega_{\lim} \longrightarrow 0,$$

where we've written ω for the line bundle $\sigma^*(\mathcal{I}/\mathcal{I}^2) = \sigma^*(\omega_{C/B})$. We have a natural map

$$\omega_{\lim} \longrightarrow \omega_{\lim} \otimes \mathcal{O}_C / \mathcal{I}^2,$$

which induces an "evaluation map"

$$\varphi : \mathcal{E} \longrightarrow \mathcal{F}.$$

The description of Weierstrass points as ramification points of (limit) canonical series tells us that the degeneracy locus of φ is W. Thus $W = c_1(\mathcal{F}) - c_1(\mathcal{E})$, and it remains to compute $c_1(\mathcal{E})$ and $c_1(\mathcal{F})$.

For $\mathcal{F}$ this is immediate: we obviously have $\sigma^* \omega_{\lim} = \omega(-\delta_1)$, so from the exact sequence above

$$c_1(\mathcal{F}) = 3\omega - 2\delta_1.$$

To evaluate $c_1(\mathcal{E})$ we use the sequence

$$0 \longrightarrow \omega_{\lim} \longrightarrow \omega_{C/B} \longrightarrow \sum_{b \in R} (\omega_{C/B})|_{Z_b} \longrightarrow 0,$$

which pushes forward on B to an exact sequence

$$0 \longrightarrow \mathcal{E} \longrightarrow \pi_* \omega_{C/B} \longrightarrow \mathcal{O}_{\delta_1}|_B$$

since

$$(\omega_{C/B})|_{Z_b} = \mathcal{O}_{Z_b}(q)$$

and

$$H^0(\mathcal{O}_{Z_b}(q)) \cong \mathbb{C}.$$

We claim that for each b lying in δ_1 (i.e., with reducible fiber), the map

$$\pi_* \omega_{C/B} \longrightarrow H^0((\omega_{C/B})|_{Z_b}) = \mathbb{C}$$

is onto — that is, that a nonzero section of $(\omega_{C/B})|_{Z_b}$ extends to a neighborhood of C_b in $\mathcal{C}$. This holds since $\pi_* \omega_{C/B}$ is a vector bundle with fiber

$$H^0((\omega_{C/B})|_{Y_b}) \oplus H^0((\omega_{C/B})|_{Z_b})$$

over b. Thus we get an exact sequence

$$0 \longrightarrow \mathcal{E} \longrightarrow \pi_* \omega_{C/B} \longrightarrow \mathcal{O}_{\delta_1}|_B \longrightarrow 0,$$

which shows that $c_1(\mathcal{E}) = \lambda - \delta_1$ whence the desired relation

$$W = c_1(\mathcal{F}) - c_1(\mathcal{E}) = 3\omega - \lambda - \delta_1.$$

Divisors on $\mathcal{M}_g$ that miss $i(\overline{\mathcal{M}}_{0,g})$

We have now reached the final step: verifying the second part of Theorem (6.65). This amounts to finding relations on the classes $i^*\lambda$ and $i^*\delta_i$. The first is easy. On any family $\pi : \mathcal{X} \to B$ of curves of genus g formed by attaching fixed elliptic tails to curves in $\overline{\mathcal{M}}_{0,g}$ at the marked points, the vector bundle $\pi_* \omega_{C/B}$ is trivial, so $i^*\lambda = 0$. Also, since $i(\overline{\mathcal{M}}_{0,g})$ misses δ_0 we have $i^*\delta_0 = 0$. To obtain relations amongst the higher δ_i's we express these classes in terms of certain classes ε_i on $\overline{\mathcal{M}}_{0,g}$ defined as follows.

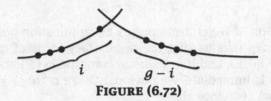

i $g - i$

FIGURE (6.72)

For $i = 2,\ldots,\lfloor g/2 \rfloor$, we take ε_i to be the class of the divisor that is the closure in $\overline{\mathcal{M}}_{0,g}$ of the set of two-component curves with exactly i of the g marked points on one of the components as illustrated schematically in Figure (6.72).

EXERCISE (6.73) Consider the birational map $\overline{\mathcal{M}}_{0,g} \to \mathbb{P}^{g-3}$ taking a smooth curve to the moduli of the marked points. Show that the divisors ε_i contract to distinct lower-dimensional subvarieties under this map and deduce that they are all independent.

For $i \geq 2$, we have $i^*\delta_i = \varepsilon_i$. We compute $i^*\delta_1$. by showing that

$$\textbf{(6.74)} \qquad i^*\delta_1 = - \sum_{i=2}^{\lfloor g/2 \rfloor} \frac{i(g-i)}{(g-1)} \varepsilon_i .$$

If, then, D is any divisor, given in terms of standard classes as in Theorem (6.65), it will pull back on $\overline{\mathcal{M}}_{0,g}$ to

$$-b_1 \left(- \sum_{i=2}^{\lfloor g/2 \rfloor} \frac{i(g-i)}{(g-1)} \varepsilon_i \right) - \sum_{i=2}^{\lfloor g/2 \rfloor} b_i \varepsilon_i .$$

If, in addition, D misses $i(\overline{\mathcal{M}}_{0,g})$ this pullback must be 0 and equating coefficients immediately gives the claimed relations on the coefficients b_i.

As usual, it suffices to check (6.74) after restricting to families

$$\pi : \mathcal{C} \to B, \quad \sigma_1,\ldots,\sigma_g : B \to \mathcal{C}$$

of stable rational g-pointed curves, where B is a smooth curve missing any inconvenient codimension 1 loci in $\overline{\mathcal{M}}_{0,g}$, and transverse to relevant codimension 1 loci in $\overline{\mathcal{M}}_{0,g}$. We can thus assume that all reducible fibers of $\mathcal{C}$ have exactly two components, the general fiber is a smooth curve, and the total space $\mathcal{C}$ is a smooth surface. Fix g pointed elliptic curves (E_i, p_i), and let $\mathcal{C}' \rightarrow B$ be the family obtained by attaching a copy of $B \times E_i$ along σ_i and $B \times p_i$. The family $\mathcal{C}' \rightarrow B$ lies in the g-fold self-intersection of the normal crossing divisor δ_1, and, by Proposition (3.32), $i^* \delta_1$ is thus the sum of the pullbacks of the normal bundles to the branches.

At the point of Δ_1 corresponding to a fiber C'_b of $\mathcal{C}'$, the branch corresponding to the ith node has normal bundle equal to $T_{\sigma_i(b), C_b} \otimes T_{p_i, E}$. Thus it pulls back on B to the normal bundle to the section $\sigma_i(B)$, which we may rewrite as $\pi_*(\sigma_i(B))^2$. Thus

$$i^* \delta_1 = \pi_* \Big(\sum_{i=1}^{g} \sigma_i(B)^2 \Big).$$

We may contract the component of each reducible fiber meeting the smaller number of sections (or either component if both components meet $g/2$ sections) to obtain a $\mathbb{P}^1$-bundle $\tilde{\pi} : \tilde{\mathcal{C}} \rightarrow B$ with g sections $\tilde{\sigma}_i : B \rightarrow \overline{\mathcal{C}}$. These sections meet transversely in groups of i over points of ε_i, and are otherwise disjoint. Thus,

$$\tilde{\pi}_* \Big(\sum_{i=1}^{g} \tilde{\sigma}_i(B)^2 \Big) = \pi_* \Big(\sum_{i=1}^{g} \sigma_i(B)^2 \Big) + \sum_{i=2}^{\lfloor g/2 \rfloor} i \varepsilon_i.$$

On any $\mathbb{P}^1$-bundle the difference of two sections is a linear combination of fibers, and thus has self-intersection 0. Applying this remark to $\tilde{\sigma}_i(B) - \tilde{\sigma}_j(B)$ gives the relation $\tilde{\sigma}_i(B)^2 + \tilde{\sigma}_j(B)^2 = 2\tilde{\sigma}_i(B) \cdot \tilde{\sigma}_j(B)$. Summing over all pairs with $i < j$, we get

$$(g-1)\tilde{\pi}_* \Big(\sum_{i=1}^{g} \tilde{\sigma}_i(B)^2 \Big) = 2\tilde{\pi}_* \Big(\sum_{i<j} (\tilde{\sigma}_i(B) \cdot \tilde{\sigma}_j(B)) \Big) = \sum_{i=2}^{\lfloor g/2 \rfloor} (i(i-1)\varepsilon_i).$$

Putting the last three formulas together yields

$$i^* \delta_1 = \sum_{i=2}^{\lfloor g/2 \rfloor} \Big(\frac{i(i-1)}{g-1} \Big) \varepsilon_i - \sum_{i=2}^{\lfloor g/2 \rfloor} (i) \varepsilon_i = - \sum_{i=2}^{\lfloor g/2 \rfloor} \Big(\frac{i(g-i)}{g-1} \Big) \varepsilon_i.$$

We have thus verified (6.74) and completed the proof of Theorem (6.59).

Further divisor class calculations

Here are some further calculations of classes of divisors on $\overline{\mathcal{M}}_g$ you may wish to try. They call for the use of a variety of the techniques that have appeared in this book.

EXERCISE (6.75) In terms of the generators λ, ω and σ_i of the Picard group of $\overline{\mathcal{C}}_g$ as described on page 62 of Section 2.D, find the class of the closure $\mathcal{W} \subset \overline{\mathcal{C}}_g$ of the locus of pairs (C, p) where C is a smooth curve of genus 3 and p is a Weierstrass point of C.

See Cukierman [28] for the answer to this and related questions.

EXERCISE (6.76) 1) Consider the closure in $\overline{\mathcal{M}}_g$ of the locus of smooth curves C of genus g possessing a point p with Weierstrass semigroup $\{g - 1, g + 2, g + 3, \ldots\}$. Show that this locus is indeed a divisor, and find its class.

2) Repeat part 1) for the semigroup $\{g, g + 1, g + 3, g + 4, \ldots\}$.

3) For extra credit, show that the loci described above are the only codimension 1 components of the locus of curves with nonsimple Weierstrass points.

See Diaz [30] for a full treatment of this question.

EXERCISE (6.77) Let $\mathcal{W} \subset \overline{\mathcal{C}}_g$ be as in Exercise (6.75). What is the branch divisor of the projection $\mathcal{W} \rightarrow \overline{\mathcal{M}}_g$, and what does this have to do with the answer to Exercise (6.76)?

EXERCISE (6.78) Find the class of the closure in $\overline{\mathcal{M}}_g$ of the locus of curves C with a semicanonical pencil, that is, a line bundle L with $L^{\otimes 2} \cong \omega_C$ and $h^0(C, L) \geq 2$.

Curves defined over $\mathbb{Q}$

Among the consequences of Theorem (6.59) are (at least conjecturally) some that are arithmetic in nature, and we'll summarize them here. To begin with, for any number field K we'll say that a curve C *may be defined over* K if it can be realized as the zero locus of polynomials with coefficients in K. The moduli space $\overline{\mathcal{M}}_g$ itself may be given by equations with coefficients in $\mathbb{Q}$ in such a way that for any K the set of curves that may be defined over K is (if we exclude curves with automorphisms) just the set $\overline{\mathcal{M}}_g(K)$ of K-rational points of $\overline{\mathcal{M}}_g$.

Now, the fact that we can for small values of g write down a family of curves, parametrized by an open subset of an affine space, that includes the general curve of genus g, implies in particular that the subset $\overline{\mathcal{M}}_g(\mathbb{Q}) \subset \overline{\mathcal{M}}_g$ of curves that may be defined over $\mathbb{Q}$ is dense in $\mathcal{M}_g$. Is this the case in general? Conjecturally not: generalizing the Mordell Conjecture, we have the:

CONJECTURE (6.79) (WEAK LANG CONJECTURE) *If X is a variety of general type defined over a number field K, then the set $X(K)$ of K-rational points of X isn't Zariski dense.*

This, if true, would in conjunction with Theorem (6.59) imply that for $g \geq 24$, all the K-rational points of $\overline{\mathcal{M}}_g$ lie in a proper subvariety! But wait, there's a further conjecture:

CONJECTURE (6.80) (STRONG LANG CONJECTURE) *Let X be a variety of general type, defined over a number field K. There exists a proper closed subvariety $\Sigma \subset X$ such that for any number field L containing K, the set of L-rational points of X lying outside Σ is finite.*

If we believe this, there is for each $g \geq 24$ a subvariety $\Sigma \subset \overline{\mathcal{M}}_g$ such that, for any number field K, all but finitely many curves of genus g defined over K lie in Σ. This raises a host of intriguing questions. First, can we disprove the Lang Conjectures by exhibiting a Zariski-dense collection of curves of genus g defined over $\mathbb{Q}$? Second, if we do believe the Strong Lang Conjecture, what could the minimal such subvariety Σ be? Clearly, it has to contain the hyperelliptic locus, since the rational points are dense there; and a little more effort shows that it'll also contain the trigonal and tetragonal locus, as well as loci of plane curves with small numbers of nodes, complete intersection curves and the like.

At this point, no one has any idea what Σ might look like, if indeed it exists. There are two guesses we might make, though. The first is that Σ is contained in the locus of special curves, that is, curves possessing a linear series with negative Brill-Noether number ρ. In fact, no one has (to our knowledge) written down for large g a single curve, defined over $\mathbb{Q}$, that satisfied the Brill-Noether theorem. The second is that it's the intersection of the base loci of all the pluricanonical linear series of $\overline{\mathcal{M}}_g$: conjecturally (cf. [81]), this second locus is closely related to the first and it's known to contain some very special subloci such as the loci of hyperelliptic and trigonal curves. We leave you with the:

PROBLEM (6.81) Find better evidence that Σ equals either of the candidates above, or find a better candidate for Σ.

Bibliography

[1] ABRAMOVICH, D. & KARU, K. *Weak semistable reduction in characteristic 0.* math.AG/9707012, 1997.

[2] ALUFFI, P. & FABER, C. *Linear orbits of smooth plane curves.* J. Alg. Geom., **2**, 155–184, 1993.

[3] ARBARELLO, E. *Weierstrass points and moduli of curves.* Compositio Math., **29**, 325–342, 1974.

[4] ARBARELLO, E. & CORNALBA, M. *Footnotes to a paper of Beniamino Segre.* Math. Annalen, **256**, 341–362, 1981.

[5] ARBARELLO, E. & CORNALBA, M. *The Picard groups of the moduli spaces of curves.* Topology, **26**, 153–171, 1987.

[6] ARBARELLO, E. & CORNALBA, M. *Calculating cohomology groups of moduli spaces of curves via algebraic geometry.* math.AG/9803001, 1998.

[7] ARBARELLO, E., CORNALBA, M., GRIFFITHS, P. & HARRIS, J. *Geometry of algebraic curves, I.* Number 267 in Grundlehren der mathematischen Wissenschaften (New York: Springer-Verlag), 1985.

[8] ARBARELLO, E. & DECONCINI, C. *On a set of equations characterizing Riemann matrices.* Ann. of Math. (2), **120**, 119–140, 1984.

[9] ARBARELLO, E., DECONCINI, C., KAC, V. & PROCESI, C. *Moduli spaces of curves and representation theory.* Comm. Math. Phys., **117**, 1–36, 1988.

[10] ARTIN, M. *Algebraic spaces.* Number 3 in Yale Mathematical Monographs (New Haven: Yale University Press), 1971.

[11] ASH, A., MUMFORD, D., RAPOPORT, M. & TAI, Y. *Smooth compactification of locally symmetric varieties* (Brookline: Math-Sci. Press), 1975.

[12] BAILY, W. & BOREL, A. *Compactification of arithmetic quotients of bounded symmetric domains.* Ann. of Math. (2), **84**, 442–528, 1966.

[13] BAYER, D. *The division algorithm and the Hilbert scheme*. Ph.D. thesis, Harvard University, 1992.

[14] BEHREND, K., FANTECHI, L., FULTON, W., GOETTSCHE, L. & KRESCH, A. *Introduction to stacks*. In preparation.

[15] BERS, L. *Spaces of Riemann surfaces*. In J. A. Todd, editor, *Proc. Int. Congress of Math., Cambridge, 1958* (Cambridge: Cambridge Univ. Press), 1960.

[16] CAPORASO, L. *A compactification of the universal Picard variety over the moduli space of stable curves*. J. Amer. Math. Soc., 7, 589–660, 1994.

[17] CAPORASO, L. & HARRIS, J. *Counting plane curves of any genus*. Invent. Math., **131**, 345–392, 1998.

[18] CAPORASO, L. & HARRIS, J. *Enumerating curves on rational surfaces: the rational fibration method*. math.AG/9608023, to appear in Compositio Math., 1998.

[19] CAPORASO, L. & HARRIS, J. *Parameter spaces for curves on surfaces and enumeration of rational curves*. math.AG/9608024, to appear in Compositio Math., 1998.

[20] CASTELNUOVO, G. *Numero delle involuzioni razionali gaicenti sopra una curva di dato genere*. Rendi. R. Accad. Lincei (4), **5**, 130–133, 1889.

[21] CASTELNUOVO, G. *Ricerche di geometria sulle curve algebriche*. Atti R. Accad. Sci. Torino, **24**, 196–223, 1889.

[22] CHANG, M.-C. & RAN, Z. *Unirationality of the moduli space of curves of genus 11, 13 (and 12)*. Invent. Math., **76**, 41–54, 1984.

[23] CHANG, M.-C. & RAN, Z. *Closed families of smooth space curves*. Duke Math. J., **52**, 707–713, 1985.

[24] CHANG, M.-C. & RAN, Z. *Dimensions of families of space curves*. Compositio Math., **90**, 53–57, 1994.

[25] CILIBERTO, C. *On rationally determined line bundles on a family of projective curves with general moduli*. Duke Math. J., **55**, 909–917, 1987.

[26] CLEBSCH, A. *Zur Theorie der Riemann'schen Flachen*. Math. Ann., **6**, 216–230, 1872.

[27] CORNALBA, M. & HARRIS, J. *Divisor classes associated to families of stable varieties, with applications to the moduli space of curves*. Ann. Sci. École Norm. Sup. (4), **21**, 455–475, 1988.

[28] CUKIERMAN, F. *Families of Weierstrass points*. Duke Math. J., **58**, 317–346, 1989.

[29] DELIGNE, P. & MUMFORD, D. *The irreducibility of the space of curves of given genus.* Publ. Math. I.H.E.S., **36**, 75-110, 1969.

[30] DIAZ, S. *Exceptional Weierstrass points and the divisor on moduli that they define.* Mem. Amer. Math. Soc., **56**, 1985.

[31] DIAZ, S. & EDIDIN, D. *Towards the homology of Hurwitz spaces.* J. Diff. Geom., **43**, 66-98, 1996.

[32] DIAZ, S. & HARRIS, J. *Geometry of Severi varieties, II: Independence of divisor classes and examples.* In *Algebraic Geometry, 1986 (Sundance, Utah)*, Number 1311 in Lecture Notes in Math., 23-50 (New York: Springer-Verlag), 1988.

[33] DIAZ, S. & HARRIS, J. *Geometry of the Severi variety.* Trans. Amer. Math. Soc., **309**, 1-34, 1988.

[34] DIJKGRAAF, R., VERLINDE, E. & VERLINDE, H. *Loop equations and Virasoro constraints in nonperturbative two-dimensional quantum gravity.* Nuclear Phys. B, **348**, 435-456, 1991.

[35] DOLGACHEV, I. *Rationality of fields of invariants.* In *Algebraic Geometry Bowdoin 1985, Vol. 2*, Number 46 in Proc. Symp. Pure Math, 3-16 (Providence: Amer. Math. Soc.), 1987.

[36] EDIDIN, D. *Picard groups of Severi varieties.* Comm. Algebra, **22**, 2073-2081, 1994.

[37] EISENBUD, D. & HARRIS, J. *Divisors on general curves and cuspidal rational curves.* Invent. Math, **74**, 371-418, 1983.

[38] EISENBUD, D. & HARRIS, J. *A simpler proof of the Gieseker-Petri theorem on special divisors.* Invent. Math, **74**, 269-280, 1983.

[39] EISENBUD, D. & HARRIS, J. *Limit linear series: basic theory.* Invent. Math, **85**, 337-371, 1986.

[40] EISENBUD, D. & HARRIS, J. *Existence, decomposition and limits of certain Weierstrass points.* Invent. Math, **87**, 495-515, 1987.

[41] EISENBUD, D. & HARRIS, J. *The Kodaira dimension of the moduli space of curves of genus* $\geq$ 23. Invent. Math, **90**, 359-387, 1987.

[42] EISENBUD, D. & HARRIS, J. *When ramification points meet.* Invent. Math, **87**, 485-493, 1987.

[43] ENRIQUES, F. & CHISINI, O. *Teoria geometrica delle equazione e delle funzione algebriche*, volume III (Bologna: Zanichelle), 1924.

[44] ESTEVES, E. *Compactifying the relative Jacobian over families of reduced curves.* math.AG/9709009, 1997.

[45] FABER, C. *Chow rings of moduli spaces of curves, I: The Chow ring of* $\overline{\mathcal{M}}_3$. Ann. of Math., **132**, 331-419, 1990.

[46] FABER, C. *Chow rings of moduli spaces of curves, II: Some results on the Chow ring of* $\overline{\mathcal{M}}_4$. Ann. of Math., **132**, 421-449, 1990.

[47] FABER, C. *Algorithms for computing intersection numbers on moduli spaces of curves, with an application to the class of the locus of Jacobians.* math.AG/9706006, 1996.

[48] FABER, C. *A conjectural description of the tautological ring of the moduli space of curves.* Preprint, 1997.

[49] FOGARTY, J. *Invariant theory* (New York: W. Benjamin), 1969.

[50] FOGARTY, J., KIRWAN, F. & MUMFORD, D. *Geometric invariant theory.* Number 34 in Ergebnisse der Mathematik und ihrer Grenzgeibeite, 3. Folge, third edition (New York: Springer-Verlag), 1994.

[51] FRANCHETTA, A. *Sulle serie lineari razionalmente determinate sulla curve a moduli generali di dato genera.* Le Mathematiche, Catania, **9**, 126–147, 1954.

[52] FRIEDMAN, R. *Hodge theory, degenerations and the global Torelli problem.* Ph.D. thesis, Harvard University, 1981.

[53] FULTON, W. *Intersection theory.* Number 2 in Ergebnisse der Mathematik und ihrer Grenzgeibeite, 3. Folge (New York: Springer-Verlag), 1980.

[54] FULTON, W. & PANDHARIPANDE, R. *Notes on stable maps and quantum cohomology.* In J. Kollár, R. Lazarsfeld & D. Morrison, editors, *Algebraic Geometry Santa Cruz 1995,* Number 62.2 in Proc. of Symposia in Pure Math., 45–92 (Amer. Math. Soc.), 1997. math.AG/9608011.

[55] GETZLER, E. *Intersection theory on $\overline{\mathcal{M}}_{1,4}$ and elliptic Gromov-Witten invariants.* J. Amer. Math. Soc., **10**, 973–998, 1997.

[56] GETZLER, E. *Topological recursion relations in genus 2.* math.AG/9801003, 1998.

[57] GIESEKER, D. *Global moduli for surfaces of general type.* Invent. Math, **43**, 233–282, 1977.

[58] GIESEKER, D. *Lectures on moduli of curves.* Number 69 in Tata Institute of Fundamental Research Lecture Notes (New York: Springer-Verlag), 1982.

[59] GIESEKER, D. *Stable curves and special divisors.* Invent. Math, **66**, 251–275, 1982.

[60] GIESEKER, D. *Geometric invariant theory and applications to moduli problems.* In F. Gherardelli, editor, *Invariant theory, Proceedings of the 1st 1982 Session of the C.I.M.E (Montecatini, Italy),* Number 996 in Lecture Notes in Math., 45–73 (New York: Springer-Verlag), 1983.

[61] GIESEKER, D. *A degeneration of the moduli space of stable bundles.* J. Diff. Geom., **19**, 137–154, 1984.

[62] GIESEKER, D. & MORRISON, I. *Hilbert stability of rank two bundles on curves.* J. Diff. Geom, **19**, 1–29, 1984.

[63] GOTZMANN, G. *Eine Bedingung für die Flachheit und das Hilbert-polynom eines graduiertes Ringes.* Math. Zeit., **158**, 61–70, 1978.

[64] GRIFFITHS, P. & HARRIS, J. *Principles of algebraic geometry* (New York: Wiley-Interscience), 1978.

[65] GRIFFITHS, P. & HARRIS, J. *The dimension of the space of special linear series on a general curve.* Duke Math. J., **47**, 233–272, 1980.

[66] GROMOV, M. *Pseudo-holomorphic curves in symplectic manifolds.* Invent. Math., **82**, 307–383, 1985.

[67] GROTHENDIECK, A. *Techniques de construction et théorèmes d'existence en géometrie algébrique, IV: Les schémas de Hilbert.* In *Seminaire Bourbaki, 1960–61, Exposés 205–222*, Exposé 221, 221.1–221.28 (New York: Benjamin), 1966.

[68] GROTHENDIECK, A. *Seminaire de géometrie algébrique, 1: Revêtements étales et groupe fondamental.* Number 224 in Lecture Notes in Math. (Heidelberg: Springer-Verlag), 1973.

[69] HABOUSH, W. *Reductive groups are geometrically reductive.* Ann. of Math. (2), **102**, 67–83, 1975.

[70] HAIN, R. & LOOIJENGA, E. *Mapping class groups and moduli spaces of curves.* In J. Kollár, R. Lazarsfeld & D. Morrison, editors, *Algebraic Geometry Santa Cruz 1995*, Number 62.2 in Proc. of Symposia in Pure Math., 97–142 (Amer. Math. Soc.), 1997.

[71] HALPHEN, G. *Memoire sur la classification des courbes gauches algébriques.* J. École Polytechnique, **52**, 1–200, 1882.

[72] HARER, J. *The second homology group of the mapping class group of an orientable surface.* Invent. Math., **72**, 221–239, 1983.

[73] HARER, J. *Stability of the homology of mapping class groups of orientable surfaces.* Ann. of Math. (2), **121**, 215–251, 1985.

[74] HARER, J. *The cohomology of the moduli space of curves.* In E. Sernesi, editor, *Theory of moduli : lectures given at the 3rd 1985 session of the Centro Internazionale Matematico Estivo (C.I.M.E.) held at Montecatini Terme, Italy, June 21-29, 1985*, Number 1337 in Lecture Notes in Math., 138–221 (New York: Springer-Verlag), 1988.

[75] HARER, J. *The third homology group of the moduli space of curves.* Duke Math. J., **63**, 25–55, 1991.

[76] HARER, J. & ZAGIER, D. *The Euler characteristic of the moduli space of curves.* Invent. Math., **85**, 457–485, 1986.

[77] HARRIS, J. *Galois groups of enumerative problems.* Duke Math. J., **46**, 685–724, 1979.

[78] HARRIS, J. *The genus of space curves.* Math. Ann., **249**, 191–204, 1980.

[79] HARRIS, J. *On the Kodaira dimension of the moduli space of curves, II: The even genus case.* Invent. Math., **75**, 437–466, 1984.

[80] HARRIS, J. *On the Severi problem.* Invent. Math., **84**, 445–461, 1986.

[81] HARRIS, J. & MORRISON, I. *Slopes of effective divisors on the moduli space of curves.* Invent. Math., **99**, 1990, 321–355.

[82] HARRIS, J. & MUMFORD, D. *On the Kodaira dimension of the moduli space of curves, with an appendix by William Fulton.* Invent. Math., **67**, 1982, 23–88.

[83] HARTSHORNE, R. *Connectedness of the Hilbert scheme.* Publ. Math. I.H.E.S., **29**, 5–48, 1966.

[84] HARTSHORNE, R. *Residues and duality.* Number 20 in Lecture Notes in Math. (New York: Springer-Verlag), 1966.

[85] HASSETT, B. *Local stable reduction for plane curve singularities.* In preparation, 1998.

[86] HILBERT, D. *Lie groups: history, frontiers and applications.* Number VIII in Hilbert's invariant theory papers (Brookline: Math Sci. Press), 1978.

[87] HORIKAWA, E. *On deformations of holomorphic maps, I.* J. Math. Soc. Japan, **25**, 372–396, 1973.

[88] HORIKAWA, E. *On deformations of holomorphic maps, II.* J. Math. Soc. Japan, **26**, 647–667, 1974.

[89] KAC, V. & SCHWARZ, A. *Geometric interpretation of the partition function of 2D gravity.* Phys. Letters B, **257**, 329–334, 1991.

[90] KARU, K. *Minimal models and boundedness of stable varieties.* math.AG/9804049, 1998.

[91] KEEL, S. *Intersection theory of the moduli space of stable N-pointed curves of genus 0.* Trans. Amer. Math. Soc., **330**, 545–574, 1992.

[92] KEMPF, G. *Schubert methods with an application to algebraic geometry* (Amsterdam: Publ. Math. Centrum), 1971.

[93] KEMPF, G. *Instability in invariant theory.* Ann. of Math. (2), **108**, 299–316, 1978.

[94] KLEIMAN, S. r-special subschemes and an argument of Severi. Adv. of Math., **22**, 1–23, 1976.

[95] KLEIMAN, S. & LAKSOV, D. *On the existence of special divisors.* Amer. J. Math., **94**, 431–436, 1972.

[96] KLEIMAN, S. & LAKSOV, D. *Another proof of the existence of special divisors.* Acta Math., **132**, 163–176, 1974.

[97] KNUDSEN, F. *The projectivity of the moduli space of stable curves, I.* Math. Scand., **39**, 19–66, 1976. *II*, Math. Scand., **52** 161–199, 1983. *III*, Math. Scand., **52** 200–212, 1983.

[98] KODAIRA, K. *Complex manifolds and deformations of complex structures* (New York: Springer-Verlag), 1986.

[99] KOLLÁR, J. *Projectivity of complete moduli.* J. Diff. Geom., **32**, 235–268, 1990.

[100] KOLLÁR, J. *Rational curves on algebraic varieties.* Number 32 in Ergebnisse der Mathematik und ihrer Grenzgeibeite, 3. Folge (New York: Springer-Verlag), 1996.

[101] KOLLÁR, J. & SCHREYER, F. O. *The moduli of curves is stably rational for $g \leq 6$.* Duke Math. J., **51**, 239–242, 1984.

[102] KONTSEVICH, M. *Intersection theory on the moduli space of curves and the matrix Airy function.* Comm. Math. Phys., **147**, 1–23, 1992.

[103] KONTSEVICH, M. & MANIN, Y. *Gromov-Witten classes, quantum cohomology, and enumerative geometry.* Comm. Math. Phys., **164**, 525–562, 1994.

[104] KONTSEVICH, M. & MANIN, Y. *Quantum cohomology of a product.* Invent. Math., **124**, 313–339, 1996.

[105] KOUVIDAKIS, A. *The Picard group of the universal Picard varieties over the moduli space of curves.* J. Diff. Geom., **34**, 839–850, 1991.

[106] LAUMON, G. & MORET-BAILLY, L. *Champs algébriques.* Prépublication de l'Université de Paris-Sud 92-42, 1992.

[107] LAZARSFELD, R. *Brill-Noether-Petri without degenerations.* J. Diff. Geom., **23**, 299–307, 1986.

[108] LOOIJENGA, E. *Smooth Deligne-Mumford compactifications by means of Prym level structures.* J. Alg. Geom., **3**, 283–293, 1992.

[109] LOOIJENGA, E. *Intersection theory on Deligne-Mumford compactifications after Witten and Kontsevich.* In *Seminaire Bourbaki, 1992-93*, volume 216 of *Astérisque*, Exposé 768, 187–212 (Paris: Socété Mathématique de France), 1993.

[110] LOOIJENGA, E. *On the tautological ring of $\mathcal{M}_g$.* Invent. Math., **121**, 411–419, 1995.

[111] MACAULAY, F. R. *Some properties of enumeration in the theory of modular systems.* Proc. Lond. Math. Soc., **26**, 531–555, 1927.

[112] MESTRANO, N. *Conjecture de Franchetta forte.* Invent. Math., **87**, 365–376, 1987.

[113] MESTRANO, N. & RAMANAN, S. *Poincare bundles for families of curves.* J. Reine Angew. Math., **362**, 169–178, 1985.

[114] MILLER, E. *The homology of the mapping class group.* J. Diff. Geom., **24**, 1–14, 1986.

[115] MOISHEZON, B. *Stable branch curves and branch monodromies.* In *Algebraic Geometry (Proc. Univ. Ill. Chicago Circle, 1980)*, Number 862 in Lecture Notes in Math., 107–192 (New York: Springer-Verlag), 1981.

[116] MORRISON, I. *Projective stability of ruled surfaces.* Invent. Math., **50**, 269–304, 1980.

[117] MULASE, M. *Cohomological structure in soliton equations and Jacobian varieties.* J. Diff. Geom., **19**, 403–430, 1984.

[118] MUMFORD, D. *Further pathologies in algebraic geometry.* Amer. J. Math., **84**, 642–648, 1962.

[119] MUMFORD, D. *Picard groups of moduli problems.* In *Arithmetic algebraic geometry*, Proc. Conf. Purdue Univ. 1963, 33–81 (New York: Harper and Row), 1965.

[120] MUMFORD, D. *Lectures on curves on an algebraic surface.* Number 59 in Princeton Mathematical Series (Princeton: Princeton University Press), 1966.

[121] MUMFORD, D. *Stability of projective varieties.* L'Ens. Math., **23**, 39–110, 1977.

[122] MUMFORD, D. *Towards an enumerative geometry of the moduli space of curves.* In *Arithmetic and geometry, II*, Number 36 in Progress in Math., 271–326 (Boston: Birkhäuser), 1983.

[123] NAGATA, M. *On the fourteenth problem of Hilbert.* Amer. Math. J., **81**, 766–772, 1959.

[124] NAGATA, M. *Lectures on the fourteenth problem of Hilbert.* Number 31 in Tata Institute of Fundamental Research Lecture Notes (New York: Springer-Verlag), 1965.

[125] NOVIKOV, S. *The periodic problem for the Korteweg-deVries equation.* Funct. Anal. Appl., **8**, 236–246, 1974.

[126] PALAMODOV, V. *Deformations of complex spaces.* Russian Math. Surveys, **31**, 129–197, 1976.

[127] PANDHARIPANDE, R. *A compactification over $\overline{\mathcal{M}}_g$ of the universal moduli space of slope-semistable vector bundles.* J. Amer. Math. Soc., **9**, 425–471, 1996.

[128] PANDHARIPANDE, R. *A note on elliptic plane curves with fixed j-invariant.* Proc. Amer. Math. Soc., **125**, 3471–3479, 1997.

[129] PANDHARIPANDE, R. *Intersections of $\mathbb{Q}$-divisors on Kontsevich's moduli space $\overline{\mathcal{M}}_{0,n}(\mathbb{P}^R, d)$ and enumerative geometry.* math.AG/9504004, to appear in Trans. Amer. Math. Soc., 1998.

[130] PIENE, R. & SCHLESSINGER, M. *On the Hilbert scheme compactification of the space of twisted cubics.* Amer. J. Math., **107**, 761–774, 1985.

[131] POPP, H. *Classification of algebraic varieties and compact complex manifolds.* Number 412 in Lecture Notes in Math. (New York: Springer-Verlag), 1974.

[132] POPP, H. *Moduli Theory and Classification Theory of Algebraic Varieties.* Number 620 in Lecture Notes in Math. (New York: Springer-Verlag), 1977.

[133] RAN, Z. *Enumerative geometry of singular plane curves.* Invent. Math, **97**, 447–465, 1989.

[134] REID, M. *Young person's guide to canonical singularities.* In *Algebraic Geometry, Bowdoin 1985*, Number 46 in Proc. of Symposia in Pure Math., 345–414 (Providence: Amer. Math. Soc.), 1987.

[135] RUAN, Y. & TIAN, G. *A mathematical theory of quantum cohomology.* J. Diff. Geom, **45**, 259–367, 1995.

[136] SATAKE, I. *On the compactification of the Siegel space.* J. Indian Math. Soc., **20**, 259–281, 1956.

[137] SERNESI, E. *L'unirazionalità della varietà dei moduli delle curve di genere dodici.* Ann. Sc. Norm. Sup. Pisa, Cl. Sci. IV, **8**, 405–439, 1981.

[138] SERRE, J.-P. *Faisceaux algébriques cohérents.* Ann. of Math. (2), **61**, 197–278, 1955.

[139] SESHADRI, C. S. *Geometric reductivity over an arbitrary base.* Adv. in Math., **26**, 225–274, 1977.

[140] SEVERI, F. *Vorlesungen über Algebraische Geometrie, Anhang F* (Leipzig: Teubner), 1921.

[141] SHEPHERD-BARRON, N. *Rationality of moduli spaces via invariant theory.* In *Topological methods in algebraic transformation groups*, Number 80 in Progress in Math., 153–164 (Boston: Birkhäuser), 1989.

[142] SHIOTA, T. *Characterization of Jacobian varieties in terms of soliton equations.* Invent. Math., **83**, 333–382, 1986.

[143] STANKOVA-FRENKEL, Z. *Moduli of trigonal curves.* math.AG/9710015, 1997.

[144] STANLEY, R. *Hilbert functions of graded algebras.* Adv. Math., **28**, 57–83, 1978.

[145] TAI, Y. S. *On the kodaira dimension of the moduli space of abelian varieties.* Invent. Math., **68**, 425–439, 1982.

[146] VAINSENCHER, I. *Schubert calculus for complete quadrics.* In *Enumerative geometry and classical algebraic geometry (Nice, 1981)*, Number 24 in Progress in Math., 199–235 (Boston: Birkhäuser), 1982.

[147] VAKIL, R. *Enumerative geometry of rational and elliptic curves in projective space.* math.AG/9709007, 1997.

[148] VIEHWEG, E. *Quasi-projective moduli for polarized manifolds.* Number 30 in Ergebnisse der Mathematik und ihrer Grenzgeibeite, 3. Folge (Berlin: Springer-Verlag), 1995.

[149] VISTOLI, A. *Intersection theory on algebraic stacks and on their moduli spaces.* Invent. Math., **97**, 613–670, 1989.

[150] VISTOLI, A. *Deformation of local complete intersections.* math.AG/9703008, 1997.

[151] WEYL, H. *Classical groups, their invariants and representations* (Princeton: Princeton University Press), 1946.

[152] WITTEN, E. *Two-dimensional gravity and intersection theory on moduli space.* Surveys in Diff. Geom., **1**, 243–310, 1991.

[153] WOLPERT, S. *On obtaining a positive line bundle from the Weil-Petersson class.* Amer. J. Math., **107**, 1485–1507, 1985.

[154] WOLPERT, S. *On the Weil-Petersson geometry of the moduli space of curves.* Amer. J. Math., **107**, 969–997, 1985.

Index

The **Symbols** section below gives the defining occurrence of each notation. All other references to these notations are indexed in the alphabetical section "as read": e.g., for references to $\overline{\mathcal{M}}_g$, look under Mgbar.

Symbols

Graduate Texts in Mathematics

continued from page ii